Textile-Based Energy Harvesting and Storage Devices for Wearable Electronics

Textile-Based Energy Harvesting and Storage Devices for Wearable Electronics

Xing Fan
Nannan Zhang
Yi Wang

WILEY-VCH

Authors

Prof. Xing Fan
Chongqing University
Chemistry and Chemical Engineering
Room LC511, Like Building
No.55 Daxuecheng South Rd
Shapingba
401331 Chongqing
China

Dr. Nannan Zhang
Chongqing University
Chemistry and Chemical Engineering
Room LC511, Like Building
55 Daxuecheng South Rd
Shapingba
40133 Chongqing
China

Yi Wang
Chongqing City Management College
Daxuecheng South Rd
Shapingba
401331 Chongqing
China

Cover Image: © flovie/Shutterstock

Library of Congress Card No.: applied for

British Library Cataloguing-in-Publication Data
A catalogue record for this book is available from the British Library.

Bibliographic information published by the Deutsche Nationalbibliothek
The Deutsche Nationalbibliothek lists this publication in the Deutsche Nationalbibliografie; detailed bibliographic data are available on the Internet at <http://dnb.d-nb.de>.

Print ISBN: 978-3-527-34524-3
ePDF ISBN: 978-3-527-81820-4
ePub ISBN: 978-3-527-81822-8
oBook ISBN: 978-3-527-81823-5

Typesetting Straive, Chennai, India
Printing and binding CPI Group (UK) Ltd, Croydon, CR0 4YY

Printed on acid-free paper

C089690_061021

Contents

Preface

The emerging era of artificial intelligence (AI) has aroused immense interests for wearable electronic hardware, particularly body-area networks, one health, robotics, and internet of things (IoT). Global industry giants have been pursuing a better way of putting electronic equipment on the body. Many AI equipment are considered as merely able-to-wear medical therapeutic measures or function modules, rather than a favor-to-wear daily habit. Especially when people feel healthy, many equipment are psychologically regarded as added burden for long-term wearing.

Textiles have played an important role in the human civilization for thousands of years as a cover for the human body for warmth, beauty, and protection, as a habit. The electronic textile has been developed with electronic functions or as an electronic device in the form of a fiber in interwoven textile in recent years as a solution to issues in wearing AI equipment.

Among the various types of electronic textiles, textile-type energy harvesting and storage devices are of special interests, because they can serve as sustainable power supply for on-body electronic devices. Efforts have been devoted into energy textiles to efficiently generate electricity from near-body energy sources of solar light, body movements, wind blowing, and body heat. Not only energy harvesting devices, such as photovoltaic cell, nanogenerator, and thermoelectric cell, but also energy storage devices, such as supercapacitor, battery and phase-transit devices, have been reshaped into fiber or textile, or woven together as hybrid energy textiles, to sustainably power various on-body electronic equipment, for example, a self-powered integrated electronic system can accomplish tasks relating to healthcare, safety protection, and human–machine interaction. Electronic textiles inspire new intelligent styles of human life in the future.

This book has summarized the advancement of various fiber-type and textile-type energy harvesting and storage devices, focusing on the function performance, device fabrication, integration, and function expansions, such as stretchability, hydrophobicity, endurability, air permeability, and color-change ability, which would lead to a self-powered intelligent electronic system for comfortably wearing as an everyday habit. This book is intended for scientists, engineers, graduate students, and undergraduate students majoring in physical chemistry, material chemistry, or energy chemistry who are interested in energy materials and devices. It may be also useful to many non-specialists in the industry who are devoted to promoting the wearable and related technologies. As textile-based energy devices have been studied only recently, there may be many imperfections and omissions in this book also due to the limit of my ability. I will greatly appreciate critical and suggestive comments from the readers.

This book was co-edited by Dr. Nannan Zhang and Ms. Yi Wang and written mainly based on our continuous research works since 2007. I enthusiastically acknowledge many postdoctoral fellows and graduate students who joined in the writing of this book at my lab: Hongmei Tao for Chapter 1, Jie Xue for Chapter 2, Qian Wen for Chapter 3, Xiaofei Wei and Yuxin Yang for Chapter 4, Jiaye Wu for Chapter 5, Chang Xu and Fang Huang for Chapter 6, Huang Zhou and Qin Zhang for Chapter 7, Siwei Xiang for Chapter 8, Jie Wu and Han Li for Chapter 9, and Nannan Zhang for Chapter 10. In particular, Nannan Zhang and Yi Wang devoted a lot of effort for the organization of this book during writing and revising at the final stage. I sincerely thank the Editor of Wiley for kindly inviting us to write this book and for the warm support in publishing it.

Chongqing, China *Xing Fan*
2020

1

On the Basis of Fibers and Textiles

Fibers are substances composed of continuous or discontinuous filaments. They are mainly used in making textiles for human wear, thus called textile fibers. Traditionally, textile fibers can be classified as nature fibers, chemical fibers, and functional fibers.

Nature fibers are textile fibers directly obtained from plants or animals that are originally natural or artificially cultivated. Their advantages are that they can be directly obtained from nature and have good moisture absorption, air permeability, and extensibility. Nevertheless, the wide application of nature fibers is restricted by its limited source, low yield, poor chemical stability, and low abrasion resistance, thus their inability to supply the fast-growing needs of human beings.

Chemical fibers refer to the fibrous object manufactured by artificial processing, which can be divided into synthetic fiber and regenerated fiber. Compared with nature fibers, chemical fibers exhibit superior light resistance, abrasion resistance, mildew resistance, and mothproof ability. Besides, their length, thickness, whiteness, and luster can be conveniently adjusted during the production process. Therefore, they can be an important complementary for nature fibers.

Furthermore, functional fibers refer to new types of fibers that have some special functions such as flame retardant, antistatic, anti-ultraviolet, etc., in addition to common physical and mechanical properties of ordinary fibers. Nowadays, functional fibers have been developed in order to endow chemical fibers with special functions such as antiflaming, hydrophile, and antistatic.

Textiles are products made by processing and interweaving textile fibers. These can be made into clothes, which can be utilized for covering the ugliness, adorning the beauty, keeping out cold wind, and protecting the body from insects. In addition to these basic functionalities of traditional textiles, people have been exploring functional textiles by weaving with functional fibers to endow clothes with more functions, like better skin protection, body temperature maintenance, decoration and aestheticism, etc.

Textile-Based Energy Harvesting and Storage Devices for Wearable Electronics, First Edition.
Xing Fan, Nannan Zhang, and Yi Wang.

This chapter mainly discusses the basis (such as definition, classification, and application) of fibers and textiles and the evolution process from classical functional fibers to intelligent fibers and textiles.

1.1 On the Basis of Fibers

Fibers usually refer to a soft slender body with a length-to-width ratio above 10^3 and thickness of several microns to hundreds of microns. Since fibers are mostly used to make textiles, they are also called textile fibers [1]. Traditionally, textile fibers are divided into nature fibers and chemical fibers, according to their sources of raw materials. In recent years, textile fibers with some special functions (such as flame retardant, hydrophilic, antistatic, anti-ultraviolet, etc.) have emerged, which are called functional fibers. The following section will mainly introduce the definitions, classifications, characteristics, and applications of nature fibers, chemical fibers, and functional fibers.

1.1.1 Nature Fibers

Nature fibers can be directly obtained from planted plants or animal hair and secretions. They are important material sources for the textile industry, of which the history can be dated back to the prehistoric times. Humans have begun to use nature fibers even since the Neolithic. They have been used in making clothings and life production equipment (such as ropes, nets, bags, etc.). According to the source of raw materials, nature fibers can be divided into cotton fiber, hemp fiber, wool fiber, and silk fiber [1].

The first type of nature fiber is cotton fibers. They are single-cell material that grow on cotton seeds and lengthen first and then thicken deposits to mature (Figure 1.1a).[1] In addition, Figure 1.1b shows the scanning electron microscope (SEM) of cotton fibers.[2] The advantages of cotton fibers are that they have slender and soft texture and good moisture absorption, strong alkali resistance, organic solvent resistance, and heat resistance. Therefore, the applications of cotton fibers are suitable for making all kinds of clothes as shown in Figure 1.1c, furniture cloths and industrial cloths (https://site.douban.com/108216/widget/notes/181120/note/146893919).

The second type of nature fiber is hemp fibers. They refer to fibers that are obtained from a variety of hemp plants, including phloem fibers in the cortex of annual or perennial herbs, dicotyledonous plants, and leaf fibers in monocotyledonous plants (Figure 1.2a).[3] Also, Figure 1.2b shows the SEM of hemp fibers [2]. The advantages of hemp fibers are that they have good hygroscopicity, high strength, and low denaturation ability, as well as being very pleasantly cool. The applications of hemp fibers are that they can be used as textile materials to woven into various

1 http://www.xjxnw.com/c/2017-06-15/1215397.shtml
2 https://www.meipian.cn/h8g0qnf
3 http://baike.jc001.cn/words/56558.html#

Figure 1.1 Cotton fibers. (a) Photographs of cotton fibers. (b) The SEM of cotton fibers. (c) Fashion clothing made of cotton fibers. Source: From Glenn Morrison et al. (2015), Figure 00 (p.149)/with permission of Elsevier. https://doi.org/10.1016/j.atmosenv.2015.05.051.

cool linen cloths, summer cloths, spinning ropes, and sacks for packaging as shown in Figure 1.2c.[4]

The third type of nature fiber is wool fibers that are developed from cells on sheep skin (Figure 1.3a).[5] Therefore, sheep wool is the most important textile wool fibers. Figure 1.3b shows the SEM of wool fibers [3]. The advantages of wool fibers are that they have soft and flexible texture and excellent warmth retention, so the applications of these are that they can be used in making woolen cloths, woolen yarns, blankets, felts, and other textiles. In addition, because wool fibers have the characteristics of good warmth retention and comfortable wearing, they are mostly used in autumn and winter clothings as shown in Figure 1.3c.

The fourth type of nature fiber is silk. It is a kind of continuous long fiber, which is solidified by silk liquid secreted by mature silkworm (Figure 1.4a).[6] In addition, Figure 1.4b shows the SEM of silk [4]. The advantages of silk are that it is light, fit, soft, smooth, and breathable and has gorgeous color, luster, elegance, and comfort. Additionally, silk is the mildest, softest, and thinnest nature fiber. The applications of silk are that it is used for silk weaving and then processed into a variety of silk

4 http://www.pibu.com/news/show-17833.html
5 http://www.czeps.cn/nongyechangshi/10387.html
6 https://baike.baidu.com/tashuo/browse/content?id=69510a16e826351b68c4ec14&fr=vipping

Figure 1.2 Hemp fibers. (a) Photographs of hemp fibers. (b) The SEM of hemp fibers. Source: Mwaikambo and Ansell [2]. (c) Photographs of spinning ropes and sacks for packaging.

textiles with different styles and finally made into silk products, clothing as shown in (Figure 1.4c), bedding, and other commodities (http://art.cfw.cn/news/153346-1.html).

1.1.2 Chemical Fibers

With the increase of human population and clothing requirements, limited natural resources are far from meeting with people's needs. As complementary to nature fibers, people have begun to consider the use of chemical fiber to address the shortage issue of cotton, silk, and wool. As a matter of fact, the use of chemical fibers has been started from the end of the nineteenth century. In addition, the period from the end of the nineteenth century to the 1930s of the twentieth century was a fast developing stage for the innovation of chemical fibers.

Chemical fibers are fibers made from natural or synthetic polymer compounds through chemical processing. According to the different sources of raw materials and processing methods, they can be divided into synthetic fibers and regenerated fibers.

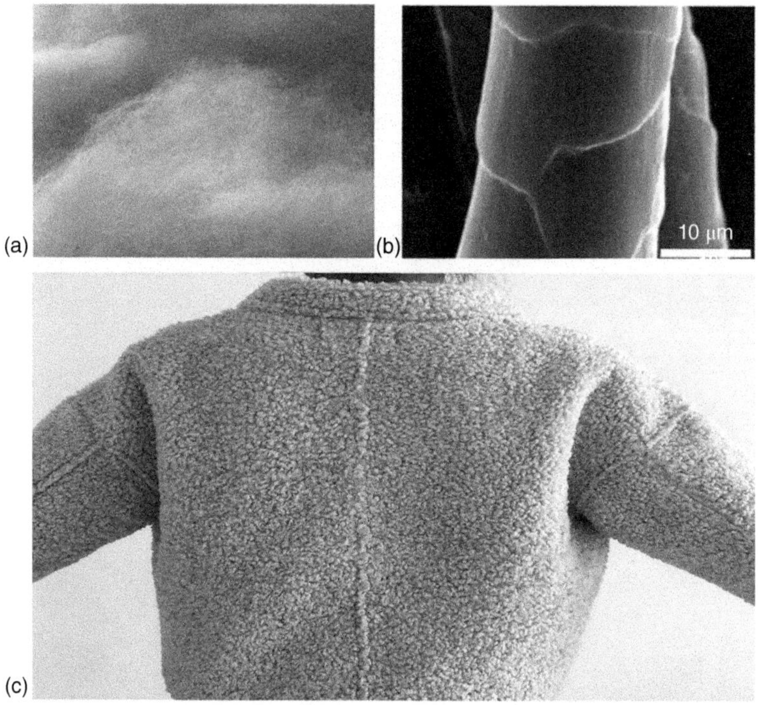

Figure 1.3 Wool fibers. (a) Photographs of wool fibers. (b) The SEM of wool fibers. Source: Long et al. [3]. (c) Photograph of wool fibers used in autumn and winter clothing.

Synthetic fibers are chemical fibers made from synthetic polymer compounds. Organic polymer compounds are synthesized by the addition polymerization reaction or condensation polymerization reaction with small molecular organic compounds as raw materials. People have developed various types of synthetic fibers for textile use, including polyester, nylon, acrylic, and vinylon.

Polyester is made from polyethylene terephthalate, which is produced by the condensation polymerization of terephthalic acid or dimethyl terephthalate and ethylene glycol (Figure 1.5a). The advantages of polyesters are that they have high strength, good elasticity, heat resistance, and density resistance. The applications of polyesters are that they are widely used in the manufacture of clothing and industrial products.

Nylon is linear structure polymer formed by amide bonds and several methylene groups (Figure 1.5b). The advantages of nylons are that they have strong abrasion resistance, ranking first among of all fibers. The applications of nylons are that they can be blended or purely spun into various medical and knitted products.

Acrylic is synthetic fiber made by copolymerization of acrylonitrile with other monomers, of which the acrylonitrile content is usually more than 85% (Figure 1.5c). The advantages of acrylics are that they have good elasticity and excellent lightfastness. Besides, their strengths will decrease by <20% in the

Figure 1.4 Silk. (a) Photographs of silk fibers. (b) The SEM of silk fibers. Source: Fink and Zha [4]. (c) Photographs of silk clothing. Source: From Andrea et al. (2016), Figure 01 (p.186)/with permission of Elsevier. https://doi.org/10.1016/j.saa.2015.10.024.

summer. Therefore, the applications of acrylics are that they can be used for curtains, tarpaulins, clothing, etc.

Vinylon is made by polymerization of vinyl alcohol acetate, which was usually alcoholyzed to obtain polyvinyl alcohol. After spinning, it can be treated with formaldehyde to obtain heat-resistant water-resistant vinylon (Figure 1.5d).[7] The advantages of vinylons are that they have good moisture absorption and air permeability, and its moisture absorption rate is the strongest among synthetic fibers, ranging from 4.5% to 5%. The applications of vinylon are that they are used to make knitted fabrics such as muslin, poplin, corduroy, work clothes, underwear, cotton sweaters, sweatshirts, etc. They can also be used for packaging materials, tarpaulins, canvas, fishing nets, surgical sutures, filter material, etc.

The second chemical fibers are regenerated fibers. They refer to the textile fibers obtained by using natural polymer compounds such as cellulose and protein as raw materials. After chemically processed into a concentrated polymer solution, they are

7 http://www.texindex.com.cn/buy/484056.html

Figure 1.5 Four kinds of synthetic fibers. Photographs of (a) polyester, (b) nylon, (c) acrylic, and (d) vinylon.

spined into fibers. There are many types of regenerated fibers, such as viscose fiber, bamboo fiber, and cuprammonium fiber.

Taking viscose fibers as an example, they are made of natural cotton linters and wood as raw materials, which are regenerated by separating natural cellulose through chemical reactions (Figure 1.6a).[8] The transverse morphological characteristics of viscose fibers are shown in Figure 1.6b [5]. Viscose fibers exhibit good moisture absorptions and spinnability, better dyeability, and antistatic resistance. The preparation process of viscose fiber can be divided to three steps. First, plant cellulose was alkalized into alkali cellulose; second, alkali cellulose was reacted with carbon disulfide to generate cellulose xanthate; and third, cellulose xanthate was dissolved in dilute alkali solution and wet-spined to obtain fibers. Viscose fibers are often blended or interwoven with cotton, wool, or various synthetic fibers and used in various clothing and decorative textiles as shown in Figure 1.6c[9] because of their good moisture absorption, comfortable wearing, and excellent spinnability.

1.1.3 Classical Functional Fibers

Functional fibers refer to a new type of fibers that they are given special functions to the fibers through special surface processing technology or adding a certain

8 https://www.sohu.com/a/395160376_527026
9 http://www.yzzao.com/news/show.php?itemid=439

(a) (b) (c)

Figure 1.6 The regenerated fibers. (a) Photographs of viscose fibers. (b) The transverse morphological characteristics of viscose fibers. Source: Chi and Chen [5]. (c) Photographs of decorative textiles.

functional additive. They have been developed in order to endow chemical fibers with special functions such as antiflaming, hydrophile, and antistatic. There are many types of functional fibers such as flame-retardant fibers, the hydrophilic fibers, and the antistatic fibers.

Firstly, flame-retardant fibers refer to fibers that they can reduce the flammability of fiber materials in the flame and slow down the spread of the flame, so that they can extinguish themselves quickly after leaving the flame and no longer smolder. The flame-retardant mechanism is aiming to prevent the thermal decomposition of fiber and inhibit the generation of combustible gas. Therefore, it will block the thermal feedback loop or the air supply, which would then eliminate or reduce the three key elements of combustion (combustible substances, temperature, oxygen) and finally achieve the purpose of flame retardancy. The advantages of flame-retardant fibers are that they have good flame-retardant properties. Furthermore, the applications of flame-retardant fibers are mainly used for high-temperature work clothes, space suits, national defense clothing, and fireproof clothing for firefighters.

The preparation of flame-retardant fibers includes two methods: adding thermal-stable function group and compositing with flame-retardant materials [6]. On the one hand, adding thermal-stable function group can be achieved by adding aromatic rings or aromatic heterocycles to the macromolecular chain, which would increase the rigidity of the molecular chain and the density and cohesion of the macromolecular chain. This method has four goals, including increasing the thermal cracking temperature, inhibiting the generation of combustible gas, increasing the degree of carbonization, and making the fiber difficult to burn. On the other hand, compositing is to add flame retardant to spinning melt or slurry and then spin to make flame-retardant fibers. For example, Jung et al. [7] reported on a fiber modification methods to use keratinous fibers as the host material for creating an effective flame retardant as shown in Figure 1.7a. A simple solution-based treatment to implant amine phosphate and phosphoric acid in the fiber through sequential monomer infiltration is found to be significantly effective for applying flame retardancy and reducing the flammability of polymeric materials. As shown in Figure 1.7b, after the flame-retardant fiber modification, polypropylene (PP) shows significantly improved flame retardancy to achieve V-0 grade and >70% reduced peak heat release rate in vertical burning and cone-calorimeter tests, respectively.

Secondly, hydrophilic fibers are highly hygroscopic fibers. Hydrophilicity refers to the ability of fibers to absorb water and transport water to neighboring fibers. The mechanism of hydrophilic fibers is that there are a certain number of strong polar groups (such as —OH, —NH, C=O, etc.) on the macromolecular chain, which can form hydrogen bonds with water molecules. The edges of the amorphous region and the crystalline region are relatively large, the molecular structure is relatively loose, and water molecules easily penetrate into the tiny gaps on the fiber surface. The advantages of hydrophilic fibers are that the textiles made of hydrophilic fibers can emit water vapor discharged from the human body in time, so that the user feels comfortable, without the bad feeling of stuffiness and airtightness. The applications of hydrophilic fibers are that they are suitable for making underwear fabrics and summer cool fabrics and are also often used to blend with hydrophobic synthetic fibers to improve the consumption of synthetic fibers.

The preparation of hydrophilic fibers includes two methods: chemical methods and physical methods. On the one hand, chemical methods include copolymerization with hydrophilic monomers, graft modification of hydrophilic monomers, and hydrophilic treatment of fiber surface. On the other hand, physical methods include endowing the fibers with a porous structure, roughening the surface, and introducing heteromorphosis fibers. For example, Song et al. [8] used one-step process that was developed to generate hydrophilic coating on polytetrafluoroethylene (PTFE) hollow fiber membranes based on co-deposition of polymerized dopamine (PDA) and poly(ethyleneimine) (PEI) from aqueous solutions as shown in Figure 1.8a. PDA and PEI were successfully deposited on PTFE membranes while the hydrophilicity and wettability of the modified membrane were greatly improved. The modified PTFE membranes showed high water permeate fluxes, good long-term stability, and durability in strong acidic aqueous solution. Figure 1.8b shows the photographs of the modified PTFE membranes with different reaction time.

Thirdly, antistatic fibers are fibers that improve the electrical conductivity of the fiber by improving the moisture absorption of the surface of the fiber. The antistatic mechanism of the antistatic fiber is to leak most of the static electricity generated by

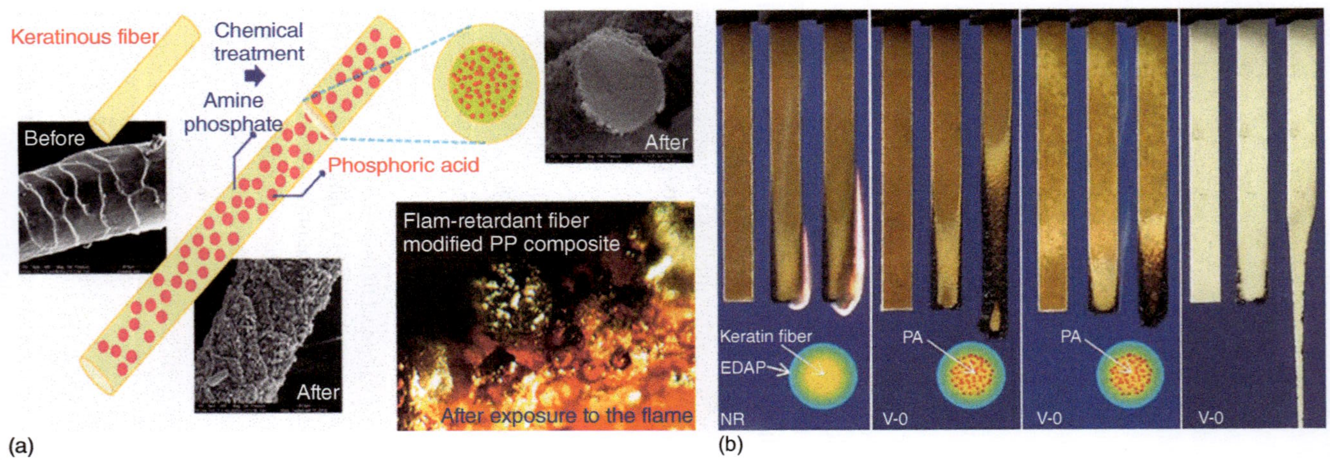

Figure 1.7 The flame retardant fibers. (a) Schematic diagram of fabrication of FR wool and SEM images of before and after wool surface treatment. (b) Captured images during the vertical burning test. Source: (a) Jung and Debes [7]. © 2018, American Chemical Society, (b) Jung and Debes [7].

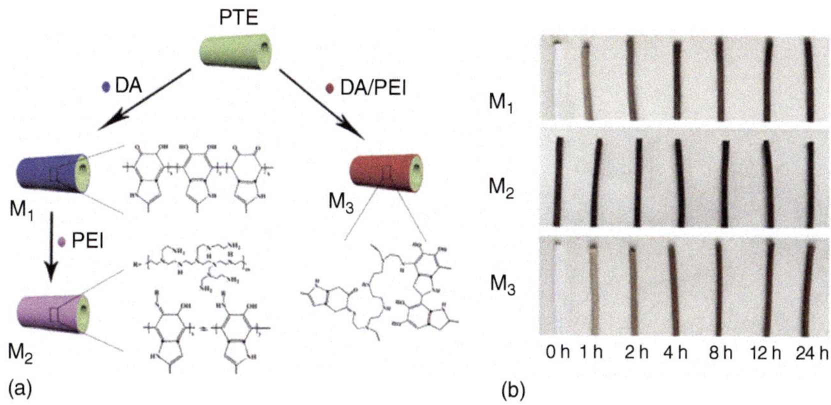

Figure 1.8 The hydrophilic fibers. (a) Schematic illustration for the modification of PTFE membranes. M_1, PTFE membranes coated by PDA; M_2, M_1 coated by PEI; and M_3, PTFE membranes co-deposited by DA/PEI. (b) Photographs of the modified PTFE membranes with different reaction time. Source: Song et al. [8]. © 2017, Elsevier.

moisture absorption, using the leakage effect. It needs to absorb moisture in the environment to increase the amount of static electricity leakage, so it is highly dependent on environmental humidity, and the general relative humidity is more than 40%. The advantages of antistatic fibers are that they are not easy to accumulate static charges, so they are often used in the processing all kinds of various clothing, carpets, mine conveyor belts, automotive interior decoration products, and special dust-free, sterile, and static-free work clothes.

The preparation of antistatic fibers includes four methods: (i) surface coating with antistatic agent, (ii) surface modifying with hydrophilic polymer finishing agent, (iii) co-spinning of ordinary polymer and antistatic polymer (such as polyethylene glycol, sodium alkyl sulfonate, etc.), and (iv) copolymering with antistatic monomer. For example, Gu et al. [9] synthesized PP/glass fiber (GF) composites showing excellent antistatic performance, which were prepared by a simple melt process blending PP with GF and a small amount of organic salts (OSs) as shown in Figure 1.9a. Also, Figure 1.9b shows schematic of the GF network and the surface absorption of OSs on the surface of GF. It was found that the GF and OSs exhibited significant synergistic effects on the antistatic performance and the mechanical properties of the composites. The GF network in the PP matrix provided perfect orbits for the movement of ions, inducing the excellent antistatic performance exhibited by the PP/GF composites at an OS loading of as low as 0.25 wt% when the GF formed a network in the PP matrix.

1.2 On the Basis of Textiles

Textiles are products made by processing and weaving textile fibers. Textiles can be divided into traditional textiles and functional textiles. Traditional textiles were basically utilized for covering the ugliness, adorning the beauty, keeping out cold

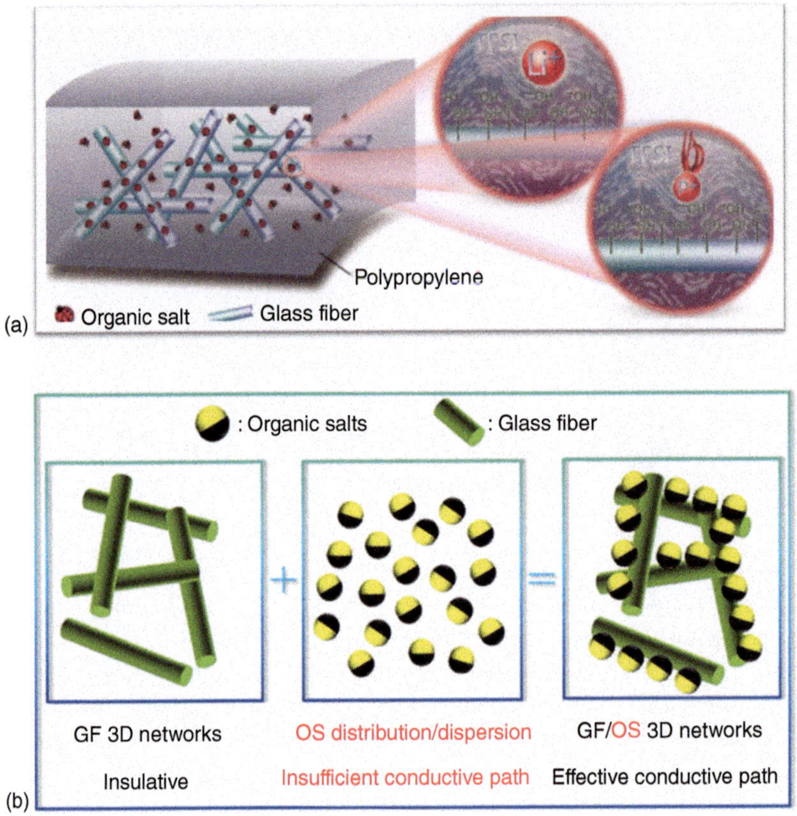

Figure 1.9 The antistatic fibers. (a) Polypropylene (PP)/glass fiber (GF) composites add a small amount of organic salts (OSs). (b) Schematic of the GF network and the surface absorption of OSs on the surface of GF. Source: Gu et al. [9]. © 2017, American Chemical Society.

wind, and protecting the body from insects. With the increase of human needs, traditional textiles could not meet with requirements of healthcare, protection, comfort, and medical and environmental protections. Therefore, various functional textiles (such as flame-retardant textiles, waterproof and moisture-permeable textiles, and antistatic textiles) begin to emerge and have become a trend in textile development.

This section will introduce the definition, classification, characteristics, and applications of both traditional textiles and functional textiles.

1.2.1 Traditional Textiles

Textiles can be seen everywhere in daily life, and their application fields include clothing textiles, household (decorative) textiles, and industrial textiles. Traditional textiles are objects that are formed by interspersing, interlacing, and entangling textile fibers and yarns in a weaving processing way. According to the production

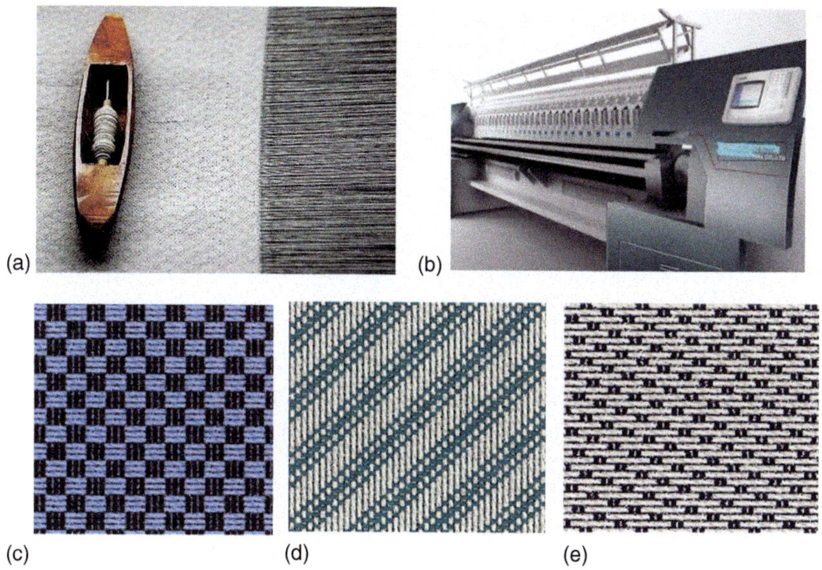

Figure 1.10 Description of woven fabric. Photographs of (a) woven fabric and (b) weaving machine of woven fabric. The organization structures of (c) plain weave, (d) twill weave and (e) satin weave.

methods, traditional textiles can be divided into woven fabrics, knitted fabrics, and nonwoven fabrics.

Firstly, woven fabrics (Figure 1.10a)[10] are made of two systems of vertical and horizontal yarns that are arranged vertically and interwoven according to certain rules on the loom (Figure 1.10b).[11] These fabrics exhibit stable structure, smooth cloth surface, and neat appearance. They have been used in all kinds of clothing, especially suitable for all kinds of outerwear and shirts. According to the organization structure, woven fabrics include plain weave, twill weave, and satin weave.

Plain weave is made up of warp and weft interlaced in a pattern of up and down (Figure 1.10c).[12] The characteristics of plain weave are that the numerator represents the warp yarn organization point and the denominator represents the weft yarn organization point. Warp and weft yarns are interwoven with other yarns, with more interweaving points and more yarn buckling points. Among them, the numerator represents the warp organization point, and the denominator represents the weft organization point. The advantages of plain weave are that the fabric is firm, wear-resistant, stiff, and flat and has good air permeability. The disadvantages of plain weave are that the elasticity is relatively small and the gloss is average. The applications of plain weave are mostly used for yarn-dyed and high-end embroidery fabrics.

10 https://ss2.baidu.com/6ON1bjeh1BF3odCf/it/u=1969820724,1531930458&fm=27&gp=0.jpg
11 https://ss0.baidu.com/6ON1bjeh1BF3odCf/it/u=2850963774,3906526589&fm=27&gp=0.jpg
12 https://xw.qq.com/cmsid/20190821A0JY2P00?f=newdc

(a) (b) (c)

Figure 1.11 Description of knitted fabric. (a) Photograph of knitting machine, (b) warp knitted fabrics, and (c) weft knitted fabrics.

Twill weave is an organization in which weave points (or weft weave points) are continuous into diagonal lines (Figure 1.10d). The characteristics of twill weave are that the numerator represents the number of warp weave points on each yarn in the tissue cycle that has at least three warp and three weft threads and the denominator represents the number of weft weave points. These include gauze, twill, serge, etc. The advantage of twill weave is that there are obvious oblique lines, soft feel, strong three-dimensional sense, and good glossiness and elasticity. The disadvantages of these are poor wear resistance and fastness. The applications of twill weave are mostly used in printing, umbrella cover, sports shoes, and sportswear clips.

Satin weave consists of separate, discontinuous warp weave points (or weft weave points) that are regularly and evenly distributed in the tissue cycle (Figure 1.10e). The characteristics of satin weave are that the separate weave points on two adjacent warp yarns are far apart and independent and discontinuous and arranged in a certain order. The numerator represents the latitude and longitude of a complete organization, and the denominator represents the flying number that refers to the number of lines separated between two adjacent lines or two adjacent tissue points. These include horizontal tribute, straight tribute, soft satin, etc. The advantages of satin weave are that they have soft texture and smooth silk surface, good gloss, and most rich and luxurious. The disadvantage is that the floating length is longer and the fastness is the worst. The applications of satin weave are used for brocade, floral soft satin, handkerchiefs, etc.

Secondly, knitted fabrics are made of a group or groups of yarns connected to each other in circles according to certain rules on a knitting machine (Figure 1.11a). The advantages of knitted fabrics are that they have soft texture, good wrinkle resistance, and air permeability, as well as greater extensibility and elasticity. They are suitable for making underwear, tights, sportswear, etc. According to different knitting methods and fabric loop structures, knitted fabrics can be divided into warp knitted fabrics and weft knitted fabrics.

Warp knitted fabrics are fabrics formed by one or more groups of parallel yarns that are formed into loops along the warp direction of the fabric and are connected in series (Figure 1.11b).[13] This fabric passes through each yarn of a set of warp yarns to form only one or two stitches in one course and then forms stitches in the next course. The stitches formed by one yarn are arranged along the warp direction of the

13 http://info.texnet.com.cn/detail-760483.html

(a) (b)

Figure 1.12 Descriptions of nonwoven fabrics. (a) Photographs of nonwoven fabrics of different colors. (b) Photographs of weaving machine of nonwoven fabric.

knitted fabric. The advantages of warp knitted fabrics are that they have small extensibility, good elasticity, and low dispersibility. These are suitable for outer clothes, mosquito nets, fishing nets, flower headbands, towels, etc.

Weft knitted fabrics are fabrics in which one (or more) yarns are sequentially bent into loops along the weft direction of the knitted fabric, and the loops are sequentially wrapped (Figure 1.11c).[14] This is formed by feeding the raw material (yarn) into the knitting needles along the weft direction and then knitting them into loops in a certain order and forming them by stringing each other. The advantages of weft knitted fabrics are that they have greater horizontal extensibility, a certain degree of elasticity, and great dispersibility. At present, the applications of these are that they are used in underwear, pants, sweaters, etc.

Thirdly, nonwoven fabric refers to a product similar to cloth made by chemical (or hot melt) bonding with chemical fiber as the basic raw material, which is made without fiber weaving (Figure 1.12a).[15] These fabrics are formed by oriented or random arrangement of textile short fibers or filaments to form a web structure, which is then reinforced by mechanical and thermal bonding or chemical methods. Figure 1.12b shows the weaving machine of nonwoven fabrics.[16] These fabrics are moisture-proof, breathable, flexible, lightweight, noncombustible, easy to decompose, nontoxic, nonirritating, rich in color, low price, and recyclable. These are used for filling in clothes as linings and making disposable garments.

1.2.2 Classical Functional Textiles

People's pursuit of new textiles not only is satisfied with the basic functions of warmth and decoration but also has an increasing demand for healthy and environmentally friendly functional textiles. Therefore, functional textiles are constantly innovating in types.

14 http://www.52bjw.cn/product-info/51429294.html
15 http://www.yzwufangbu.com
16 https://ss0.baidu.com/6ON1bjeh1BF3odCf/it/u=2995699824,2939041623&fm=15&gp=0.jpg

Functional textiles refer to textiles with special functions in addition to basic functions. In other words, these textiles not only have their own basic use values (such as comfortable wearing, conventional decoration, warmth, etc.) but also have some special functions of antibacterial, anti-mildew, antivirus, anti-mosquito, mothproof, flame retardant, and so on. According to different functional characteristics, classical functional textiles include flame-retardant textiles, waterproof and moisture-permeable textiles, and antistatic textiles.

Firstly, flame-retardant textiles are made of flame-retardant fibers or textiles that have undergone flame-retardant finishing, which have reduced flammability to varying degrees, and can significantly delay their burning rate during the combustion process, and can quickly self-extinguish after leaving the fire source. The advantages of flame-retardant textiles are that they are more washable and air permeable, as well as superior dyeing. Furthermore, the main characteristics of flame-retardant textiles are high limiting oxygen index. The applications of these textiles are that they can be used in making work clothes such as military protective combat uniforms, firefighting uniforms as shown in Figure 1.13a (https://www.51wendang.com/doc/d77bc160ad876beb14191011/2), steelmaking work clothes, welding work clothes, and medical protective clothing; curtains for public facilities, insulation cotton for exterior walls of buildings, and other flame-retardant products in industrial; and carpets, sofas, and curtains as shown in Figure 1.13b (http://www.hometex114.com/Leads/Detail_11_10604.html).

To prevent the burning process, it is sufficient to interfere with, or suppress, one or several of these subprocesses. This is the underlying principle of textile protection against fire via (i) application to the fabric of substances whose decomposition at the burning temperature releases noncombustible gases, (ii) formation on the fabric of a film isolating the oven fabric surface from atmospheric oxygen, (iii) chemical modification of the functional groups of the fiber, increasing the thermal stability of the macromolecular chains of the fiber, and (iv) use of combined methods of flame-retardant finishing of fabrics.

It is possible, in principle, to improve the fire resistance of textiles, i.e. to reduce their flammability or render them fire-resistant, via the creation of flame-retardant fiber-forming polymers or via the use of special agents, flame retardants, reducing the textile flammability (Scheme 1.1) [10].

It should be noted that fires today are much more dangerous than in the past, considering different burning temperatures and release of toxic substances. Therefore, the development of new low-toxic low-smoke substances and compositions reducing the flammability of polymeric materials will long remain an issue of much importance.

Secondly, waterproof and moisture-permeable textiles are the textile products that have integrated waterproof, moisture permeability, windproof, and warmth retention performance (Figure 1.14a) (https://www.sohu.com/a/229792964_297345). These textiles can not only meet with the wearing needs of people during activities in harsh environments such as severe cold, rain, snow, and windy weather, but also they are suitable for people's daily life requirements for raincoats. These can be

Figure 1.13 The applications of flame-retardant textiles. (a) Photograph of firefighting uniforms. Source: From Fent et al. (2014), Figure 01 (p.833)/CC BY/with permission of Oxford University Press. https://doi.org/10.1093/annhyg/meu036. (b) Photographs of curtains. Source: From Chunming Xu et al. (2011), Figure 01 (p.87)/with permission of John Wiley & Sons, Inc. https://doi.org/10.1002/fam.1089.

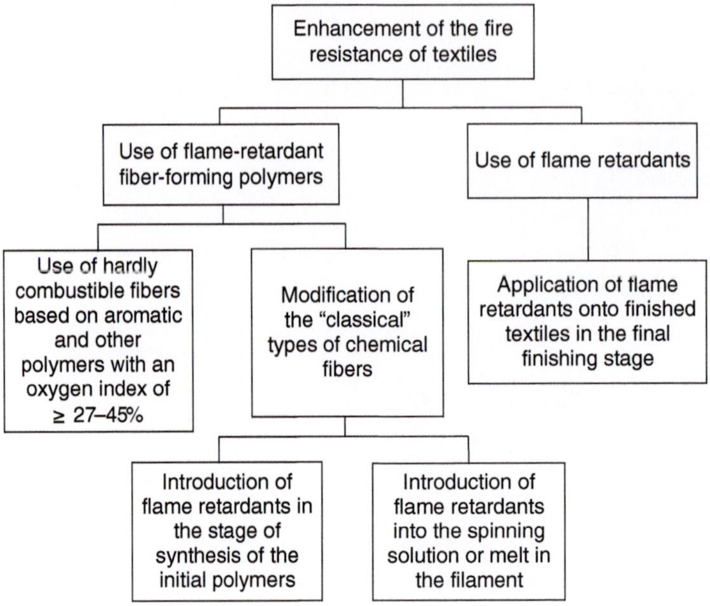

Scheme 1.1 Ways to reduce the flammability of textiles. Source: Vladimirtseva et al. [10]. © 2014, Spinger Nature.

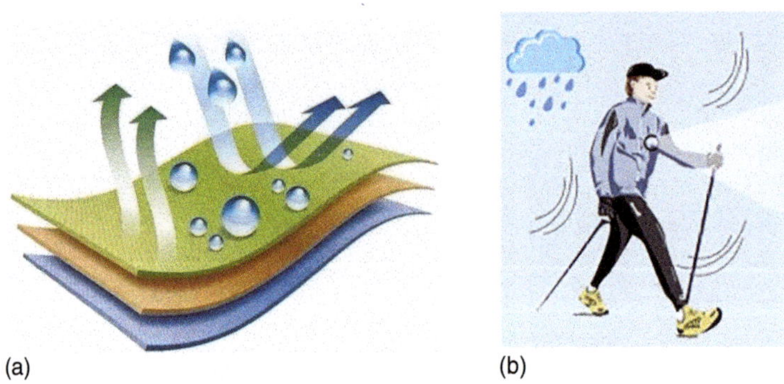

Figure 1.14 Waterproof and moisture-permeable textiles. (a) Photographs of waterproof and moisture-permeable textiles. Source: ag visuell/Adobe Stock. (b) Photographs of rainwear. Source: From Yating Wang et al. (2020), Figure 01 (p.03)/with permission of Elsevier. https://doi.org/10.1016/j.cej.2020.126222.

used to make sportswear, casual wear, rainwear as shown in Figure 1.14b,[17] sailing and water work clothes, firefighting clothing, etc.

The processing methods of waterproof and moisture-permeable textiles include the following three types:

(i) High-density fabric, which uses ultrafine cotton fiber or ultrafine synthetic fiber filament to be woven into a high-density fabric, so that the yarn gap of this fabric is small, preventing water droplets from passing.

17 https://guangdiu.com/detail.php?id=6379721

(a) (b)

Figure 1.15 Antistatic textiles. (a) Photographs of antistatic textiles. Source: https://www.sohu.com/a/229792964_297345. (b) Photographs of blankets.

(ii) Coated fabric, which throughs the use of dry or wet coating technology. Moreover, the pores on the surface of the fabric are closed or reduced to a certain extent by the coating agent, thereby obtaining water resistance.
(iii) Laminated fabric, which uses special adhesives and laminating process to laminate the microporous or hydrophilic film with waterproof and moisture-permeable function with ordinary fabrics to form a waterproof and moisture-permeable material.

Thirdly, antistatic textiles are textiles that are spun with conductive fibers or embedded in conductive filaments during weaving (Figure 1.15a) (https://www.sohu.com/a/229792964_297345). They can eliminate static electricity by themselves and can avoid electric shock and fire. These textiles can prevent the accumulation of static electricity on the surface of the textiles, and the static electricity on the surface of the human body made of antistatic textiles can be quickly released. They not only can solve the static electricity phenomenon in daily life but also can be used in work clothes or blankets (Figure 1.15b)[18] to play a role in special places to avoid safety accidents.

The preparation of antistatic textiles usually adopts the following three methods:

(i) Surface treatment method, which uses conductive metal salts or antistatic agents (such as quaternary ammonium salt, sodium alkyl sulfate, and polyol) and then through methods of spraying, dipping, and coating to carry out surface treatment (such as coating a conductive layer) of fibers and textiles.
(ii) Blending method, in which a small amount of conductive material (such as carbon black or metal oxide) or antistatic agent is added to the fiber and it is made by blending with the fiber stock solution.
(iii) Bulky chemical modification method, adding hydrophilic monomers (such as methyl acrylate) during the synthesis of fiber interpolymers and then synthesizing hydrophilic fibers through copolymerization to improve the moisture absorption of textiles and make them resistant to electrostatic performance.

18 http://379188.shop.cnlist.org/product/1509345.html

1.3 The Evolution from Classical Functional Fibers to Intelligent Fibers and Textiles

With the advancement of human civilization, people have been increasingly seeking for self-cognition, communication, and cooperation. On the one hand, human have tried to explore inward to get a better knowledge of ourselves such as memory, learning, and aging. On the other hand, we are falling into the incremental connection with others by relying on the portable electronics including smartphones and laptops. Those needs have accelerated the development of wearable electronic devices to deal with related information fast, accurately, directly, and conveniently. However, classical functional fibers cannot fulfill the new function requirements and failed to meet with the higher demands of human life, which then lead to the emerging of intelligent textiles.

Intelligent textiles can be defined as textiles device that are able to sense and respond to changes in their environment, rather than a type of fiber materials. Classical functional textiles can only play the role of protection, hygiene, and senses, but they cannot in terms of responding to their environment. Popular technologies for functionalizing textiles include vapor deposition, high energy beam processes, nanomaterial coatings, etc. Different from the classical functional textiles, intelligent textiles can sense and react to environmental conditions or stimuli according to thermal, mechanical, electrical, magnetic, or other bases [11]. Intelligent textiles can be prepared by dyeing and finishing of fabrics, weaving smart fibers into textiles, and integrating electronics into textiles. The development of intelligent fibers and intelligent textiles is an innovative combination of electronic devices and textiles, which provides a new idea for the human–machine interface in the future.

At present, many common intelligent textiles have been developed, which include shape memory textiles, intelligent temperature-regulating textiles, intelligent color-changing textiles, and electronic intelligent textiles, which will be introduced in details in the next section.

1.3.1 Shape Memory Fibers and Textiles

Shape memory fibers and textiles are a kind of special textiles with shape memory effect. The so-called shape memory effect means that the fiber or textile has a certain initial state under certain conditions, and when the external conditions change, the fiber or textile shape changes and fixes accordingly. When the external environment changes in a specific way, the fiber or textile can reversibly return to its initial state with its high ability to restore deformation. In addition, external conditions generally include thermal energy, light energy, electrical energy, external mechanical forces, and other physical factors and chemical factors such as pH, phase transition reactions, etc.

At present, most researches on shape memory materials are focused on shape memory polymers and shape memory alloys.

Firstly, shape memory polymer is a new type of functional polymer material, which includes polynorbornene, trans-polyisoprene, cross-linked polyethylene,

Figure 1.16 Photograph of shape memory fiber fashion designed by British designers. Source: From Dang et al. (2021), Figure 01 (p.03)/CC BY 3.0/with permission of IOP Publishing. https://doi.org/10.1088/1742-6596/1790/1/012084.

intercalated polyurethane, polyester, copolyimide, polyamide, etc. From the perspective of microphase structure, the shape memory polymer has a two-phase structure of "stationary phase" and "reversible phase." The "stationary phase" has the function of maintaining the shape of the solidified polymer. Supramolecular structure is manifested as the cross-linking, crystallization, glass state of macromolecules, or entanglement of polymer chains. The "reversible phase" is the variable part of the polymer. With the change of external conditions (mainly heat), the morphology changes, and the reversible phase undergoes crystal melting and formation or the reversible phase transition between glass and rubber. The advantages of shape memory polymer are that they are soft and easy to form and have better shape stability, a larger range of adjustable mechanical properties, etc.

The applications of shape memory materials in textiles are that these materials are made into filaments and then spun into yarns, and then shape memory yarn is woven into various shape memory polymer textiles. Shape memory polymer textiles are usually used in sportswear, mountaineering clothing, and tents. These kinds of clothing are easy to take care. After the clothing is wrinkled, it can be restored to the original flat state only by smoothing it by hand, and various styling treatments can be performed on the garments by manual grasping as shown in Figure 1.16 that is designed by British designers (https://ss2.baidu.com/6ON1bjeh1BF3odCf/it/u=2498880830,312056152&fm=27&gp=0.jpg).

Secondly, shape memory alloy is a special metal material, including nickel–titanium alloy, copper–zinc alloy, copper–tin alloy, nickel–titanium–copper alloy, and copper–zinc–aluminum alloy. This has the ability to deform and restore shape under certain heat treatment conditions. The basic principle of the alloy's shape memory effect is attributed to the thermal change of the metal crystal structure. Shape memory alloys have a certain transition temperature. Above the transition temperature, the metal crystal structure is stable; below the transition temperature,

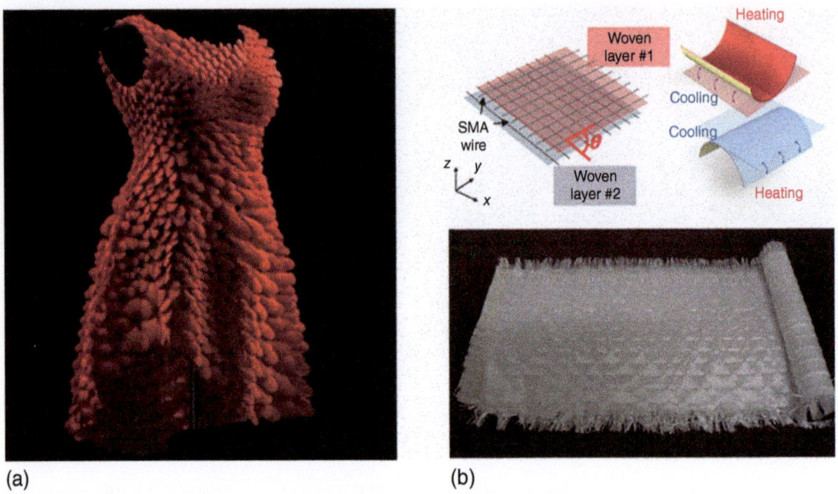

Figure 1.17 The applications of shape memory fibers and textiles. (a) Photograph of anti-wrinkle clothing. Source: Biswas et al. [12]. (b) Photograph of anti-scald clothing. Source: Han et al. [13].

the crystal is in an unstable structure state and deforms when subjected to external forces. When the heating temperature rises above the transformation temperature, the metal crystal will return to the shape of the stable structure state, which will cause the deformation and deformation recovery of the alloy fiber.

Shape memory alloy fiber or textile is made of shape memory alloy. At present, nickel–titanium alloys are the most used shape memory alloys in textiles because its fibers have a large deformation recovery rate, good stability, and good biocompatibility. For example, an Italian company recently launched a "smart shirt" that uses a fabric interwoven with nickel–titanium alloy wires and nylon and a woven fabric with five nylon wires and one nickel–titanium alloy wire. This shirt is used in the hot summer, the shape memory fiber is activated, and then the sleeves of the shirt will be automatically rolled up immediately. Furthermore, the sleeves can automatically recover when the temperature drops. In addition, this shirt is not afraid of wrinkling, even if it is crumpled into a wrinkled ball, the hot air blown by a hair dryer or the human body temperature will immediately restore it (Figure 1.17a) [12]. And the British textile agency processed and fixed nickel-titanium alloy fibers inside the garments when they developed anti-scald garments. Once they come into contact with high-temperature shape memory fibers, they will be excited to achieve the purpose of anti-scalding as shown in Figure 1.17b [13] (https://www.sohu.com/a/237181827_783093).

1.3.2 Intelligent Temperature-Regulating Fibers and Textiles

Intelligent temperature-regulating textiles are high-tech products developed by combining the phase change materials with the fiber and textile manufacturing technology. The advantages of intelligent temperature-regulating textiles are that

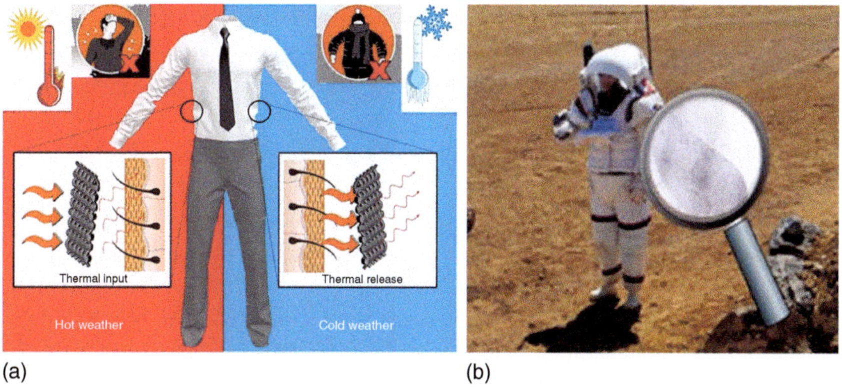

Figure 1.18 The applications of intelligent temperature-regulating fibers and textiles. (a) Photograph of protective outerwear. Source: From Linghui Peng et al. (2019), Figure 01 (p.6416)/with permission of Springer Nature. https://doi.org/10.1007/s10570-019-02534-6. (b) Photograph of astronaut clothing. Source: From Malgorzata Holyńska et al. (2020), Figure 18 (p.24)/with permission of John Wiley and sons. https://doi.org/10.1002/admt.202000028.

they can adjust the micro-temperature between human body and external environment and improve the dress comfort of human body. The applications of intelligent temperature-regulating fibers are that they are used in textile industries such as protective outerwear shown in Figure 1.18a (https://www.sohu.com/a/203630756_368281) and astronaut clothing shown in Figure 1.18b (https://www.sohu.com/a/237181827_783093). These not only are a technological innovation in textile industry but also provide a guarantee for people working at extreme environmental temperatures.

The phase change material (such as simple salt, alkali metal and alloy, paraffin, fatty acid, and eutectic) is an important component of intelligent temperature-regulating textiles. Through specific technology, phase change materials are combined with textiles or fibers to realize the intelligent temperature regulation of textiles. When a substance undergoes a phase change, which is accompanied by heat absorption or heat release, the energy absorbed or released is called phase change heat. The mechanism of heat storage and temperature adjustment of phase change materials is that the solid–liquid reversible change occurs in the phase change material of the textiles according to the change of the ambient temperature of the outside. When the ambient temperature is increased, the stored heat is absorbed, and the phase change material is changed from the solid state to a liquid state. Also, when the temperature is reduced, the stored heat is discharged from a liquid state to a solid state. Therefore, the two-way temperature regulation of the textile is realized, and a comfortable temperature environment is created.

There are three commonly used manufacturing methods for intelligent temperature-regulating textiles:

(i) Hollow fiber filling method, which makes hollow fibers first and then immerse them in the phase change material solution to fill the hollow part of the fibers with the phase change material and finally close the ends of the fibers after drying. For example, the hollow fiber is immersed in polyethylene glycol solution to obtain a temperature-regulating fiber containing polyethylene glycol in the hollow part.

(ii) Spinning method, which adds phase change materials and additives to the spinning solution. For example, *n*-octadecane microcapsules are mixed with a polyacrylonitrile/vinylidene chloride copolymer solution, and the temperature-regulating fiber is prepared by a solution spinning process.

(iii) Fabric finishing method, which uses the padding method and coating method to finish the phase change material on the textile. Moreover, this method is simple and easy. For example, a blend of energy storage microcapsules and coating glue is coated on the surface of the fabric to make smart temperature-regulating textiles.

1.3.3 Intelligent Color-Changing Fibers and Textiles

Intelligent color-changing fibers and textiles refer to the fibers and textiles that can show different color with the change of external environment condition (such as light, temperature, electricity, etc.). The advantages of intelligent color-changing textiles are that they have good washing resistance and long-lasting color-changing effect. The applications of these are that they are used in textile, military, entertainment, anti-counterfeiting, and other fields because these are kinds of intelligent products with high added value and high efficiency. For example, they can be used in the production of fashion discoloration clothing and decorative fabrics and anti-counterfeiting field, which can be used as anti-counterfeiting materials and widely used in bills, documents, and trademarks and so on.

According to the external stimulus conditions, the color-changing textiles are mainly divided into photochromic textiles, thermochromic textiles, and electrochromic textiles.

Firstly, photochromic textiles refer to the textiles whose color change with the wavelength of light or the intensity of light. Photochromism is observed when there is a change in sunlight (UV radiation) exposure to a photochromic material. The chemical structure of a photochromic material is temporarily altered as a result of UV irradiation. This change in chemical structure leads to a shift in the absorption of electromagnetic waves to the visible part of the spectrum, upon which the color changes from colorless to colored. A reverse change in its chemical structure, and consequently in its electromagnetic wave absorption spectrum, can take place in the absence of UV rays. As a result, the material returns to its original colorless state.

The application of photochromic textiles adds new content to clothing fashion. More recently, British designer Amy Winters has created a sun-reactive dress by printing photochromic inks directly onto the fabric. The rainforest dress is white indoors but becomes purple when exposed to sunlight (Figure 1.19a) [14]. Photochromic textiles have also been used for military purposes to provide protection by camouflage and, by monitoring UV radiation, can be applied for solar protection.

Figure 1.19 The applications of photochromic textiles. (a) Amy Winter's photochromic "rainforest" dress expressing the effects without (left) and with (right) UV exposure. Source: Kelly and Cochrane [14]. (b) Photograph of photochromic textiles in fashion. Source: From Zidan Gong et al. (2019), Figure 02 (p.03)/CC BY/with permission of MDPI. https://doi.org/10.3390/ma12203311.

Figure 1.19b shows the photochromic textiles in fashion (https://m.sohu.com/a/374940893_559321).

Secondly, thermochromic textiles are made of a special thermal discoloration material. Thermochromism is observed when there is a change in heat exposure to a thermochromic material. Thermochromic materials change their molecular or supramolecular structure and absorption spectrum as a result of the variation in environmental temperature. The structural changes of thermochromic compounds may include rearrangement of molecules (e.g. cleavage of covalent bonds or changes in the spatial configuration of a molecule) and changes in crystalline structures.

The applications of thermochromic textiles have predominately been used for decorations and fashions. For example, a US kick-starter company by the name of

(a) (b)

Figure 1.20 The applications of thermochromic textiles. (a) Heat-sensitive color-changing textiles of Radiate Athletics' heat-sensitive sportswear and Shi Yuan's blossoming wallpaper. Source: Kelly and Cochrane [14]. (b) Photograph of Stone Island thermal response jacket. Source: From Zidan Gong et al. (2019), Figure 02 (p.03)/CC BY/with permission of MDPI. (https://doi.org/10.3390/ma12203311).

Radiate Athletics has applied thermochromic materials to sports clothes in order to monitor the body temperature of a person. When a person is working out, their body heats up, and consequently the fabric of the T-shirt, or other item of clothing, will change color (Figure 1.20a). Certain parts of the shirt will change color before others, therefore making it possible to determine which parts of your body are being exercised the most. As a result, the color-changing function of the sports clothing offers the ability to monitor and maximize your workout [14]. Figure 1.20b shows the Stone Island thermal response jacket (https://m.sohu.com/a/374940893_559321).

Thirdly, electrochromic textiles are kinds of textile materials whose color changes reversibly with the change of external electric field. Electrochromism is induced by a change in the electrical potential applied to an electrochromic material. Electrochromic materials experience a reversible change in color as a result of an electrochemical redox (i.e. oxidation and reduction) reaction. These require only a short electrical pulse in order to effect the color change. Due to the memory effect of electrochromic color changes, little or no input of power is required to retain the new color. Electrochromic displays are therefore far more economical in the consumption of power when compared with their light-emitting analogs. Additionally, there is no visual dependence on the viewing angle, and multiple colors, with high contrast, are available. The advantages of electrochromic textiles are that the light transmittance can be adjusted freely in a large range and have multi-color continuous change, storage memory functions, low driving color change voltage, simple power supply, power saving, less environmental impact, etc.

The applications of electrochromic textiles have great prospects in some special fields such as safety warnings, fashion clothing, decorations, and adaptive camouflage. For example, Thermochromic textiles are used for cloth of different colors

(a) (b)

Figure 1.21 The applications of thermochromic textiles: (a) Photograph of electrochromic cloth of different colors. Source: From Zidan Gong et al. (2019), Figure 02 (p.03)/CC BY/with permission of MDPI. https://doi.org/10.3390/ma12203311. (b) Digital photographs of electrochromic fibers or textiles in the colored/bleached states for proof-of-concept military camouflage. Source: (b) Fan et al. [15].

(Figure 1.21a) (https://doi.org/10.3390/ma12203311). Furthermore, in military stealth technology, human used electrochromic fibers and textiles to make the covering of color-changing dragon suit or gunsuit, achieving good results as shown in Figure 1.21b [15].

1.3.4 Wearable Electronic Intelligent Fibers and Textiles

Traditionally, the collection, analysis, storage, and transmission of signals are completed by an integrated electronic system based on printed circuit board (PCB), which has been implanted with rigid electronic components, such as sensors, didoes, transistor, switches, etc. However, integrated circuit (IC) boards, even though bendable or stretchable, still have body-adaptation problems for many people and were usually regarded as special medical measures, rather than the daily wearing habit, especially when people feel healthy.

Emerging wearable electronic intelligent textiles refer to highly intelligent textiles that can also function as the integrated electronic systems, information processing, computing, and other electronic functions. Until now, there are three technical solutions of integrating electronic components based on textiles: (i) embedding small electronic components or equipment on ordinary fabrics or fabric-type power sources, (ii) directly assembling flexible electronic devices based on fabric substrate, and (iii) weaving various fiber-shaped electronic devices into one textile.

In fact, the first two technical solutions are very similar to the PCB-based solution for conventional ICs.

For example, for the first technical solution, Fan and coworkers [16] wove fiber-shaped batteries with interwoven textile-type solar cell into one textile by

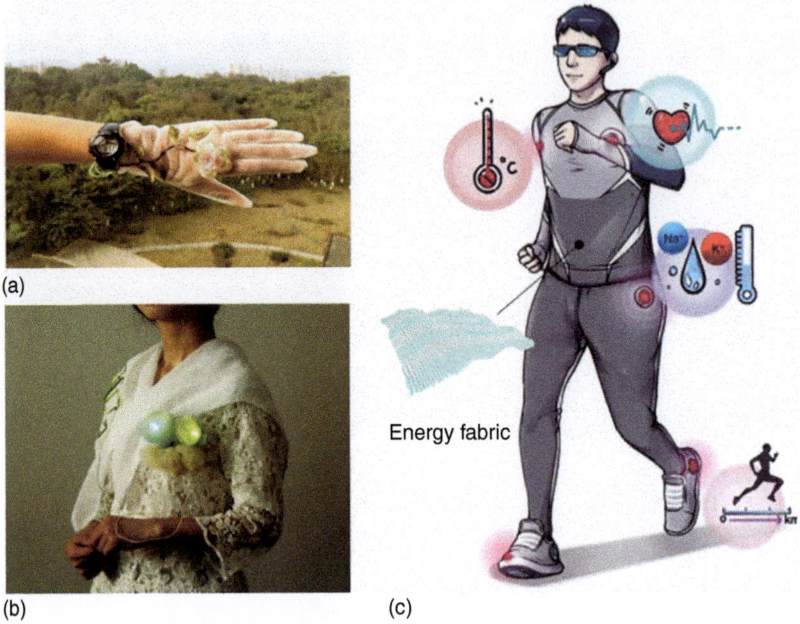

Figure 1.22 Integration of electronic devices into textiles. Demonstration of the energy fabric to serve as the power source for various portable electronics. (a) Electronic watch and (b) light-emitting diode (LED) lights. Source: Huang et al. [16]. (c) Illustration of an autonomous body area sensor network. Source: Zhang et al. [17]. © 2020, Elsevier.

couching embroidery or weaving. This energy textile can be reformed as a gauze glove, which can serve as both a fabric substrate and a portable power source for embedding with wearable electronics, such as sensors, LEDs, and electric watch (Figure 1.22a). With the appropriate appearance and pattern design, it can also be designed to a gauze kerchief, which can be worn around the body and power colorful light-emitting diode (LED) lights (Figure 1.22b) or decorations in stage performances. With a collection of compelling features, the wearable photo-rechargeable fabric paves a new way to sustainably power body area sensor networks for personalized healthcare [17]. As shown in Figure 1.22c, a body area sensor network is a collection of wearable/implantable biosensors that can be employed to continuously monitor human physiological signals for personalized healthcare. Given such features as lightweight, robustness, and cotton-like softness, the photo-rechargeable fabric integrated with sensor and communication components can act as autonomous textile sensor network for continuous human physiological, biomechanical, and ambient environmental monitoring, with the possibility of succeeding as a future medical application in the era of the Internet of Things.

For example, for the second technical solution, Peng and coworkers [18] integrated of the electronic components into textiles by a screen-printed process on the textile directly to form the patterned circuits (Figure 1.23a, b). Different electronic

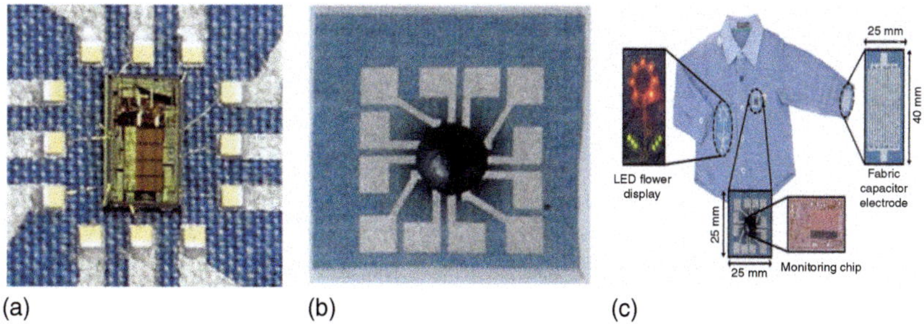

Figure 1.23 The screen-printed process. (a) Electronic component bonding on a P-FCB. (b) Electronic component packaging on a P-FCB. (c) System integration on P-FCB with capacitive sensor, chip, and LED display. Source: Adapted with permission from Wang et al. [18]. Copyright (2019) by WILEY-VCH Verlag GmbH & Co. KGaA, Weinheim.

components, e.g. system-on-a-chip, chip capacitors, resistors, and LEDs, can be bonded directly on the textile lead frame (Figure 1.23c). Liquid molding epoxy was then coated on the chip after bonding process, which provided robust protection against stretching.

Compared with the first two technical solutions, intelligent textile woven from fiber-shaped electronic devices are soft and deformable, and the structure of textile can be designed into various patterns with high integration degrees. By integration of fiber-shaped electronic devices into textiles, it shows unique and attractive application prospects in the field of wearable/portable electronic products.

During the weaving process, there are two main assembling strategies for device units. The first strategy is to make complete fiber-shaped electronic devices, such as sensor, solar cells, lithium ion batteries, and light-emitting devices, and then integrate them into textiles through weaving or embroidery technologies. The second strategy is to weave fiber-shaped cathode and anode with an interlaced structure to directly construct textile-type electronic devices. Comparing with the first strategy, the second strategy has been verified by successfully fabricating solar cell textiles, in which the warp and weft yarns serve as the anode and modified cathode, respectively. As a matter of fact, Fan et al. have already presented a cloth-like 24/7 private AI nursing system based on fully interwoven IC fabric, for which both devices (Figure 1.24a) assembling and circuit wiring (Figure 1.24b) were completely accomplished along polymer wires or at their cross-nodes during weaving, rather than on a PCB. Figure 1.24c and 1.24d shows a designed interwoven IC textile and the practical interwoven IC textile, respectively. Such a body-fitted cloth has demonstrated its capabilities of uninterrupted physiological monitoring, signal amplifying, logic computing, and wireless communication, as an independent mobile electronic system for routine monitoring and emergency assistance.

Starting with the lightweight, low-cost polymer fibers and taking advantage of industrial weaving technology, flexible wearable electronic intelligent textiles are an increasingly expanding field that has raising value for both academicians and industry, in areas such as energy harvesting and storage, environment monitoring,

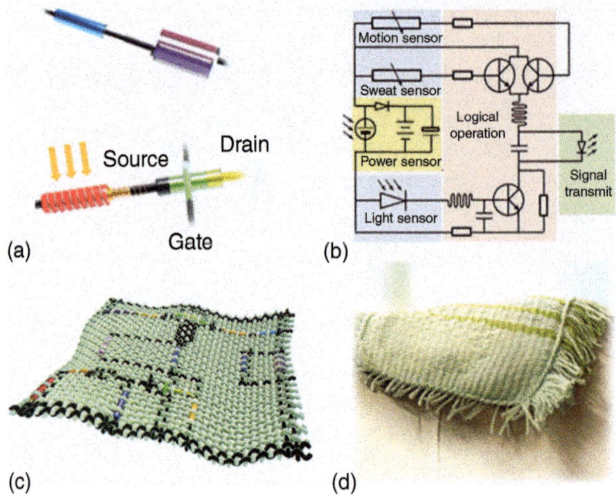

Figure 1.24 Integration of fiber-shaped electronic devices into textiles by interwoven IC. (a) Fiber-shaped field-effect transistor. (b) Design drawing of IC. (c) A designed interwoven IC textile. (d) The practical interwoven IC textile. Source: Based on Fan et al. [15].

healthcare systems, and entertainment products. Clothing is a feature of all human societies. Merging electronics and textiles is becoming increasingly important. Global industry giants have been exploring an efficient way of merging electronics with common cloth by pursuing an ultimate dream of "fabric computer" [19].

Although the integration of different functions of fibers has been widely demonstrated, the mismatch in material, structure, and fabrication requirement among the different kinds of devices still remains to be a critical barrier for practical applications. It is vital to develop efficient processing techniques to integrate different fiber electronic devices into textiles with a better and more comfortable human/textile interface design. Moreover, reliable connecting technique and IC are desired to achieve the higher integration of electronic textile. Apart from the problems mentioned above, safety issues are important for practical applications. Some fiber-shaped electronic devices such as batteries often require the use of flammable and toxic organic electrolytes, and they suffer from risks of fires and explosions induced by short circuit during deformation.

1.4 Conclusions

This chapter has discussed the definition, classification, characteristics, and applications of fibers, textiles, and intelligent textiles.

Fibers are closely related to every human life. Nature fibers such as silk and cotton were woven into textiles by human, which can keep human warm and comfortable. Because nature fibers are limited by natural conditions, they cannot meet with human needs, so people began to synthesize chemical fibers by processing methods, which appeared same as nature fibers but greatly enhanced human lives.

However, traditional textiles only have the basic functionalities, which can cover the ugliness and adorn the beauty, keep out the cold wind, and protect the body from insects. They cannot meet with human demands for healthcare, medicine, etc. Then, various functional fibers and textiles with special functions are constantly emerging, which are obtained through special processing of the original fibers or textiles.

Recently, textiles have faced a new challenge with the advancement of electronics and the Internet. Therefore, textiles are now expected to exhibit intelligent functionalities besides healthcare, protection, and medicine. Intelligent textiles are the product of the effective combination of high technology and traditional textile technology. They have broad market prospects and have great application potential in clothing, construction, military, etc., even though many new varieties are still under development.

In the near future, varieties of intelligent fibers and textiles will become more and more abundant, and the scope of application will become more and more extensive. The intelligent textiles can lead an important development direction and become a new economic growth point of the textile industry.

References

1 Yu, W. (2006). *Textile Materials Science.* Beijing: China Textile Press.

2 Mwaikambo, L.Y. and Ansell, M.P. (2002). Chemical modification of hemp, sisal, jute, and kapok fibers by alkalization. *Journal of Applied Polymer Science* 84 (12): 2222–2234.

3 Long, L., Lei, S., and Guiqin, J. (2008). Properties of wool fibers treated under high-temperature water and steam. *Journal of Applied Polymer Science* 109 (2): 852–858.

4 Fink, T.D. and Zha, R.H. (2018). Silk and silk-like supramolecular materials. *Macromolecular Rapid Communications* 39 (17): e1700834.

5 Chi, F. and Chen, H. (2010). Fabrication and characterization of zein/viscose textile fibers. *Journal of Applied Polymer Science* 118 (6): 3364–3370.

6 Zhu, P. (2016). *Functional Fiber and Functional Textile.* Beijing: China Textile Press.

7 Jung, D. and Debes, B. (2018). Keratinous fiber based intumescent flame retardant with controllable functional compound loading. *ACS Sustainable Chemical Engineering* 6 (10): 13177–13184.

8 Song, H., Yu, H., Zhu, L. et al. (2017). Durable hydrophilic surface modification for PTFE hollow fiber membranes. *Reactive and Functional Polymers* 114: 110–117.

9 Gu, S., Zhu, L., Mercier, C., and Li, Y. (2017). Glass-fiber networks as an orbit for ions: fabrication of excellent antistatic PP/GF composites with extremely low organic salt loadings. *ACS Applied Materials & Interfaces* 9 (21): 18305–18313.

10 Vladimirtseva, E.L., Smirnova, S.V., Odintsova, O.I., and Vinokurov, M.V. (2016). Flame-retardant finishing of different textiles. *Russian Journal of General Chemistry* 86 (2): 460–469.

11 Ismar, E., Kurşun Bahadir, S., Kalaoglu, F., and Koncar, V. (2020). Futuristic clothes: electronic textiles and wearable technologies. *Global Challenges* 4: 1900092.

12 Biswas, M.C., Chakraborty, S., Bhattacharjee, A., and Mohammed, Z. (2021). 4D printing of shape memory materials for textiles: mechanism, mathematical modeling, and challenges. *Advanced Functional Materials*, 2100257.

13 Han, M.W., Kim, M.S., and Ahn, S.H. (2020). Shape memory textile composites with multi-mode actuations for soft morphing skins. *Composites Part B: Engineering* 198: 108170.

14 Kelly, F.M. and Cochrane, C. (2015). Color-changing textiles and electrochromism. In: *Handbook of Smart Textiles* (ed. X. Tao), 859–889. Singapore: Springer.

15 Fan, H., Li, K., Liu, X. et al. (2020). Continuously-processed, long electrochromic fibers with multi-environmental stability. *ACS Applied Materials & Interfaces* 12: 28451–28460.

16 Huang, F., Yu, H., Xiang, S. et al. (2019). Embroidering a filmsy photorechargeable energy fabric with wide weather adaptability. *ACS Applied Materials & Interfaces* 12: 3654–3660.

17 Zhang, N., Huang, F., Zhao, S. et al. (2020). Photo-rechargeable fabrics as sustainable and robust power sources for wearable bioelectronics. *Matter* 2: 1260–1269.

18 Wang, L., Fu, X., He, J. et al. (2019). Application challenges in fiber and textile electronics. *Advanced Materials* 32 (5): 1901971.

19 Chen, J., Huang, Y., Zhang, N. et al. (2016). Micro-cable structured textile for simultaneously harvesting solar and mechanical energy. *Nature Energy* 1 (10): 16138.

2

A Brief Introduction to Typical Film Deposition Technologies

More than a thousand years ago, people began to make color glaze on the surface of ceramic ware, which is the first example of the preparation and application of precious metal film. Nowadays, the thin film technique has been involved in many disciplines and have a wide range of applications, especially the electronic equipment industry.

In the electronic equipment industry, wearable flexible devices have attracted much attention due to their advantages of low cost, lightweight, flexibility, and so on. The flexible device structures include sheet, network, fiber, etc. Methods for making flexible devices into wearable fabrics include laminated and woven fabrics. Therefore, the requirements for the functional film on the device are very high, especially the fiber structure device with large curvature and high flexibility.

From the point of the film preparation process on the fiber device, it is different from the typical film-forming technology: the diameter of the fiber substrate can reach a few tenths of a millimeter, which is much smaller than the typical film substrate, the curvature of the fiber substrate is also much higher than that of the typical film substrate, and the continuous production of fiber substrates is also much more difficult than typical film substrates.

However, the typical thin film deposition technology has the characteristics of uniform film layer, high purity, and good step coverage, which also can be well applied in wearable flexible devices. What needs to be solved is how to deposit thin films on flexible substrates. Because of these production process requirements, typical thin film technology has also developed a lot.

To meet with different scenarios and application requirements, there are many methods for coating. These are divided into three categories, namely, dry-process film deposition, wet-process film deposition, and patterning technique, which can be applied to different purposes and applications. These advances will also be more conducive to coating functional layers on fiber substrates. In addition, after forming a film on a fibrous device, the characterization of the film performance is also extremely important. In this section, the ways to monitor the morphology and characterization methods of thin films are discussed.

Textile-Based Energy Harvesting and Storage Devices for Wearable Electronics, First Edition.
Xing Fan, Nannan Zhang, and Yi Wang.

2.1 Dry-Process Film Deposition Technologies

Dry-process film deposition is mainly a vapor deposition method under vacuum. The vacuum refers to a gas state that is at a lower than one atmosphere pressure. To obtain the vacuum state, it can be obtained by means such as mechanical pumps and diffusion pumps. At the meanwhile, dry-process film deposition can be divided into physical vapor deposition (PVD) method and chemical vapor deposition (CVD) method.

Among them, PVD is the use of evaporation or sputtering means to make solid materials into vapor and then condense and deposit on the surface of the silicon wafer. It is a purely physical process, does not include chemical reactions, and is also the most important way to prepare metal films. The advantages of PVD are simple equipment, easy operation, high film purity, and fast film-forming rate. However, the disadvantages are small adhesion between the film and the substrate and poor step coverage.

CVD refers to a process in which gaseous substances are chemically reacted on a solid surface and deposited on the surface by using heat energy, plasma discharge, ultraviolet irradiation, etc., individually or comprehensively, to form a stable solid film. It has the following characteristics: the reactants and by-products are gases, the film formation speed is fast, the composition is precisely controllable, the film structure is complete and dense, the adhesion to the substrate is good, and a smooth deposition surface can be obtained, as well as high purity and complete crystallization film. Its disadvantage is particle contamination. Sometimes, some applications are not satisfied with the simple form of the film, so it need to process the film purposefully to form a specific pattern.

Micromachining the surface layer (thin film) of the material, that is, patterning, requires the use of surface etching technology. Corresponding to the CVD and PVD technologies for preparing thin film materials, there are chemical and physical surface etching methods. In a sense, the etching process is the reverse process of the thin film deposition process.

The following section will introduce the specific method content and application of PVD and CVD and patterning of thin films.

2.1.1 Physical Vapor Deposition for Film Deposition

The process of PVD is divided into the following three steps. First, the film-forming material is vaporized. Second, the film-forming atoms, molecules, or ions are transferred from the evaporation source to the substrate. Third, the film-forming atoms adsorb, accumulate, and grow to form a film on the surface of the substrate. The main methods of PVD are vacuum evaporation, sputter coating, molecular beam epitaxy (MBE), and pulse laser deposition (PLD).

The first PVD method is vacuum evaporation. The vacuum evaporation coating method uses an evaporator to heat the evaporation material, so that the film particles evaporate from the evaporation source in atomic form, fly to the workpiece in a high vacuum, and pass through the surface of the workpiece. Figure 2.1a is a

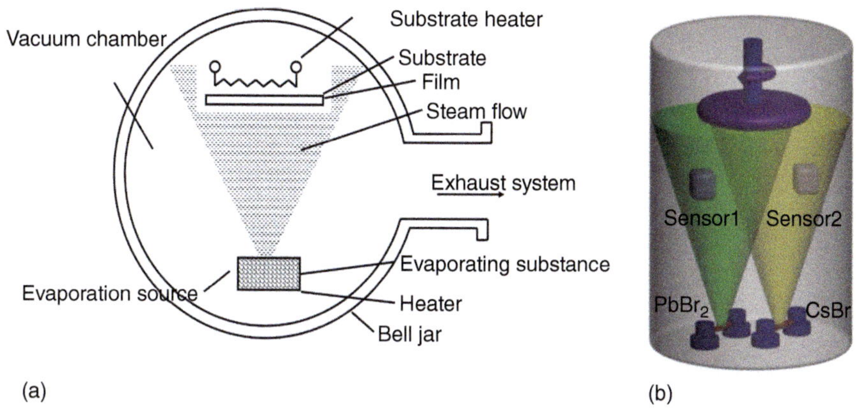

Figure 2.1 Device schematic and an application example of vacuum evaporation. (a) Schematic diagram of the vacuum evaporation. (b) Schematic diagram of the vacuum chamber for the dual-source vacuum coevaporation method. Source: (a) Based on Chen et al. [1], (b) Chen et al. [1]. © 2018, John Wiley & Sons.

schematic diagram of the vacuum evaporation. The advantages of this method are the following: the film particles reach the workpiece directly, a high-purity film can be prepared, and the structure and performance of the film according to requirements can be controlled. Its disadvantages are the following: the compound will decompose by bombardment, the structure is complicated, the equipment is expensive, and the ionization of gas molecules affects the quality of the film. This method is widely used in industrial fields such as optics, electronics, light industry, and decoration. For example, Li and co-workers [1] prepared the inorganic perovskite active layer by vacuum evaporation, so that the photoelectric conversion efficiency of the laminated battery reaches 14.03%, which is also the highest efficiency of the inorganic perovskite/organic laminated battery (Figure 2.1b).

The second method of PVD is sputter coating. The principle of this method is to let particles with enough energy bombard the surface of the solid target in the vacuum chamber, sputter the atoms in the target, and deposit them on the workpiece to form a film. Compared with vacuum evaporation, sputter coating has the following characteristics: it can sputter any substance; the adhesion between the sputtered film and the substrate is better; sputter coating has higher density, less pinholes, less pollution, and higher film purity; and it can deposit a film of uniform thickness over a large area. Disadvantages of sputter coating are the following: complex sputtering equipment, low deposition rate of sputter deposition, and high substrate temperature. This method has been widely used in high-tech fields such as display components, electronic components, large-scale integrated circuits, optical components, and magnetic components. For example, Barsoum and co-workers [2] made a conducting and hydrophilic Ti_3C_2 films by sputtering, which can be applied in electronic, photonic, and sensing applications (Figure 2.2a). Chen et al. [3] prepared field-effect transistor by sputtering gold nanoparticles onto the surface of an object, which showed both high sensitivity and selectivity toward human immunoglobulin G (Figure 2.2b).

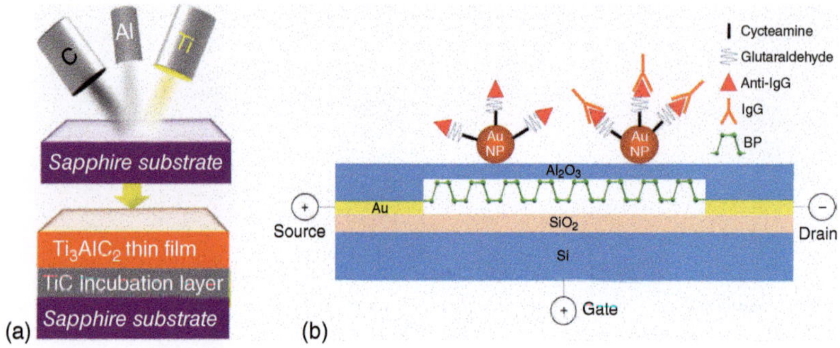

Figure 2.2 Application examples of sputter coating. (a) Magnetron sputtering of Ti, Al, and C forming a few-nanometer TiC incubation layer on a (0001) sapphire substrate, followed by the deposition of Ti_3AlC_2. Source: Halim et al. [2]. Licensed under CC BY 4.0. (b) Schematic of the BPFET sensor. Source: Chen et al. [3]. © 2017, Elsevier.

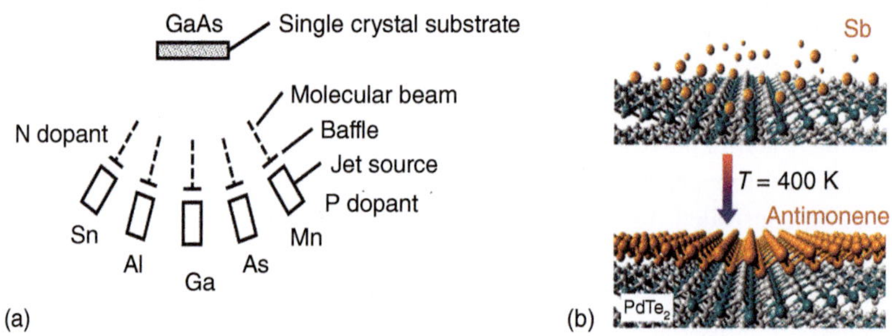

Figure 2.3 Device schematic and an application example of molecular beam epitaxy. (a) Schematic diagram of molecular beam epitaxy. (b) Preparation of monolayer antimony on $PdTe_2$ by MBE. Source: Xu et al. [4]. © 2017, WILEY-VCH Verlag GmbH & Co. KGaA, Weinheim.

The third method of PVD is MBE. It is a technology of spraying molecular beams of thin film component elements directly onto a substrate surface under an ultra-vacuum condition, thereby forming an epitaxial layer thereon. The schematic diagram of the MBE technology is shown in Figure 2.3a. Compared with other methods such as vacuum evaporation, MBE has the following advantages: it can form a monolayer with controllable thickness, it can control accurately the growth process and growth rate, it can control the composition and doping concentration of the thin film, it can reduce self-doping diffusion effect, and it is particularly suitable for growing superlattice materials because of low growth rate. Its disadvantage is harsh growth conditions. It has very high requirements for the flatness of the surface of the substrate loaded into the system before growth and needs to remove organic impurities and certain metal impurities on the surface and at the same time remove the damaged layer of the wafer surface and surface polishing. MBE has been widely used in the research of solid-state microwave devices, optoelectronic devices, multilayer periodic structure devices, and monolayer films. For example, Xu et al. [4] successfully fabricated devices with the singular characteristics of next-generation information nanodevices by MBE (Figure 2.3b).

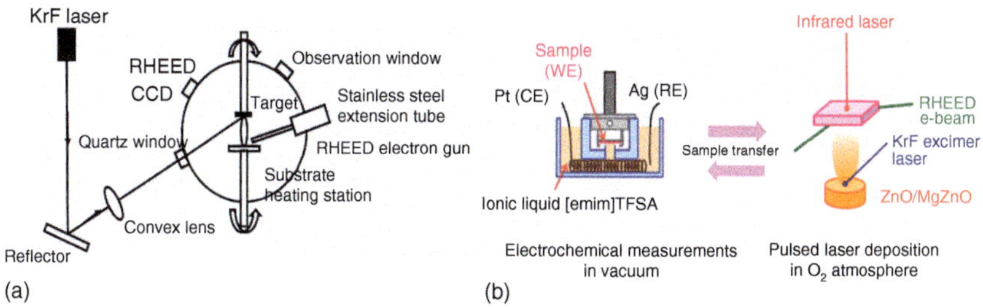

Figure 2.4 Device schematic and an application example of pulse laser deposition. (a) Schematic diagram of the basic structure of a PLD device. (b) Schematic of an in situ pulsed laser deposition-electrochemical cell (PLD-EC) system. Source: (a) Based on Kanai et al. [5], (b) Kanai et al. [5]. © 2019, Royal Society of Chemistry.

The fourth method of PVD is PLD. It focuses the high-power pulsed laser generated by the pulsed laser on the surface of the target material, causing the surface to generate high temperature and ablation and further generate high-temperature and high-pressure plasma. In addition, the plasma is oriented and locally expanded to deposit a film on the substrate. Figure 2.4a is a schematic diagram of the basic structure of a PLD device. Compared with other continuous PVD methods, one of the most important advantages of PLD is vaporization-assisted deposition of some things that are not easy to vaporize. Another significant advantage of PLD is that the instantaneous deposition rate can reach 10^4 nm/s. At the same time, there are other advantages, such as extremely high temperatures due to the laser. PLD also has its own disadvantages, such as the easy formation of small particles on the film surface and the uneven thickness of the film, which limits its application to a certain extent. At present, PLD method can be used to deposit diamond-like carbon films, high-temperature superconducting films, various nitride films, complex multicomponent oxide films, ferroelectric films, nonlinear waveguide films, and synthetic nanocrystalline quantum dot films. For example, Matsumoto and co-workers [5] prepared epitaxial ZnO and MgZnO/IL heterointerfaces by depositing ZnO or MgZnO by PLD, which can be used in the electric double-layer capacitance (Figure 2.4b).

2.1.2 Chemical Vapor Deposition for Film Deposition

The process of CVD is divided into the following steps: reactive gas transportation, gas-phase chemical reaction, initial product diffusion, initial product adsorption and diffusion, heterogeneous reaction with the matrix, and by-product dissociation and desorption. The main methods of CVD are atmospheric-pressure CVD (APCVD), low-pressure CVD (LPCVD), low-temperature CVD, high-temperature CVD, plasma-enhanced CVD (PECVD), metal–organic CVD (MOCVD), and atomic layer deposition (ALD).

The first CVD method can be divided into APCVD and LPCVD by pressure. On the one hand, APCVD is a system at near atmospheric pressure. The steps of APCVD are the following: the reaction gas is transported to the reactor for gasification, the

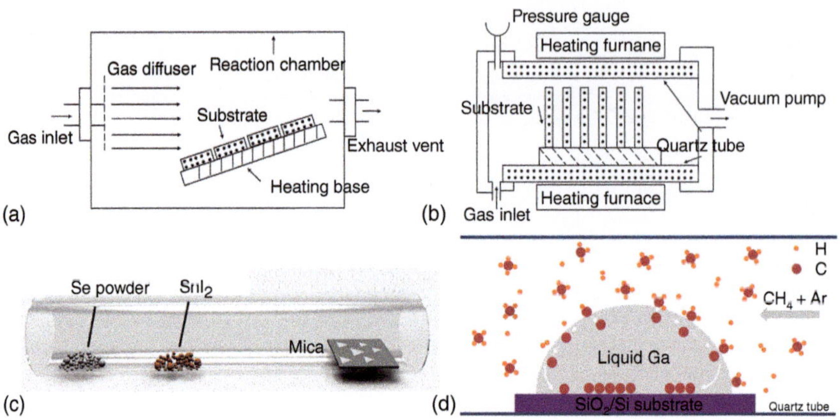

Figure 2.5 Device schematics and application examples of atmospheric-pressure CVD and low-pressure CVD. (a) Atmospheric-pressure CVD apparatus diagram. (b) Low-pressure CVD apparatus diagram. Source: (a, b) Based on Zhou [6]. (c) Schematic diagram of the chemical vapor deposition method. Source: (c) Zhou et al. [6]. © 2015, John Wiley & Sons. (d) Schematic diagram of the synthesis of high-quality and single-layer graphene film by low-pressure CVD. Source: Zhang et al. [7]. © 2020, Springer Nature.

reaction gas material undergoes a chemical reaction, the product is adsorbed on the substrate and the film is formed, and the reaction gas is recovered. The APCVD system unit is shown in Figure 2.5a. The advantages of APCVD are high deposition rate and good film uniformity. However, the disadvantage of APCVD is that it requires a faster steam flow. Otherwise, gas-phase chemical reactions are prone to occur. In addition, dust will get stuck on the walls of the deposition chamber, and the deposition chamber must be cleaned frequently. On the other hand, LPCVD is a system that increases the diffusion of reactive gases to obtain higher gas mass transmission under reduced pressure conditions and no longer affects the CVD reaction speed. Strict temperature control can deposit a uniform film on the surface of silicon wafers. The typical device structure is shown in Figure 2.5b. Compared with APCVD, the main advantage of LPCVD lies in the intentional film uniformity and better step coverage. However its disadvantage includes the lower deposition rate and the frequent use of toxic, corrosive, and flammable gases. For example, Zhai and co-workers [6] used APCVD to synthesize high-quality ultrathin single-crystalline $SnSe_2$ flakes for photodetector (Figure 2.5c). Chen and co-workers [7] realized the synthesis of high-quality and single-layer graphene film by LPCVD, which proved that it is conductive to form a controllable channel pattern based on the current silicon-based process (Figure 2.5d).

The second CVD method can be divided into low-temperature CVD (200–500 °C) and high-temperature CVD (1000–1300 °C). On the one hand, when attention is paid to the low-temperature preparation conditions of thin film materials, low-temperature CVD systems are often used. The synthesis steps of low-temperature CVD are the following: the cleaned substrate is put into the deposition chamber and connected the power supply; when the temperature of the workpiece rises to the specified deposition temperature, the reactant gas undergoes a chemical

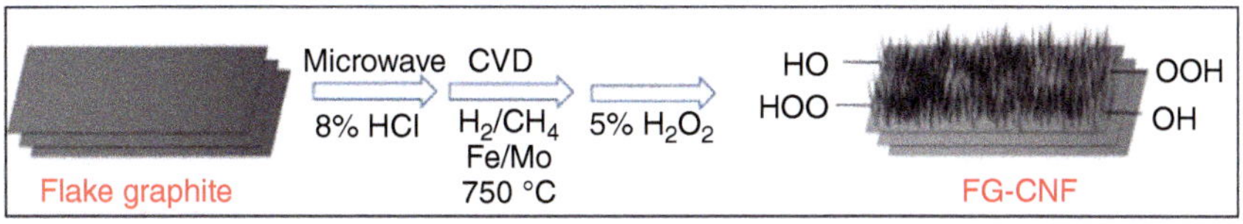

Figure 2.6 Scheme of preparation route of supporting material. Source: Li et al. [8]. © 2019, Elsevier.

reaction, and then the surface of the object is bombarded and deposited by a direct current pulse power supply. In the semiconductor industry, low-temperature CVD devices are widely used in various types of insulating dielectric films. On the other hand, if the integrity of the material is emphasized in thin film deposition, a high-temperature CVD system is often used. This type of device is characterized by using an external heater to heat the reaction chamber to a higher temperature. High-temperature CVD systems are widely used in the preparation of semiconductor epitaxial films to ensure the preparation quality of materials. In addition to the epitaxial growth of semiconductor materials, high-temperature CVD devices are also widely used in the preparation of wear-resistant coatings on metal parts. For example, Long and co-workers [8] used high-temperature CVD to prepare mineral-coupled supporting material to further enhance thermal efficiency of composites (Figure 2.6).

The third CVD method is PECVD. It uses microwave or radio frequency to ionize the gas containing the film constituent atoms to form a plasma locally. The plasma has a strong chemical activity and is easy to react to deposit the desired film on the substrate. The schematic diagram is shown in Figure 2.7a. The main advantage of PECVD is that it has a lower deposition temperature. However, its disadvantages are low yield, easy to be polluted by particles, and the film often containing a lot of hydrogen atoms. For example, Ardaheh and co-workers [9] synthesized high-quality diamond nanoparticles by PECVD method on steal 304 substrates and

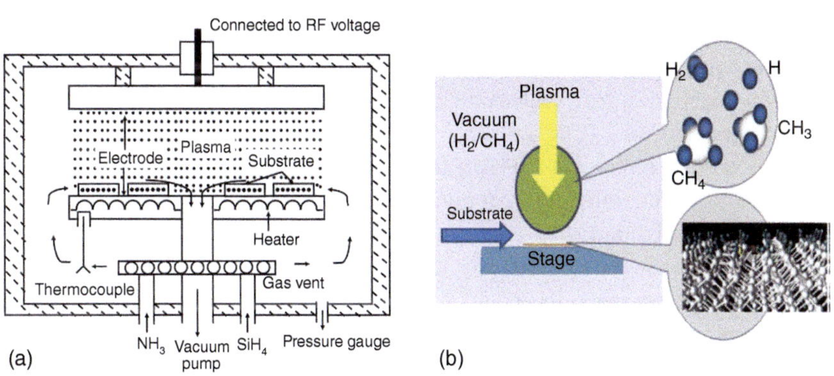

Figure 2.7 Device schematic and an application example of plasma enhanced CVD. (a) Schematic diagram of plasma-enhanced CVD apparatus (capacitive coupling type). (b) Deposition device diagram of high-quality diamond nanoparticles. Source: Asgary et al. [9]. © 2019, Springer Nature.

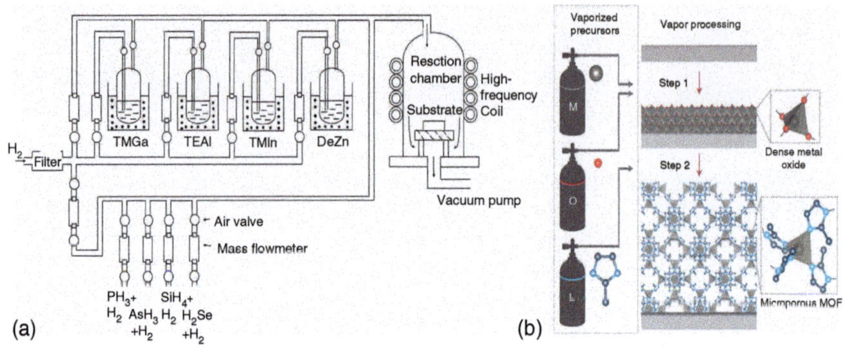

Figure 2.8 Device schematic and an application example of metal organic CVD. (a) Schematic diagram of metal organic CVD thin film growth device. (b) Chemical vapour deposition of ZIF-8 thin films. Source: Stassen et al. [10]. © 2016, Springer Nature.

revealed changes in its physical properties by studying deposition under different conditions (Figure 2.7b).

The fourth CVD method is MOCVD. It uses organic compounds of group III and group II elements and hydrides of group V and group VI elements as crystal growth source materials and then conducts vapor-phase epitaxy on the substrate by means of thermal decomposition reaction to grow various III–V groups and II–VI group compound semiconductors and their multiple solid solution thin-layer single-crystal materials. Figure 2.8a shows a schematic diagram of MOCVD. This method has the following characteristics: the device has a simple structure and is easy to realize mass production design; the growth rate of the film is easy to control, and the thickness of the film can be precisely controlled to the atomic level; the composition, conductivity type, carrier concentration, and the like of the epitaxial film layer can be controlled; it can be plated with a large area of uniform film; and it is easy to prepare a multilayer film and a nonuniform film by performing multiple depositions by controlling the deposition components. The disadvantages are that the level of residual impurities is still high and the flammable and explosive and highly toxic gas components in the reaction gas are relatively high. Thin films prepared by MOCVD have been used in mixing diodes, Hall sensors, field-effect transistor photocathodes, and LED light-emitting elements. For example, Ameloot and co-workers [10] demonstrate a CVD approach for producing ZIF-8, zinc-(2-methylimidazolate)$_2$, and this method greatly facilitate MOF integration in microelectronics (Figure 2.8b).

The fifth CVD method is ALD. It refers to a method of forming a thin film by alternately pulsed gas-phase precursors into the reaction chamber, and a gas/solid-phase chemical adsorption reaction occurs on the surface of the deposition substrate [11]. The advantages of ALD are the successive deposition of monoatomic layers, extremely uniform thickness, and excellent consistency. Its disadvantage is the slow deposition rate. ALD thin film technology has many applications in transistor material preparation, energy storage devices, catalysis, and so on. For example, Chen et al. [3] used ALD for surface passivation as a biosensor (Figure 2.9a). Fang et al. [12] used ALD deposition of Al_2O_3 on the surface of the substrate on which

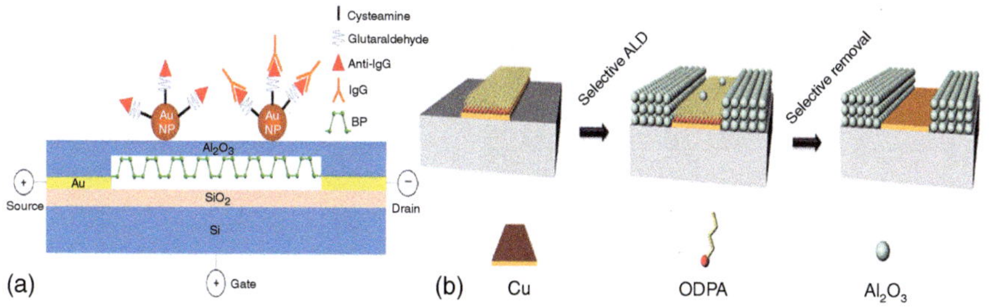

Figure 2.9 Application examples of atomic layer deposition. (a) Schematic of the BPFET sensor. Source: Chen et al. [3]. © 2017, Elsevier. (b) Illustrative schematic demonstrating the hybrid AS-ALD process, which combines the selective deposition and selective removal steps. Source: Fang and Ho [12]. © 2015, American Chemical Society.

octadecylphosphonic acid (ODPA) molecules were previously adsorbed effectively prevented Al_2O_3 from being deposited on the Cu surface, thereby obtaining high insulation (Figure 2.9b).

In conclusion, the essential difference between PVD and CVD is that PVD has no chemical reaction and is purely a physical process, while CVD includes five basic chemical reaction processes (pyrolysis, photolysis, reduction, oxidation, and redox). The coating prepared by CVD has the advantages of high crystallinity, strong film-based bonding force, and good winding properties. However, due to the high deposition temperature, high gas cost, large film stress, and environmental pollution, the CVD technology seriously restricts its application and development. The PVD technology uses a physical method to first convert a solid substance into a gas phase by "evaporation or sputtering" and then deposit it on the surface of the substrate to form a new solid material coating. Compared with CVD, PVD technology has the following advantages: first, the coating structure is dense, uniform, and high in quality, and second, it can deposit various kinds of coatings such as nitride, oxide, and carbide. Furthermore, the target has wide applicability. The materials that can be used to make thin films in the PVD method are all solid, halide, and thermally stable compounds, while the materials that can be used to make thin films in the CVD method are metals, carbides, nitrides, borides, oxides, metals other than alkali and alkaline earth sulfides, selenides, tellurides, metal compounds, and alloys. From the specific application point of view, PVD can be used for decoration, electronic materials, optics, and other purposes, while CVD can be used for material refining, decoration, surface protection, electronic materials, and other purposes.

2.1.3 Morphology and Pattern Design

Film preparation has important applications as a functional layer in flexible electronics. However, the development of thin film materials still faces a key problem: the functionalization of the thin film surface. Thin film patterning refers to the generation of regular surface structures on the order of nanometers or micrometers in

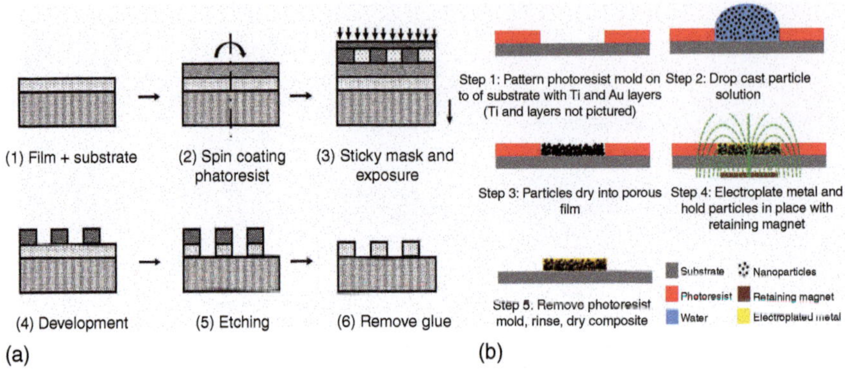

Figure 2.10 Device schematic and an application example of photolithography. (a) Schematic diagram of general photolithography process. (b) Schematic process of nickel/iron-oxide nanocomposites. Source: Smith et al. [13]. © 2020, Elsevier.

at least one dimension. With the development of modern science and technology, the microprocessing or patterning of micro- and nanoscale surfaces has attracted widespread attention. At present, there are many methods to control the microstructure and properties on the solid surface, including photolithography, wet etching, dry etching, and ion etching.

The first method of patterning is photolithography, which is a major process in the production of planar transistors and integrated circuits. It requires the use of photosensitive materials called photoresists. The general process of photolithography is shown in Figure 2.10a, which can be divided into six steps: coating on the substrate and processing it into the desired shape, dropping the photoresist onto the substrate through the nozzle of the spin coater, a reticle adhered on the surface of the photoresist film and exposed, development, etching, and the result of thin film etching obtained by removing the photoresist. The advantage of photolithography is that the etching process has a strong selectivity. Its disadvantage is that the erosion of SiO_2 is an isotropic process, which will form a "champagne cup"-shaped profile. This method has been widely used in the fields of microelectronics, binary optics, photonic crystals, integrated optics, and nanotechnology. For example, Arnold and co-workers [13] combined nickel/iron-oxide nanocomposites with photolithography for potential applications in power electronics (Figure 2.10b).

The second method of patterning is wet etching, which is a technique in which an etching material is immersed in an etching solution. Because the reaction products usually appear in the etching solvent, wet etching consists of three steps: the etchant moves to the surface of the silicon wafer, chemical reactions with the surface film generate soluble by-products, and reaction products are removed from the surface of the silicon wafer. Wet etching is a pure chemical reaction process with excellent selectivity. However, it has serious disadvantages: lack of anisotropy, poor process control, and excessive particle contamination. For example, Lou and co-workers [14] first developed a method for assembling two-dimensional (2D) nanosheets into layered three-dimensional (3D) nanostructures after the continuous etching of the template and the phosphating process (Figure 2.11).

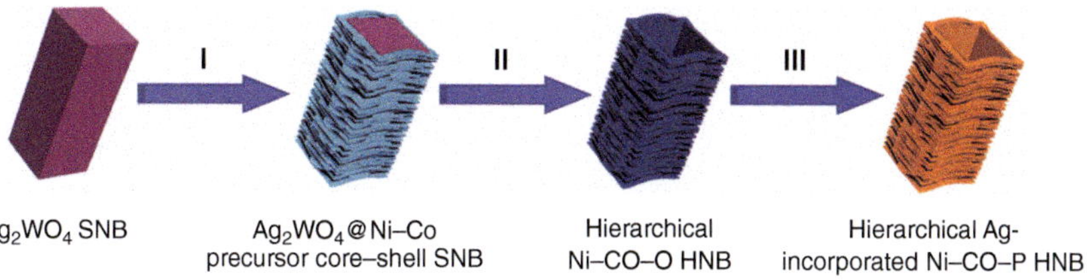

Figure 2.11 Schematic illustration of the construction of hierarchical Ni–Co–P hollow nanobricks. Source: Hu et al. [14]. © 2018, Royal Society of Chemistry.

The third method of patterning is dry etching, which is a technique for thin film etching using plasma. It can be divided into two methods: physical etching and chemical etching. The steps of physical etching are following: first, it uses glow discharge to ionize gas (such as argon) into positively charged ions; second, it uses bias to accelerate the ions and splash them on the surface of the etched object; and third, it knocks out the atoms of the etched object. The process is completely physical energy transfer. Its characteristic is that it has very good directivity and can obtain a nearly vertical etch profile. However, because the ions are sputtered on the chip uniformly, all the photoresist and the etched material are etched at the same time, resulting in low etching selectivity. The steps of chemical etching are following: first, it uses plasma to ionize the etching gas and form charged ions, molecules, and highly reactive atomic groups; second, they diffuse to the surface of the etched film and interact with the etched film; third, the surface atoms react to produce volatile reaction products, which are drawn out of the reaction chamber by vacuum equipment. Compared with wet etching, the advantages of dry etching are good side wall profile control, low chemical use cost, and good in-chip, inter-chip, and batch-to-chip etching uniformities. Its disadvantages are the following: poor choice of etching of underlying materials, plasma-induced damage, and expensive equipment. For example, Kajiya [15] demonstrated a facile means of removing transparent conducting oxides from a glass substrate in selected areas and provides a useful means to easily fabricate patterned substrates for use in electronics and related technologies (Figure 2.12).

The fourth method of patterning is ion etching, which can be divided into plasma etching and ion beam etching. Plasma etching uses a low-temperature plasma chemical reaction. The schematic diagram of the device is shown in Figure 2.13a. Under the action of a radio frequency electric field, the introduced gas is ionized to generate a plasma with a small ionization degree. When the substrate is placed in the plasma, the atoms on the film surface react with the active atoms and molecules in the plasma, thereby vaporizing the exposed part of the film. The principle of ion beam etching is that the sample (film) to be etched is placed on the target, and the ion beam directly bombards the surface of the sample for etching. A great advantage of ion beam etching is that the sample is placed in a high vacuum chamber, and the process can be free from any pollution, ensuring a high clean environment. Its disadvantage is that the sputtering speed is low, and the ion beam source equipment is more complicated. Ion etching has many applications in various aspects in recent years. For example, Wang and co-workers [16] etched Co_3O_4 with Ar plasma

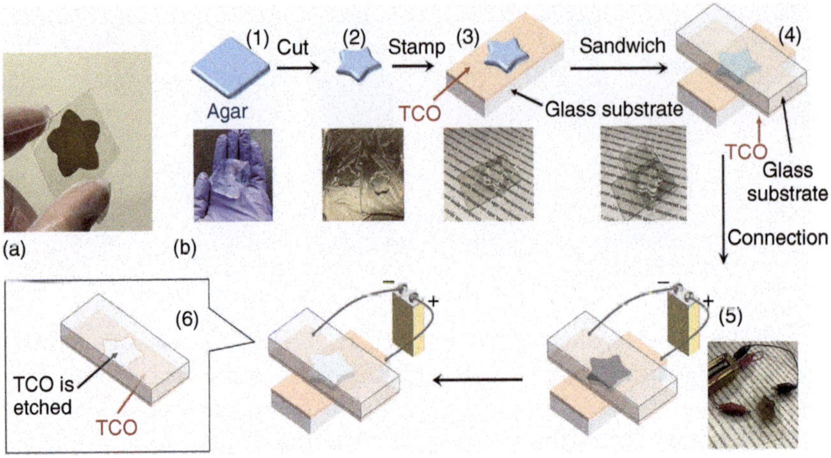

Figure 2.12 An application example of dry etching. (a) Photographs and (b) schematic diagrams of the agar sandwich procedure. Source: (a, b) Kajiya [15]. © 2019, Springer Nature.

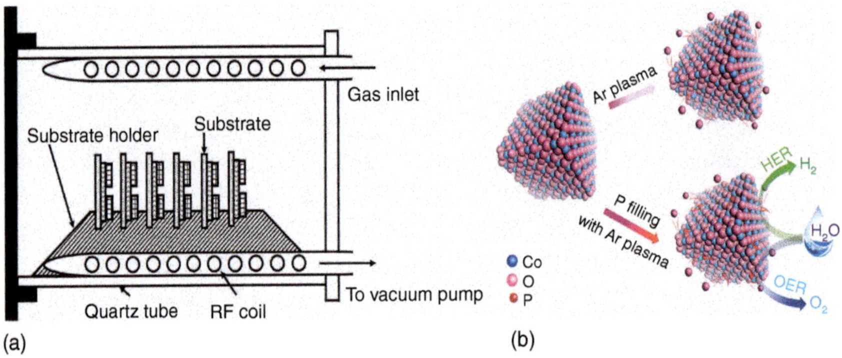

Figure 2.13 Device schematic and an application example of ion etching. (a) Schematic diagram of plasma etching device. (b) Schematic diagram of P-Co_3O_4 and VO-Co_3O_4 prepared by plasma etching. Source: (a) Based on Xiao et al. [16]. (b) Xiao et al. [16]. © 2017, Royal Society of Chemistry.

containing phosphorus precursor and successfully filled Co_3O_4 with phosphorus to obtain P-Co_3O_4 (Figure 2.13b).

2.2 Typical Wet-Process Technologies for Roll-to-Roll Device Fabrication

Wet-process deposition technology is a technique for depositing a film on the substrate surface by a chemical reaction or an electrochemical reaction. The advantages of wet-process technologies are the following: it does not require a vacuum environment, the equipment is simple, it can be formed on various substrate surfaces, and its application involves electronic components, surface coating, and decoration. The disadvantages of wet-process technologies are that there are many factors that can influence film's quality.

Wet-process deposition technology can be divided into chemical reaction coating and electrochemical reaction coating. Chemical reaction coating refers to the use of chemical reactions to deposit films on the substrate. Its advantages are as follows: the thickness of the plating layer is uniform, the material to be plated can be unlimited, the power supply is not required and the equipment is simple, and the plating solution can generally be regenerated. Electrochemical reaction coating refers to the deposition of a thin film on the surface of a metal or conductive substrate under the action of an electric field. Its advantages are the following: the thickness of the film is controllable, and the film is smooth and meticulous. But it has many influencing factors, such as current density and electrolyte concentration.

In addition, film deposition methods include spray pyrolysis (SP) and Langmuir–Blodgett (LB) technique. SP can be described as the preparation of various metal salts into precursor solutions, which are atomized by an atomizer and carried by a carrier gas into a high-temperature reaction furnace. In the furnace, a series of physical and chemical processes such as solvent evaporation, solute precipitation to form solid particles, particle drying, particle thermal decomposition, and sintering molding are completed instantly, and finally ultrafine powder is formed. The SP requires high-temperature and high-vacuum conditions and requires high equipment and operation, but it is easy to produce powder with small particle size and good dispersibility. LB film forms a monolayer of substances on the liquid surface that are incompatible with the liquid phase. When the substrate is immersed in or out of the liquid phase, the monolayer can be transferred from the liquid phase to the solid surface to form a thin film. Although the combination of the film and the substrate relies on the intermolecular force and the mechanical properties are poor, the thickness of the film can reach nanometer level and is controllable.

The following section will introduce the specific method content and application of wet-process technologies.

2.2.1 Chemical Reaction Coating for Thin Film Preparation

Chemical reaction coating is a processing method in which a chemical reducing agent is added to the plating solution containing a metal salt solution, and the metal ions in the plating solution are reduced and deposited on the surface of the processed part. The main methods of chemical reaction coating are electroless plating, dip plating, sol–gel method, and self-assembled monolayer (SAM) film.

The first method of chemical reaction coating is electroless plating. This is a method of obtaining a film layer by using a reducing agent to reduce metal ions in a metal salt to an atomic state and depositing it on the surface of a substrate. Metals such as nickel, iron, copper, etc. have a catalytic effect on the reduction reaction, and the reaction will not stop until the film thickness reaches the required thickness and the plating is removed. This type of electroless plating that relies on the catalysis of the plated metal itself is called autocatalytic electroless plating.

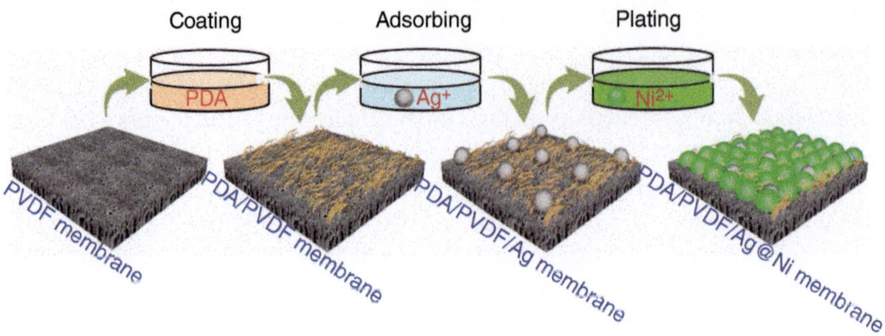

Figure 2.14 Diagram of electroless Ni plating of the polyvinylidene fluoride (PVDF). Source: Zhang et al. [17]. © 2019, Elsevier.

Autocatalytic chemical plating has the following characteristics: it can form a uniform coating on the surface of complex plating; the porosity of the film layer is low; it can be directly coated on non-conductors such as plastic, ceramics, and glass; and it does not need power supply and electrodes. There are many advantages of electroless plating but also disadvantages. For example, the complexing agent added during electroless nickel plating will have a greater impact on wastewater. Although network breaking can also be solved, this operation will increase the cost. Electroless plating technology has been widely used in the automotive industry, mechatronics, textile, printing, aerospace, and military industries. For example, Wu and co-workers [17] prepared electroless Ni plating of the polyvinylidene fluoride (PVDF) by electroless plating for dyeing wastewater treatment (Figure 2.14).

The second method of chemical reaction coating is dip plating. The principle of this method is to deposit the metal on the substrate by chemical substitution in the solution of the metal salt to be plated, without the need for an external power source. Under these conditions, the deposited metal forms a coating on the surface of the base metal. The advantages of immersion plating are the following: simple process, low investment, low cost, and good performance. For example, Tang and co-workers [18] prepared CdS QD-functionalized mesoporous TiO_2 (CdS QDs/TiO_2) on the basis of the displacement reaction method for the determination of human serum specimens and newborn calf serum-diluted PSA standards (Figure 2.15a). Rezaei and co-workers [19] fabricated a bending artificial muscle by chemical synthesis of polypyrrole (PPy) onto nanofibrous polyurethane (PU) structures for conductive polymer-based artificial muscles (Figure 2.15b).

The third method of chemical reaction coating is sol–gel method. The principle of this method is to add a certain concentration of metal alkoxide to an organic solvent, then add a catalyst and water to make a sol, and finally hydrolyze and polymerize at a certain temperature to form a gel. The substrate to be plated is placed in the gel. After soaking, the gel is coated on the surface of the substrate. After extraction, it can be heated and baked to form a film on the substrate. The thin film prepared by the sol–gel method has the advantages of uniform mixing of multiple components, easy control of components, uniform film formation, thinner films on a larger area, low cost, short cycle, and easy industrial production. It also has some disadvantages: the

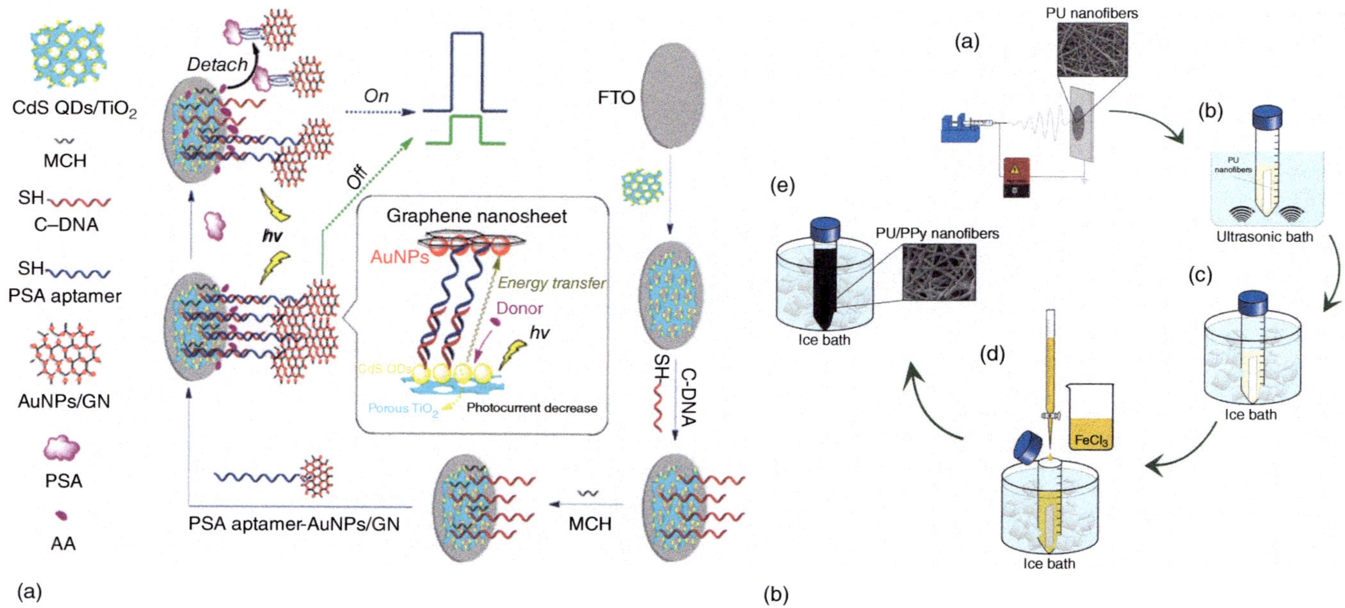

Figure 2.15 Application examples of dip plating. (a) CdS QDs/TiO_2, CdS quantum dot-functionalized mesoporous TiO_2; MCH, 6-mercapto-1-hexanol; C-DNA, capture DNA; AA, ascorbic acid; and FTO, fluorine-doped tin oxide. Source: Cai et al. [18]. © 2018, American Chemical Society. (b) Schematic representation of fabricating PU/PPy-TFSI nanofibers. Source: Adapted with permission from Ebadi et al. [19]. © 2019, Elsevier.

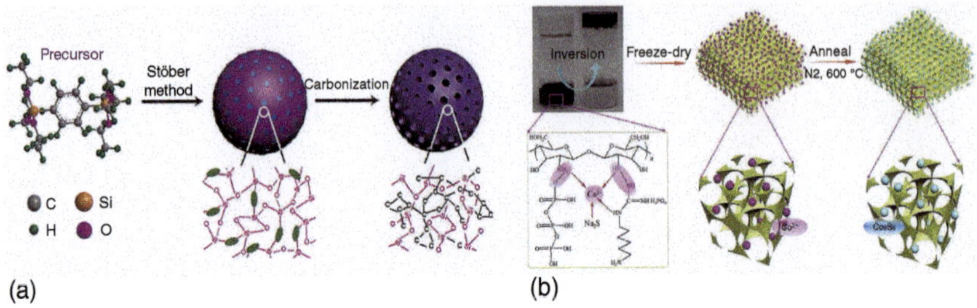

Figure 2.16 Device schematic and an application example of sol–gel method. (a) Preparation of the porous silicon-based nanocomposite consisting of the homogeneous atomic-scale distribution of carbon. Source: Zhu et al. [20]. © 2019, John Wiley & Sons. (b) Schematic illustration of the preparation of Co_9S_8/N, P-APC. Source: Adapted with permission from Hu et al. [21]. © 2019, Elsevier.

raw materials used are relatively expensive, some of the raw materials are organic that are harmful to health, and it usually takes several days or weeks. At present, it has been used in the preparation of glass, ceramics, films, fibers, composite materials, etc. For example, Yang and co-workers [20] reported a porous silicon-based nanocomposite anode derived from phenylene-bridged mesoporous organosilicas (PBMOs) through a facile sol–gel method for the application of high-performance silicon-based anodes (Figure 2.16a). Tang and co-workers [21] developed a novel hybrid hydrogel method for the preparation of Co_9S_8-doped alveolate carbon aerogels (Figure 2.16b).

The fourth method of chemical reaction coating is SAM film, which is a kind of densely organized self-assembled film formed on the substrate by surface-active substances. It is formed by immersing a suitable substrate in the solution or atmosphere of the molecules to be assembled, and the molecules spontaneously and firmly adhere to the solid surface through chemical bonds to form an ordered molecular assembly. Synthesis process of SAMs is shown in Figure 2.17a. The SAM has the following characteristics: spontaneous formation in situ, thermodynamic stability, high coverage and few defects, and orderly arrangement of molecules. For

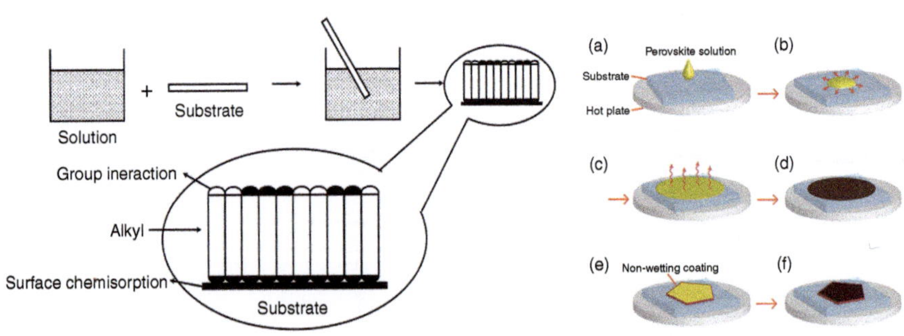

Figure 2.17 Device schematic and an application example of self-assembled monolayer film. (a) Synthesis of self-assembled monolayer film. (b) Illustration of the formation process for a drop-cast 2D perovskite film. Source: Zuo et al. [22]. © 2019, John Wiley & Sons.

example, Gao and co-workers [22] reported a 2D perovskite films by SAMs in air flexible perovskite solar cells that has the 8.0% power conversion efficiencies (PCEs) (Figure 2.17b).

2.2.2 Electrochemical Reaction Method for Thin Film Preparation

Electrochemical reaction coating refers to the process in which the cathode and anode form a loop in a certain electrolyte under the action of an electric field, and the ions in the solution are deposited on the surface of the cathode or anode through oxidation–reduction reactions to obtain the desired coating. This process can be uniformly deposited on a variety of complex substrates. The main methods of electrochemical reaction coating are anodizing method and electrodeposition.

The first method of electrochemical reaction coating is anodizing method. The idea of this method is to use certain metals or alloys as anodes in the corresponding electrolyte and graphite or metal itself as the cathode, and then an oxide film is formed on the surface of these metals. The schematic diagram is shown in Figure 2.18a. The metal oxide film formed by the anodizing method can change the state and performance of the surface, such as surface coloring, improving corrosion resistance, enhancing wear resistance and hardness, and protecting the metal surface. But it has many influencing factors, such as current density, voltage, temperature, oxidation time, and sulfuric acid concentration. For example, Li and co-workers [23] presented a novel method for preparing TiO_2/ZrO_2 nanotube/nanoporous (TiO_2 NTs/ZrO_2 NPs) heterofilm by directly anodizing Ti/Zr/Ti multilayer films deposited on silicon substrates (Figure 2.18b).

The second method of electrochemical reaction coating is electrodeposition. The chemical reaction caused by the current passing through the electrolytic salt solution is called electrolysis, and the process of using the electrolytic reaction to coat the substrate on the negative electrode is called electrodeposition. The specific process is to apply a direct current in an aqueous solution containing metal ions to be plated to discharge positive ions on the surface of the cathode to obtain a metal film. Figure 2.19a shows a schematic diagram of a plating device. In the electrodeposition

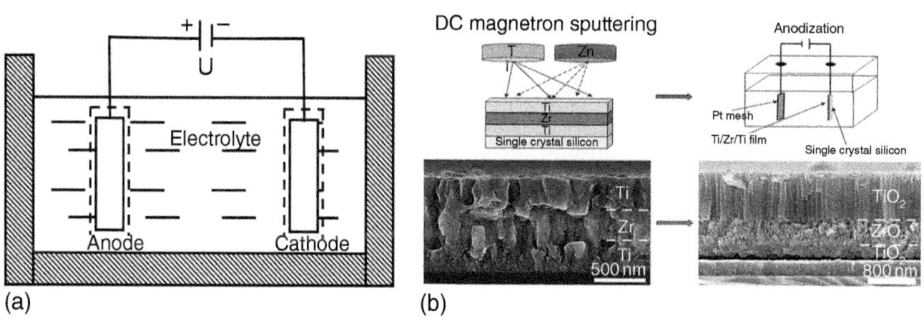

Figure 2.18 Device schematic and an application example of anodizing method. (a) Schematic diagram of anodizing method. (b) Schematic of the fabrication process of TiO_2 NTs/ZrO_2 NPs heterofilm. Source: Adapted with permission from Yu et al. [23]. © 2020, Elsevier.

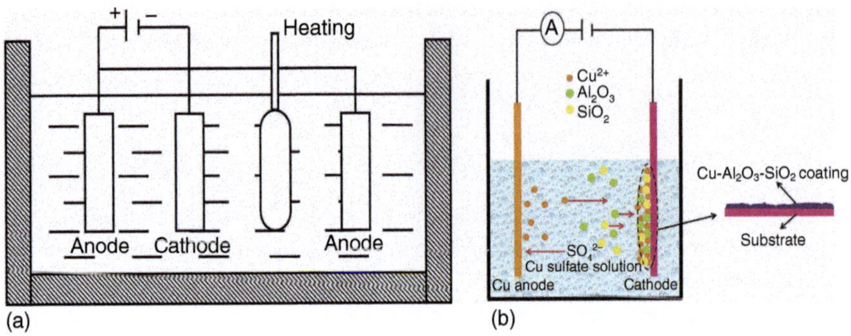

Figure 2.19 Device schematic and an application example of electrodeposition. (a) Schematic of electroplating. (b) Schematic of the electroplating process for depositing reinforced coatings with ceramic nanoparticles. Source: Mersagh and Sabzi [24]. © 2019, Elsevier.

process, the basic requirements for the electrodeposition layer are the following: fine crystals, flat plating, smooth and firm, and no pinholes. Since electrodeposition is performed at normal temperature, the plating layer has the advantages of fine and compact, flat, smooth, no pinholes, etc. Furthermore, the thickness is easy to control, so it has been widely used in the electronics industry. Although the process of electrodeposition is simple, the influencing factors are relatively complicated. The film performance not only depends on the current, voltage, temperature, solvent, pH value, and concentration of the solution but also is affected by the ionic strength of the solution and the surface state of the electrode. In addition, the generation and growth rate of crystal nuclei on the surface of the substrate cannot be controlled, the compound semiconductor thin films prepared are mostly amorphous or polycrystalline, and the performance is not good. In recent years, electroplating has also been used to prepare functional films of various metal alloys. For example, Sabzi and co-worker [24] studied the mechanisms and factors affecting the formation of the film and coating by electroplating process (Figure 2.19b).

2.2.3 Spray Pyrolysis

The SP uses an organic solution or an aqueous solution similar to the sol–gel method as a gas, atomizes the gas solution into droplets, uses a CVD-like method to send the droplets into the reaction chamber with a carrier gas, and reacts on a heated substrate and form film. The SP method for preparing thin films has the following advantages: the process equipment is simple; it can deposit films on a large area or a 3D surface, with a high deposition rate and easy industrial realization production; there are many precursors to choose; the thickness of the film can be controlled; and the deposition temperature is mostly below 600 °C, which is relatively low.

According to the different atomization methods, the thin film preparation technology of SP method can be divided into pressure atomization deposition, ultrasonic atomization deposition, and electrostatic atomization deposition. However, no matter which method is used for atomization, the thin film prepared by the SP method is grown from the deposition of droplets or fine powder particles. The surface of the prepared film is somewhat lacking in smoothness, and the porosity in the film is also high. In addition, the process conditions have a great effect on the surface morphology of the thin film prepared by the SP method. Therefore, SP is not easy to prepare smooth and dense films. During the deposition process, foreign impurities are easily brought into the film, and it is mainly limited to the preparation of materials such as oxides and sulfides. In order to address the above problems of the SP method, the following three aspects are considered: improving the atomization technology of the gas, selecting the source material and the solvent, and using a new heating deposition method.

SP is used in metal materials, optical materials, magnetic materials, and film materials. For example, Kang and co-workers [25] prepared a novel anode material for sodium-ion batteries by a one-pot SP process (Figure 2.20a). Tang and co-workers [26] used $Zn(NO_3)_2$ aqueous solution to produce ZnO buffer solution by SP (Figure 2.20b).

2.2.4 Langmuir–Blodgett Technique

LB technique is a film-forming technology that arranges insoluble molecules in a tight and orderly arrangement at the water–air interface to form a single molecular film and then transfers it to a solid surface. This technology allows precise control of film thickness and molecular structure. The specific steps of LB film are divided into two processes (Figure 2.21a). First, by applying appropriate pressure to the water surface, the disorderly oriented molecules dispersed on the water surface are compacted so that the molecules are arranged in an orderly direction perpendicular to the long axis of the molecule, which is called Langmuir film. Because the ordered molecules are easy to "collapse," it is necessary to pay attention to the size of the surface pressure and the compression method in order to maintain the uniformity of the original molecular arrangement. In addition, the vibration of the entire system will also cause the ordered molecules to be disordered, so vibration reduction is necessary. Second, when inserting and inserting the substrate in the water vertically, a single-layer or multilayer molecular film can be formed on the surface of the substrate, and the film transferred to the substrate is called Langmuir–Blodgett film.

The LB film preparation method has the following advantages: the film thickness is nanoscale, it can prepare monomolecular films but also can be accumulated layer by layer to form a multilayer film, and different polymer materials can be artificially selected to have multiple functions. However, it also has some disadvantages: the adhesion force of the LB film transferred to the substrate is the intermolecular force, so the mechanical properties of the film are poor; in order to obtain a neat and orderly LB film, the material must contain amphoteric groups, which will bring difficulties to the design of LB film-forming materials; and film-making equipment is relatively expensive, and the film-making technology requirements are very high. To obtain better film formation and low costs, the technology can be improved by changing the type of spreading agent and using

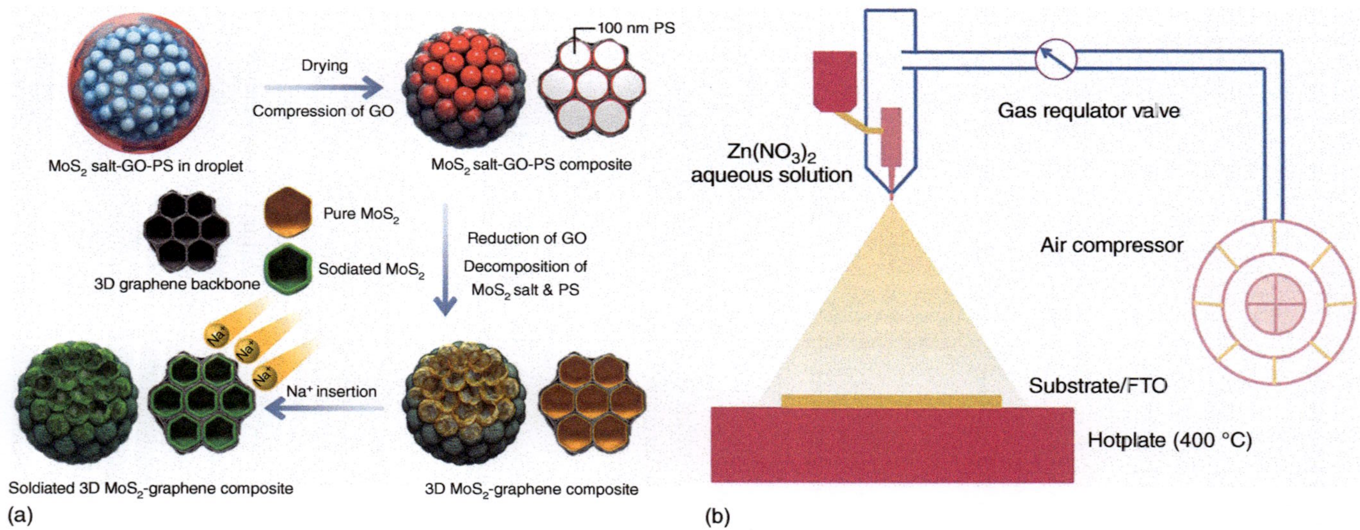

Figure 2.20 Application examples of spray pyrolysis. (a) Schematic diagram for the formation mechanism of the 3D MoS_2/graphene composite microsphere by the one-pot spray pyrolysis and description of Na^+ insertion process. Source: Choi et al. [25]. © 2015, John Wiley & Sons. (b) Schematic of deposition of ZnO and Sb_2Se_3 layer. Source: Wang et al. [26]. © 2017, Springer Nature.

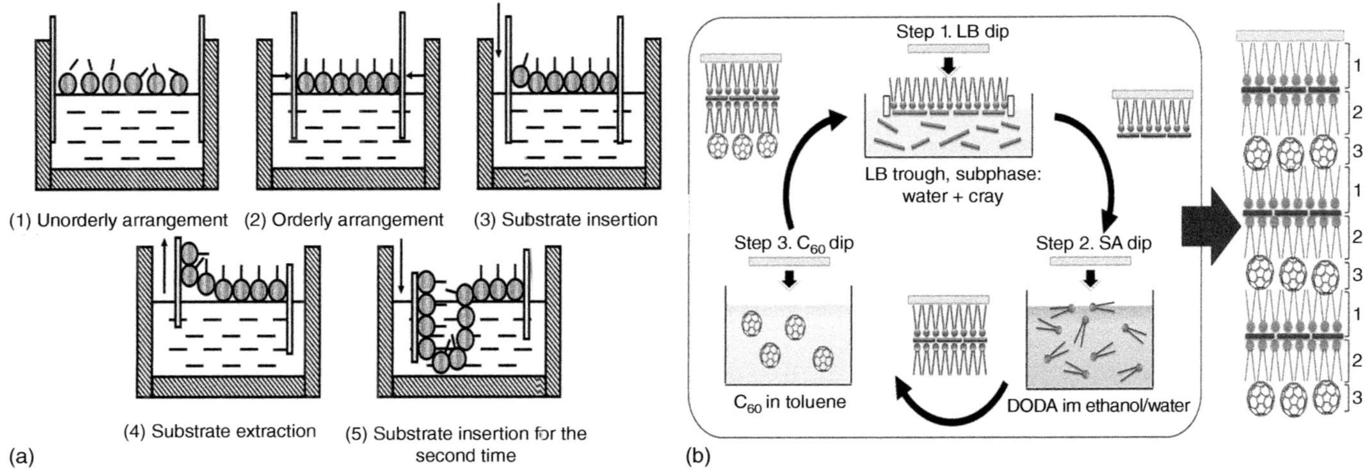

Figure 2.21 Device schematic and an application example of Langmuir–Blodgett technique. (a) Orientation of molecules on the water surface. (b) A strategy for incorporation of C60 molecules within molecular layers with the aid of clays. Source: (a) Based on Ariga et al. [27], (b) Ariga et al. [27]. © 2013, John Wiley & Sons.

electrospray technology to avoid the mutual dissolution of the spreading agent and the solvent.

In the 1980s, LB film preparation technology has attracted the attention of scholars in the fields of physics, biology, electronics, optics, chemistry, materials science, and tribology, and the research results have been applied in many fields. In recent years, it has become one of the effective methods to achieve molecular assembly and nanoscale lubrication and has received more and more attention. For example, Ariga et al. [27] proposed a strategy for incorporation of C60 molecules within molecular layers with the aid of clays (Figure 2.21b).

2.3 Typical Film Structure Characterization Technologies

From the view point of device function, film structure properties, including the crystal structure properties of the film (such as crystal phase), morphological properties (such as film thickness and uniformity) and chemical composition and structure properties (such as chemical bonds, functional groups), have great influence. Thus, characterization technologies for in situ film structure properties are essential to the process control of the film deposition toward highly efficient function devices, especially flexible devices.

There are many characterization techniques for different film characteristics. For the crystal structure properties of the film needed to be characterized, methods such as X-ray diffraction (XRD), X-ray photoelectron spectroscopy (XPS), reflective high-energy electron diffractometer (RHEED), and low-energy electron diffraction (LEED) can be used. For the morphological properties of the film, methods such as transmission electron microscope (TEM), scanning electron microscope (SEM), scanning tunneling microscope (STM), and atomic force microscope (AFM) can be used. Furthermore, for the chemical composition and structure properties of the film, methods such as auger electron spectroscopy (AES), secondary ion mass spectrum, ultraviolet–visible (UV–Vis) spectrophotometry, infrared spectroscopy, and atomic fluorescence spectrometry (AFS) can be used.

The following sections will introduce the film analysis methods for crystal structure properties, morphology properties, and chemical composition and structure properties.

2.3.1 Thin Film Analysis Method: Crystal Structure Properties

The crystal structure properties of film (such as crystal phase) are critical to the function of the devices, which should be finely controlled. For example, the formation of the crystal structure depends on the deposition conditions. When defects occur in the lattice arrangement, the film will collapse, resulting in performance degradation. For in situ characterizing the crystal structure and chemical structure properties of films in devices, the diffusion analysis method and energy spectrum are mostly used.

The existing testing methods of chemical structure properties can basically meet with the process requirements, but there are still some limitations. For example, XRD is mostly measured by powder method, thin film materials need to be ground and non-recoverable samples, while in situ XRD methods focused on a specific and smaller area are in need for device applications. Besides, XPS samples are required to be free of magnetism, radioactivity, and toxicity, which limits its application for various function materials.

Currently, the main methods for crystal structure properties are XRD, XPS, RHEED, and LEED.

The first method for crystal structure and chemical structure properties is XRD. When a beam of monochromatic X-rays is incident on a crystal, since the crystal is composed of a unit cell arranged regularly, the distance between these regularly arranged atoms is the same order of magnitude as the wavelength of the incident X-ray. Therefore, X-rays scattered by different atoms are mutually interference producing strong XRD in some special directions. The orientation and intensity of the diffraction lines in space are closely related to the crystal structure. The basic schematic of XRD is shown in Figure 2.22a. The powder method is the most convenient method for all diffraction methods and provides most of the information about the crystal structure. XRD technology has become the most basic and important structural testing method. Its main applications are as follows: quantitative analyte phase, qualitative analyte phase, measuring the crystallinity, measuring the macroscopic stress, determining the grain size, determining the crystal lattice parameter, and measuring the thickness of the film. For example, Wu and co-workers [17] found that the immobilized layer on the conductive membrane is composed of Ni and NiO through XRD patterns

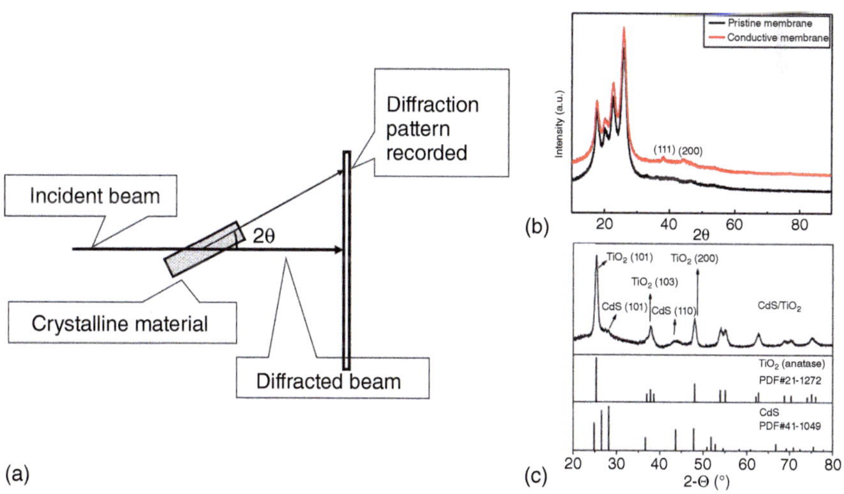

Figure 2.22 Basic schematic and application examples of XRD. (a) Simple schematic of XRD. (b) XRD curves of the pristine and conductive membranes. Source: (a) Based on Zhang et al. [17], (b) Zhang et al. [17]. © 2019, Elsevier. (c) XRD pattern data of CdS QDs/TiO_2. Source: Cai et al. [18]. © 2018, American Chemical Society.

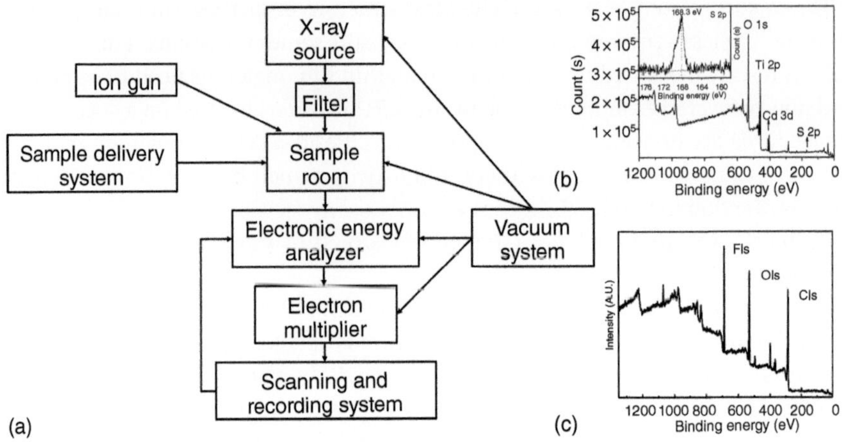

Figure 2.23 Basic schematic and application examples of XPS. (a) Simple schematic of XPS. (b) XPS spectra (inset: high-resolution XPS spectra of S element) of CdS QDs/TiO_2. Source: Cai et al. [18]. © 2018, American Chemical Society. (c) XRD curves of the pristine and conductive membranes. Source: Zhang et al. [17]. © 2019, Elsevier.

(Figure 2.22b). Tang and co-workers [18] proved the presentence of anatase TiO_2 with XRD (Figure 2.22c).

The second method for crystal structure and chemical structure properties is XPS, which is the use of X-rays to irradiate a sample so that the inner or valence electrons of an atom or molecule are stimulated to be emitted. The energy of the photoelectrons can be measured. Using the kinetic energy of the photoelectron as the abscissa and the relative intensity as the ordinate, a photoelectron spectrum can be made to obtain the composition of the test object. The kinetic energy of this electron can be measured to obtain the relevant electronic structure in the sample information. The basic schematic of XPS is shown in Figure 2.23a. Using XPS technology, information such as qualitative analysis of elements in solid samples, quantitative analysis of elements, solid surface analysis, and structure of compounds can be measured. For example, Tang and co-workers [18] proved the existence of four elements including Ti, Cd, S, and O in the hybrid nanostructures with XPS (Figure 2.23b). Wu and co-workers [17] proved the existence of C, F, O, Ag, and Ni elements in the conductive membrane with XPS (Figure 2.23c).

The third method for crystal structure and chemical structure properties is RHEED, which is one of the most important real-time monitoring tools for observing crystal growth. It can graze a single-energy electron with an energy of 10–30 KeV to the surface of the crystal through a very small grazing angle and obtain important information. Its significant advantage is that it can be conveniently loaded into an MBE device for in situ observation of epitaxial growth processes and growth modes. Figure 2.24a shows a simplified diagram of such a device. With RHEED images, we can get a lot of useful information: structural analysis of clean sample surface and ordered adsorption layer, difference between single crystal and polycrystalline, and

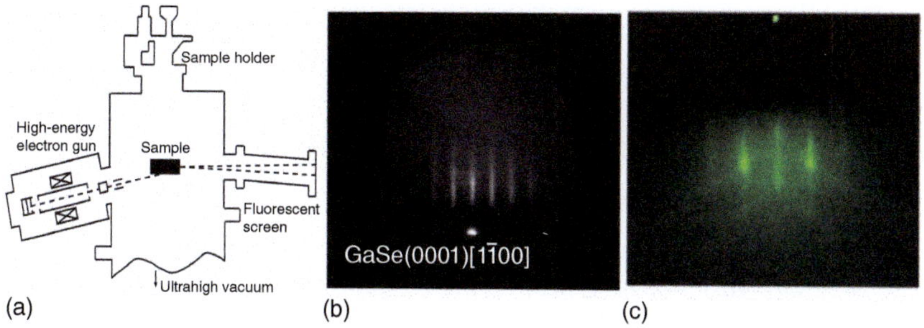

Figure 2.24 Device schematic and application examples of reflective high-energy electron diffractometer. (a) Simple schematic of RHEED. (b) RHEED image along the [1$\bar{1}$00] crystallographic axis of the GaSe base layer on GaAs(111)B. Source: (a) Based on Kanai et al. [5], (b) Adapted with permission from Kanai et al. [5]. © 2019, Royal Society of Chemistry. (c) RHEED patterns of ZnO electrodes before the electrochemical measurements. Source: Adapted with permission from O'Hara et al. [28]. © 2018, American Chemical Society.

determination of surface remodeling of the sample and the composition of the film [5, 28].

The fourth method for crystal structure and chemical structure properties is LEED, which is an electron incidence using energy of 10–500 eV. Moreover, a diffraction pattern is generated by mutual interference of elastic backscattered electron waves. LEED device generally consists of an electron gun and a spherical detector (Figure 2.25a). LEED has the following two applications: qualitative analysis, obtaining the symmetry of the surface structure and determining the relative size and direction of the unit cell of the adsorbate and the base unit lattice, and quantitative analysis, obtaining the exact positional information of the surface

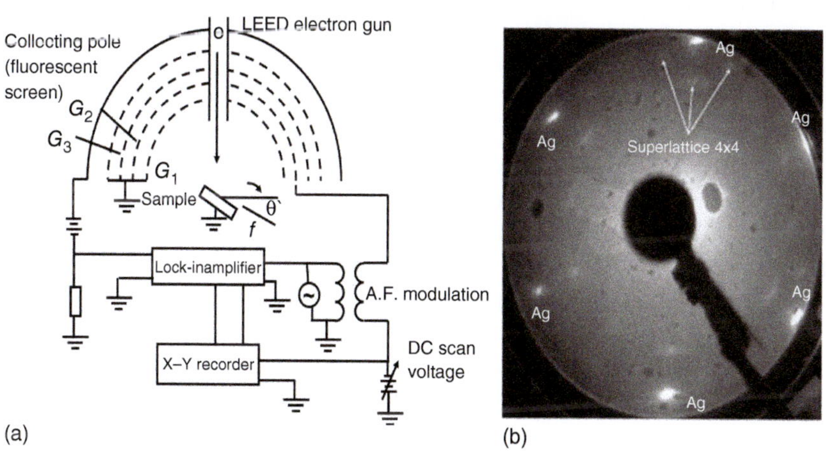

Figure 2.25 Device schematic and application examples of low-energy electron diffraction. (a) LEED device schematic. (b) LEED diagrams obtained after the deposition of about one silicon monolayer on Ag (111) maintained. Source: (a) Meng et al. [29]. © 2013, American Chemical Society, (b) Adapted with permission from Meng et al. [29]. © 2013, American Chemical Society.

atoms. For example, as shown in Figure 2.25b, the presence of silver can be observed [29, 30].

2.3.2 Thin Film Analysis Method: Morphology Properties

The morphology properties of film (such as film thickness and uniformity) are also critical to the efficiency and performance of devices. For example, when the film is deposited unevenly, certain points of the film with special thickness may act as defects or cause short circuits and even lead to cracks and peeling offs. For in situ characterizing the morphology properties of the film in devices or on electrodes, microscope observation methods are mostly used.

The existing morphological property testing methods can basically meet with the process requirements, but there are still some limitations. For example, the required detecting area for the thin film material of specific position may be smaller than the probe of microscopic topography methods, like STM, and lead to misunderstanding of certain device mechanism. Besides, SEM cannot accurately focus on samples with undulations.

Currently, the main methods of morphology properties include TEM, SEM, STM, and AFM.

The first method for morphology properties is TEM. It is a high-resolution, high-magnification electro-optical instrument that uses an extremely short wavelength electron beam as an illumination source and focuses imaging with an electromagnetic lens. Figure 2.26a shows a schematic diagram of the structure of a modern TEM and an optical path diagram of the imaging and diffraction modes of operation. The lens barrel of a TEM is mainly composed of three parts: the light source, the lens group, an observation room and a camera. The main uses of TEM in the research of materials science are general morphology observation of samples using thick contrast images; phase analysis of samples using techniques such as electron diffraction, micro-area electron diffraction, and convergent beam electron diffraction; determining the crystal structure; observing the structural defects in the crystal, determining the type of the defect, and estimating the density of the defect; analyzing the chemical composition of the micro area of the sample; and distribution of medium elements analyzed to determine whether there is component segregation in the sample. For example, Nakamura and co-workers [31] showed the difference of deposition of Au nanoparticles with TEM (Figure 2.26b), and Lee and co-workers [32] judged whether the CS-4 NS was deposited completely with the help of TEM lattice spacing (Figure 2.26c).

The second method for morphology properties is SEM. The principle of SEM is to use an electron gun to heat a tungsten filament with a vacuum to generate a thermoelectron beam. The working principle is shown in Figure 2.27a. Using SEM, it is possible to analyze and study the microstructure morphology, surface morphology, micro-area chemical composition analysis, microstructure, and ultrafine size of the film. For example, Fan and co-workers [33] showed the Ni dendritic cables with SEM (Figure 2.27b), and Alami and co-workers [34] showed the thickness of different angle of TiAlN coating (Figure 2.27c).

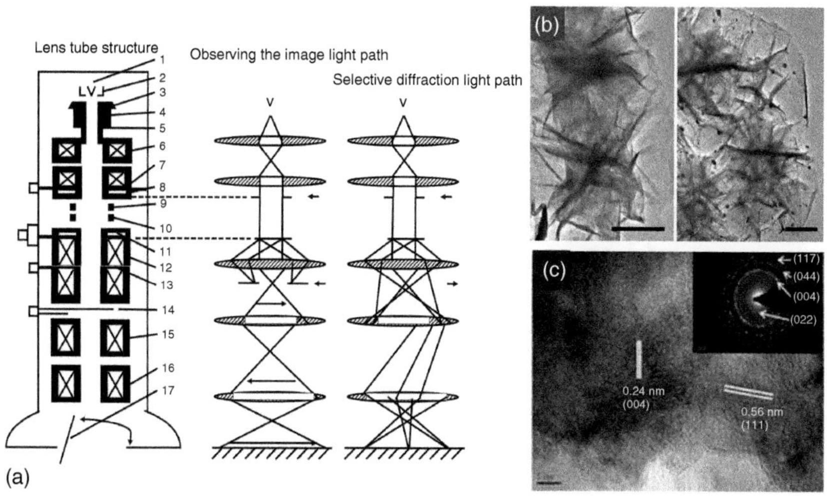

Figure 2.26 Device schematic and application examples of transmission electron microscope. (a) TEM working principle. (b) TEM images of δ-MnO_2 (left) and δ-MnO_2/AuNP (right) catalysts. Source: (a) Li et al. [31]. © 2019, John Wiley & Sons, (b) Adapted with permission from Li et al. [31]. © 2019, WILEY-VCH Verlag GmbH & Co. KGaA, Weinheim. (c) HR-TEM image and SAED pattern of the CS-4 NS. Source: Adapted with permission from Palanichamy et al. [32]. © 2015, American Chemical Society. 1, filament; 2, grid; 3, anode; 4, gun tilt; 5, gun panning; 6, first condenser; 7, secondary condenser; 8, condenser diaphragm; 9, light tilt; 10, light translation; 11, sample stand; 12, objective lens; 13, objective diaphragm; 14, selection diaphragm; 15, middle mirror; 16, projection mirror; 17, fluorescent screen.

The third method for morphology properties is STM. The principle of STM is based on the tunneling effect in quantum mechanics. In quantum mechanics, the motion of a particle is described by the Schrödinger equation. The probability of a particle at any position in space is determined by the wave function of its motion. STM has the following applications: surface structure research – high-resolution real-time observation of conductor and semiconductor surface structures in a variety of experimental environments, discrimination of individual atoms on metal surfaces with stacked structures, and observation of main features of metal surfaces – and biological research. For example, Chang et al. [35] observed the main feature of α/β-SnTe phase (Figure 2.28a), and Xu et al. [4] showed that the antimonene islands are clean and smooth with no obvious impurity on the islands (Figure 2.28b).

The fourth method for morphology properties is AFM. It uses a force-sensitive probe tip to interact with the sample atoms to achieve surface imaging. AFM mainly includes probes and samples, scanners, scanning and feedback control circuits, deflection detection, and computer data processing, as shown in Figure 2.29a. In actual application, as shown in Figure 2.29b,c, the AFM chart can not only obtain the surface morphology and roughness of the object but also determine whether other substances have an effect on the object based on the difference between the shape before and after the reaction. It is not limited to the determination of properties such as the surface of inorganic substances, but can also be used to determine properties such as biological morphology and shape [36–38].

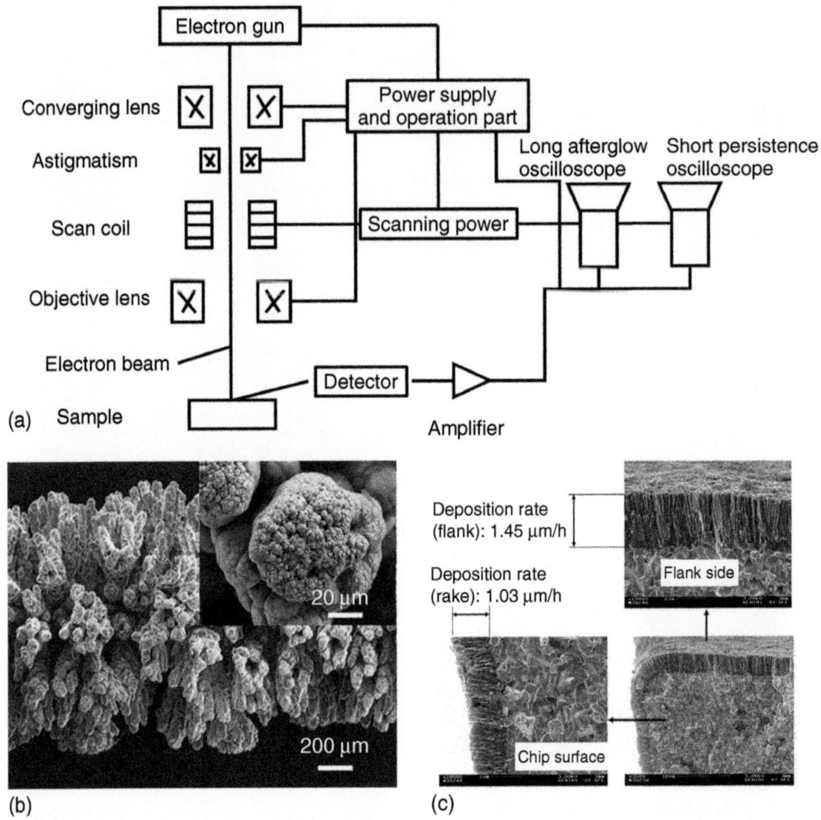

Figure 2.27 Device schematic and application examples of scanning electron microscope. (a) Working principle of scanning electron microscope instrument structure. SEM image of (b) Ni dendritic cables. Source: (a) Yang et al. [33]. © 2017, American Chemical Society, (b) Adapted with permission from Yang et al. [33]. © 2017, American Chemical Society. (c) TiAlN coating deposited on the flank and the rake side of a cutting insert. Source: Adapted with permission from Bobzin et al. [34]. © 2009, Elsevier.

2.3.3 Thin Film Analysis Method: Chemical Composition and Structure Properties

The chemical composition and structure properties (such as chemical compositions and interaction between functional groups) are critical to the fabrication control and performance of devices. For example, the type of device function, such as sensor or energy devices, basically depends on the chemical compositions and interaction between functional groups. For in situ characterizing the chemical properties of the film in devices or on electrodes, AES, UV–Vis spectrophotometry, infrared spectroscopy, and AFS are mostly used.

The existing chemical composition and structure property testing methods can basically meet with the process requirements, but there are still some limitations. For example, the aggregation of clusters would affect spectrophotometry of material and lead to additional spectral peaks, for either UV–Vis spectroscopy or infrared spectroscopy.

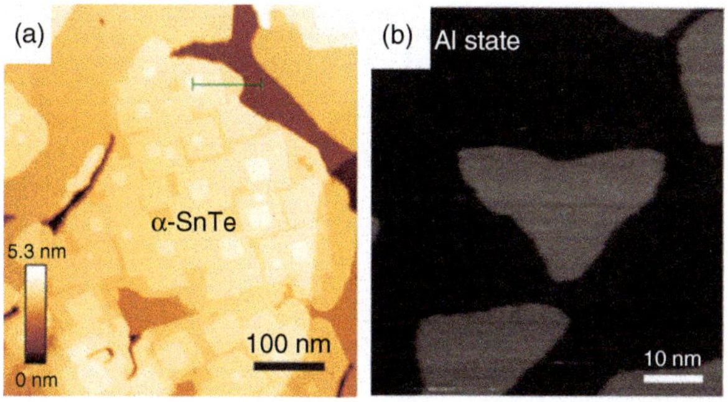

Figure 2.28 Application examples of scanning tunneling microscope. (a) STM topography of the areas in α-SnTe. Source: Xu et al. [4]. © 2017, John Wiley & Sons. (b) Typical STM image of antimonene islands on $PdTe_2$ substrate before air exposure. The surface of the islands is smooth without impurities. Source: Adapted with permission from Chang et al. [35].

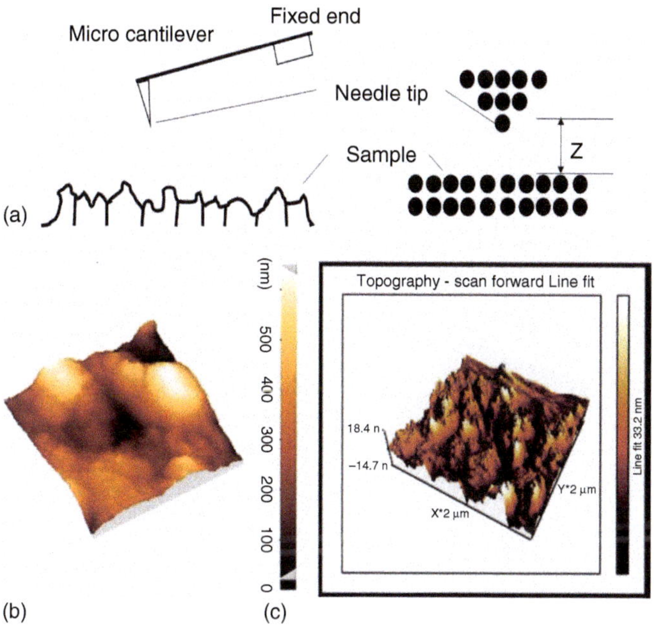

Figure 2.29 Device schematic and application examples of atomic force microscope. (a) AFM imaging principle. (b) AFM micrographs for mild steel in 1 M HCl with 0 ppm of *Citrullus lanatus* fruit extract. Source: Dehghani et al. [36]. © 2019, Elsevier. (c) AFM image of Bare CPE. Source: Adapted with permission from Shetti et al. [37]. © 2019, Elsevier.

Currently, the main methods for chemical composition and structure properties includes AES, UV–Vis spectrophotometry, infrared spectroscopy, and AFS.

The first method of chemical composition and structure properties is AES. It is mainly used for the identification of all elements except hydrogen and helium in the surface layer with a thickness of less than 2 nm. It consists of electron gun, sputtering ion gun, electron energy analyzer, vacuum system, and operation display system. AES has the following applications: qualitative analysis, identifying the element type; quantitative analysis, determining the element concentration on the surface; chemistry effect, a reference for chemical state analysis; and depth analysis, a 3D analysis of the changes in the composition and content of the sample elements with depth.

The second method of chemical composition and structure properties is UV–Vis spectrophotometry. It belongs to molecular absorption spectroscopy and is a method

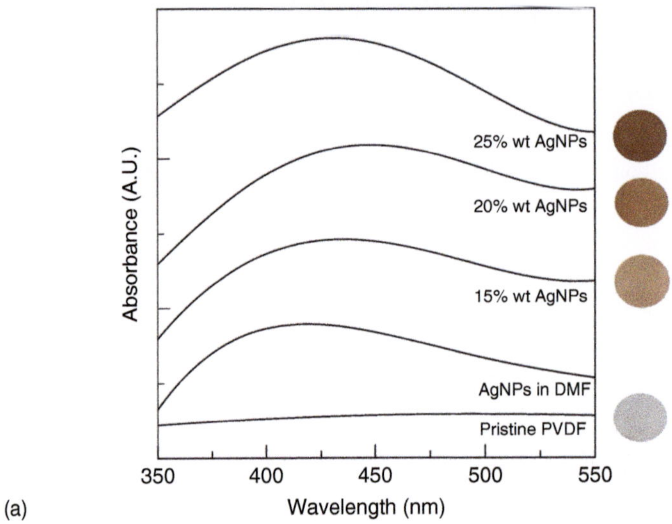

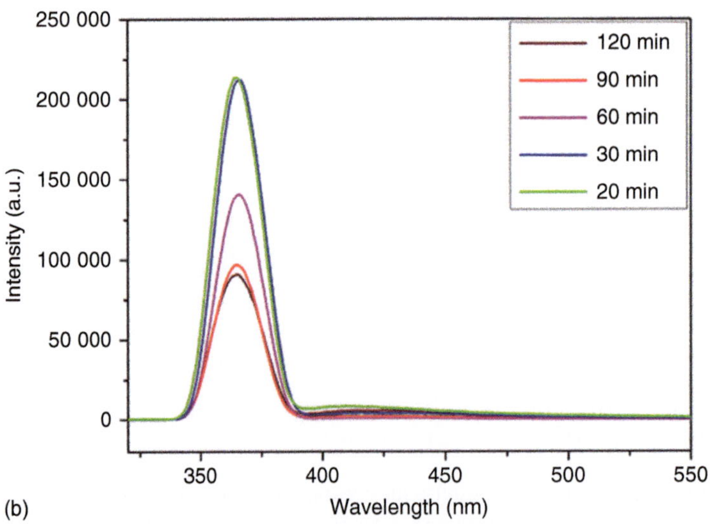

Figure 2.30 Application examples of ultraviolet–visible spectrophotometry. (a) UV–Vis characterization of mixed matrix PVDF membranes incorporating different amounts of Ag NPs. Source: Zhou et al. [39]. © 2019, Elsevier. (b) UV emission characterization of graphene/ZnO samples with different sputtering time. Source: Chen et al. [40]. © 2020, Elsevier.

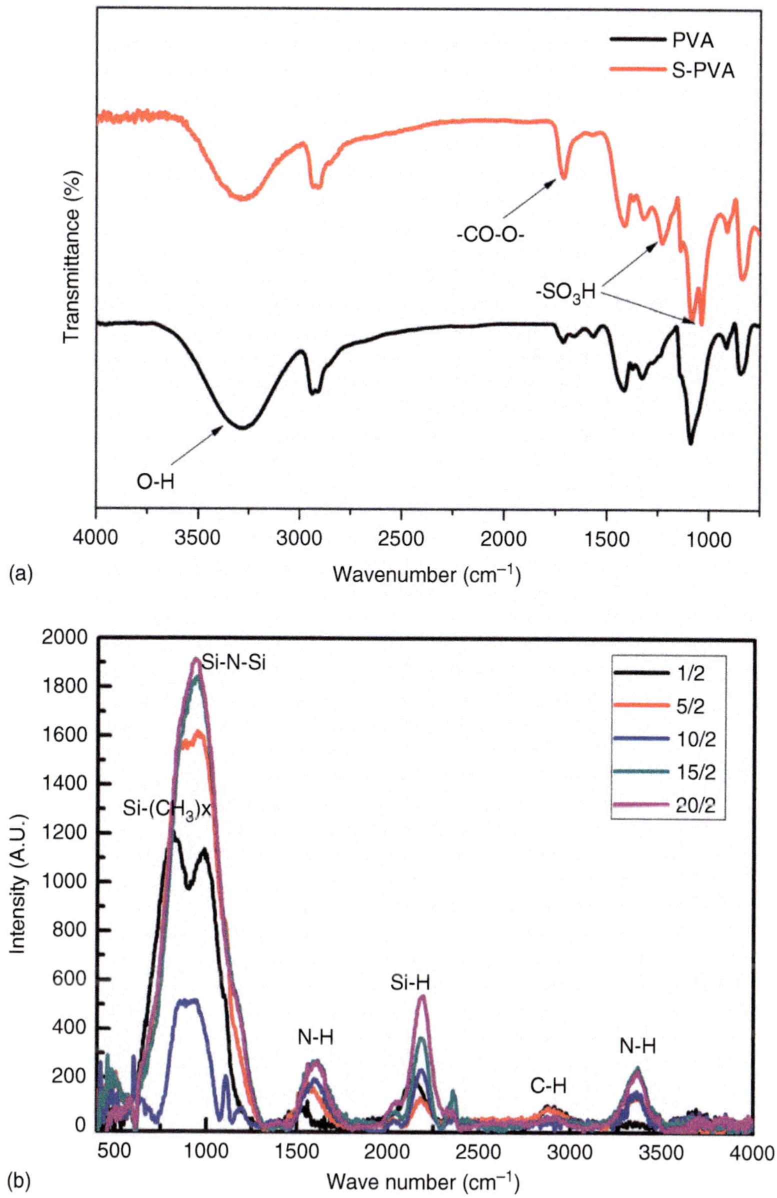

Figure 2.31 Application examples of infrared spectroscopy. (a) ATR–FTIR spectra of the PVA and S-PVA samples. Source: Zhang et al. [41]. © 2019, Taylor & Francis. (b) IR spectra obtained for different N_2/HMDSO flow ratio, as indicated. Source: Amri et al. [42]. © 2020, Elsevier.

for analyzing and measuring the absorption of radiation in the 200–800 nm spectral region using certain substances. UV–Vis absorption spectroscopy mainly results from the transition of molecular valence electrons between electron energy levels. This method is widely used for qualitative analysis, structural analysis, quantitative analysis, and spectral analysis of certain compounds [39, 40] (Figure 2.30).

The third method for chemical composition and structure properties is infrared spectroscopy. The principle of infrared spectroscopy is the absorption of infrared radiation by material molecules and the net change of the dipole moment caused by its vibration or rotational motion, resulting in the transition of molecular vibration and rotational energy levels from the ground state to the excited state. It is a commonly used analytical method to identify compounds and determine the molecular structure of substances. This analysis technique can perform not only qualitative analysis of compounds but also quantitative analysis of individual components or individual components in a mixture, especially for samples that are difficult to separate and cannot find obvious characteristic peaks in the ultraviolet and visible regions. For example, Zhang and co-workers [41] confirmed that the sulfonic acid groups are incorporated and the cross-linking network forms (Figure 2.31a); Clin and co-workers [43] proved the presence of the same absorption bands for the different ratio, but with different intensities under different ratio (Figure 2.31b).

The fourth method of chemical composition and structure properties is AFS. It is an emission spectrum analysis method for quantitative analysis by measuring the emission intensity of the fluorescence generated by the atomic vapor of the element to be measured under the excitation of radiant energy. After a gaseous free atom absorbs the characteristic radiation of a light source, the outer electrons of the atom transition to a higher energy level and then return to the ground state or a lower energy level. The main application of AFS is to measure element content.

2.4 Conclusions

This chapter provides an overview of typical thin film deposition techniques and characterization techniques. Nowadays, the film structure in devices, especially wearable electronic devices, has been evolving from the traditional rigid flat panel structure to the flexible structure like fibrous and flexible sheet. While the device structure is changing, their functional film will also be deposited with a more complex structure to achieve higher performance and diversified functions.

At present, thin film deposition technology can be divided into dry-process technology, wet-process technology, and patterning technology. Among them, by reaction type, dry-process technology can be divided into PVD and CVD, and wet-process technology can be divided into chemical reaction coating and electrochemical reaction coating. At the meanwhile, the micro/nano-assembly of thin film also requires patterning technology for support. However, there are still a lot of challenges for thin film deposition and related characterization technology.

First, devices with a more complex structure will require a complex morphology of the functional film. Film deposition technologies should be adapted for curved interface like the surface of fiber or textile. In response to this change, film deposition technologies with new principles and new setups should be designed to realize a well-controlled fluid field and electro field during film deposition process, which would also require a controlled motion of the electrode substrated, and new widely adapted micro- or nano-pattern-making methods, not only within a 2D plane but also toward a 3D space.

Second, to realize a higher device performance, like higher energy density and sensitivity, the film structure inside the device should be optimized with a higher precision or even endowed with a newly designed micro/nanostructure, which would propose a requirement for deposition technologies aiming for a multidimensional assembly. In response to this change, thin film assembly and characterization methods need to be developed with much higher precision, even to the nano- or picoscale. In terms of film assembly, it should be controlled with a layer-by-layer or even molecular-by-molecular precision.

Third, the developing of integrated circuit system would require more devices with a more diversified function. For example, we need more sensors for detecting different pollutes and physical indexes for human. We need more energy devices to harvest more energy from nature environments with weak lights or lower heats. We also need more electronic devices to construct flexible circuit for more complicated logical computing and wireless communication. Therefore, we should explore more possibility from more materials and assemble them together into a film of nano-thickness with higher endurabilities, which also requires innovations of deposition and characterization technology for a wide range of materials and fabrication conditions.

Besides, current in situ characterization methods for the film properties have been developed to adapt for various structure properties, like crystal structure property, morphological property, and chemical composition and structure properties. However, new challenges have been emerging. For example, spectroscopy methods should be expanded for new chemical properties related with the aggregation states of material. Smaller probe for microscopic topography methods should be designed to detect specific area in flexible device. Besides, characterization equipment should also be developed with wider adaptability for magnetism, radioactivity, and toxicity function materials and devices with complex structure.

In a word, we have achieved a lot of success based on current film deposition and characterization methods to make flexible devices into wearable fabrics including laminated and woven fabrics. However, we have to face more challenges to make devices with higher performance, more diversified functions, and more complex structures to realize the commercialized application in electronic industry.

References

1 Chen, W., Zhang, J., Xu, G. et al. (2018). A semitransparent inorganic perovskite film for overcoming ultraviolet light instability of organic solar cells and achieving 14.03% efficiency. *Advanced Materials* 30 (21): 1800855.

2 Halim, J., Lukatskaya, M.R., Cook, K.M. et al. (2014). Transparent conductive two-dimensional titanium carbide epitaxial thin films. *Chemistry of Materials* 26 (7): 2374–2381.

3 Chen, Y., Ren, R., Pu, H. et al. (2017). Field-effect transistor biosensors with two-dimensional black phosphorus nanosheets. *Biosensors and Bioelectronics* 89: 505–510.

4 Xu, W., Yan, S., Hang, L. et al. (2017). Epitaxial growth and air-stability of monolayer antimonene on $PdTe_2$. *Advanced Materials* 29: 1605407.

5 Kanai, M., Watanabe, K., Maruyama, S., and Matsumoto, Y. (2019). Ionic liquid/ZnO(000¯) single crystal and epitaxial film interfaces studied through a combination of electrochemical measurements and a pulsed laser deposition process under vacuum. *Physical Chemistry Chemical Physics* 21 (45): 25506–25512.

6 Zhou, X., Gan, L., Tian, W. et al. (2015). Ultrathin $SnSe_2$ flakes grown by chemical vapor deposition for high-performance photodetectors. *Advanced Materials* 27 (48): 8035–8041.

7 Zhang, P., Liu, X., Fang, X., and Chen, X. (2020). Synthesis of single-layer graphene film by chemical vapor deposition with molten gallium catalyst on silicon dioxide. *Journal of Materials Science* 55 (7): 2787–2795.

8 Li, C., Xie, B., Chen, J. et al. (2019). Emerging mineral-coupled composite phase change materials for thermal energy storage. *Energy Conversion and Management* 183: 633–644.

9 Asgary, S., Jafari, A., Ebrahminejad, Z., and Ardaheh, M. (2019). Evolution of physical properties of diamond nanoparticles deposited by DC-PECVD method after post deposition annealing. *Journal of Materials Science: Materials in Electronics* 30 (23): 20451–20458.

10 Stassen, I., Styles, M., Grenci, G. et al. (2016). Chemical vapour deposition of zeolitic imidazolate framework thin films. *Nature Materials* 15 (3): 304–310.

11 Miikkulainen, V., Leskelä, M., Ritala, M., and Puurunen, R.L. (2013). Crystallinity of inorganic films grown by atomic layer deposition: overview and general trends. *Journal of Applied Physics* 113 (2): 430.

12 Fang, M. and Ho, J.C. (2015). Area-selective atomic layer deposition: conformal coating, subnanometer thickness control, and smart positioning. *ACS Nano* 9 (9): 150909103635004.

13 Smith, C.S., Savliwala, S., Mills, S.C. et al. (2020). Electro-infiltrated nickel/iron-oxide and permalloy/iron-oxide nanocomposites for integrated power inductors. *Journal of Magnetism and Magnetic Materials* 493: 165718.

14 Hu, E., Feng, Y., Nai, J. et al. (2018). Construction of hierarchical Ni–Co–P hollow nanobricks with oriented nanosheets for efficient overall water splitting. *Energy & Environmental Science* 11 (4): 872–880.

15 Kajiya, D. (2019). An agar sandwich method for patterning transparent conducting oxides. *Journal of Materials Science: Materials in Electronics* 30 (23): 20734–20740.

16 Xiao, Z., Wang, Y., Huang, Y. et al. (2017). Filling the oxygen vacancies in Co_3O_4 with phosphorus: an ultra-efficient electrocatalyst for overall water splitting. *Energy & Environmental Science* 10 (12): 2563–2569.

17 Zhang, Y., Yu, W., Li, R. et al. (2019). Novel conductive membranes breaking through the selectivity-permeability trade-off for Congo red removal. *Separation and Purification Technology* 211: 368–376.

18 Cai, G., Yu, Z., Ren, R., and Tang, D. (2018). Exciton-plasmon interaction between AuNPs/graphene nanohybrids and CdS quantum dots/TiO_2 for photoelectrochemical aptasensing of prostate-specific antigen. *ACS Sensors* 3 (3): 632–639.

19 Ebadi, S.V., Semnani, D., Fashandi, H., and Rezaei, B. (2019). Highly conductive Faradaic artificial muscle based on nanostructured polypyrrole-bis(trifluoromethylsulfonyl)imide synthesized onto electrospun polyurethane nanofibers. *Sensors and Actuators B: Chemical* 297: 126736.

20 Zhu, G., Zhang, F., Li, X. et al. (2019). Engineering the distribution of carbon in silicon oxide nanospheres at the atomic level for highly stable anodes. *Angewandte Chemie International Edition* 131 (20): 6741–6745.

21 Hu, X., Chen, Y., Zhang, M. et al. (2019). Alveolate porous carbon aerogels supported Co_9S_8 derived from a novel hybrid hydrogel for bifunctional oxygen electrocatalysis. *Carbon* 144: 557–566.

22 Zuo, C., Scully, A.D., Vak, D. et al. (2019). Self-assembled 2D perovskite layers for efficient printable solar cells. *Advanced Energy Materials* 9 (4): 1803258.

23 Yu, Y., Zhao, Y., Li, K. et al. (2020). Microstructures and optical properties of TiO_2/ZrO_2 nanotube/nanoporous heterofilm prepared by anodizing of Ti/Zr/Ti multilayer films. *Applied Surface Science* 503: 144316.

24 Mersagh, D.S. and Sabzi, M. (2019). Deposition of ceramic nanocomposite coatings by electroplating process: a review of layer-deposition mechanisms and effective parameters on the formation of the coating. *Ceramics International* 45 (17): 21835–21842.

25 Choi, S.H., Ko, Y.N., Lee, J., and Kang, Y.C. (2015). 3D MoS_2-graphene microspheres consisting of multiple nanospheres with superior sodium ion storage properties. *Advanced Functional Materials* 25 (12): 1780–1788.

26 Wang, L., Li, D., Li, K. et al. (2017). Stable 6%-efficient Sb_2Se_3 solar cells with a ZnO buffer layer. *Nature Energy* 2 (4): 17046.

27 Ariga, K., Yamauchi, Y., Mori, T., and Hill, J.P. (2013). 25th anniversary article: what can be done with the Langmuir–Blodgett method? Recent developments and its critical role in materials science. *Advanced Materials* 25 (45): 6477–6512.

28 O'Hara, D.J., Zhu, T., Trout, A.H. et al. (2018). Room temperature intrinsic ferromagnetism in epitaxial manganese selenide films in the monolayer limit. *Nano Letters* 18 (5): 3125–3131.

29 Meng, L., Wang, Y.L., Zhang, L.Z. et al. (2013). Buckled silicene formation on Ir(111). *Nano Letters* 2 (13): 685–690.

30 Rous, P.J. and Pendry, J.B. (1989). The theory of tensor LEED. *Surface Science* 219 (3): 355–372.

31 Li, A.L., Ooka, H., Bonnet, N. et al. (2019). Stable potential windows for long-term electrocatalysis by manganese oxides under acidic conditions. *Angewandte Chemie International Edition* 58 (15): 5054–5058.

32 Palanichamy, S., Christy, M., Aravindan, V. et al. (2015). Two dimensional mesoporous cobalt sulfide nanosheets as superior anode for Li-ion battery and bi-functional electro-catalyst for Li–O_2 system. *Chemistry of Materials* 27: 5726–5735.

33 Yang, Y., Zhang, N., Zhang, B. et al. (2017). Highly-efficient dendritic cable electrodes for flexible supercapacitive fabric. *ACS Applied Materials & Interfaces* 9: 40207–40214.

34 Bobzin, K., Bagcivan, N., Immich, P. et al. (2009). Advantages of nanocomposite coatings deposited by high power pulse magnetron sputtering technology. *Journal of Materials Processing Technology* 209 (1): 165–170.

35 Chang, K., Kaloni, T.P., Lin, H. et al. (2019). Enhanced spontaneous polarization in ultrathin SnTe films with layered antipolar structure. *Advanced Materials* 31 (3): 1804428.

36 Dehghani, A., Bahlakeh, G., Ramezanzadeh, B., and Ramezanzadeh, M. (2019). A combined experimental and theoretical study of green corrosion inhibition of mild steel in HCl solution by aqueous Citrullus lanatus fruit (CLF) extract. *Journal of Molecular Liquids* 279: 603–624.

37 Shetti, N.P., Nayak, D.S., Malode, S.J. et al. (2019). Sensors based on ruthenium-doped TiO_2 nanoparticles loaded into multi-walled carbon nanotubes for the detection of flufenamic acid and mefenamic acid. *Analytica Chimica Acta* 1051: 58–72.

38 Pisonero, J., Bouzas-Ramos, D., Traub, H. et al. (2019). Critical evaluation of fast and highly resolved elemental distribution in single cells using LA-ICP-SFMS. *Journal of Analytical Atomic Spectrometry* 34 (4): 655–663.

39 Zhou, S., Xu, H., Hu, H. et al. (2019). High quality GaN buffer layer by isoelectronic doping and its application to 365 nm InGaN/AlGaN ultraviolet light-emitting diodes. *Applied Surface Science* 471: 231–238.

40 Chen, C., Kou, L., Hauffman, T. et al. (2020). Exploration and mechanism analysis: the maximum ultraviolet luminescence limits of ZnO/few-layer graphene composite films. *Applied Surface Science* 503: 144169.

41 Zhang, R., Liang, B., Qu, T. et al. (2019). High-performance sulfosuccinic acid cross-linked PVA composite pervaporation membrane for desalination. *Environmental Technology* 40 (3): 312–320.

42 Amri, R., Sahel, S., Gamra, D. et al. (2020). Experimental evidence of photonic band gap in hybrid Si/SiNOCH multilayer structure at the infrared range. *Optics Communications* 457: 124619.

43 Kuo, C.H., Li, W., Pahalagedara, L. et al. (2015). Understanding the role of gold nanoparticles in enhancing the catalytic activity of manganese oxides in water oxidation reactions. *Angewandte Chemie International Edition* 54 (8): 2345–2350.

3

The Fabrication Process of Intelligent Fibers and Textiles

With the improvement of people's living standards and the continuous development of scientific research, classical fibers can only meet with the wearing comfort of people's daily life without other functional characteristics. Then people begin to add more nano function materials into the fiber or do additional surface modifications, aiming to add some special functions such as flame-retardancy, hydrophilic, and antistatic, which were given the idea of functional fibers.

However, with intelligent life development, people have higher requirements for merging electronics with wearable textiles to construct an excellent on-body electronic function platform while maintaining textiles' features. Clothing is a feature and final form of wearing all human societies, giving superior wearing comfort, excellent mechanical strength, softness, low cost, and flexibility. To fabricate a fiber and textile with intelligent functions, people try to assemble the nano/micro-functional structure layer by layer on fiber, similar to assembling an electronic device, which has successfully turned the functional fiber materials into intelligent fiber devices to the intelligent textiles. In other words, the daily-wearable cloth can be endowed with capabilities, like sensing stimulations of mechanical, thermal, chemical, light, humidity, electricity, and magnetism, and respond to them.

The development of intelligent fibers and textiles is also developing in this direction, from classical functional fiber materials to intelligent textile devices. Firstly, functional fibers can be fabricated by various methods like wet spinning, electric spinning, thermal drawing, and other methods. Nano-functional materials were blended in the fiber materials or coated on the surface. Secondly, intelligent fibers can be assembled based on functional fibers by constructing much more complicated nano/micro-functional structures. They include nanowires, nano-arrays, nanoballs, nanoflowers, nanorods, or even assembling them into a multi-layered structure, which was far beyond the structure of conventional fiber materials. Lastly, intelligent textiles are further fabricated based on intelligent fibers for better wearing comfort. These assembly strategies have broken the traditional flat device's structure limitation, making better use of many special-shaped and narrow spaces on the human body to realize a more efficient on-body integration. Nevertheless, it has proposed higher requirements for the fabrication technologies on the fiber or textile.

Textile-Based Energy Harvesting and Storage Devices for Wearable Electronics, First Edition.
Xing Fan, Nannan Zhang, and Yi Wang.

This chapter will discuss the fabrication technologies from classical functional fibers to intelligent fibers and textiles successively.

3.1 The Synthesis of Classical Functional Fibers

The functional fibers are the foundation for the assembling of intelligent fibers and textiles. Before fabricating an intelligent fiber device, layers of conductors and semiconductors were usually assembled on a functional fiber as the electrode. Commonly, there are two strategies for realizing a functional fiber material, including blending into the classical raw materials during the drawing process and surface modification on a classical fiber.

For the blended drawing process, nanoparticles with functions such as flame-retardancy, hydrophilic, and antistatic are mixing into the polymer solution or melt. The functional fibers are then fabricated by drawing processes, including wet spinning, electrospinning, thermal drawing, and other methods. A variety of functions have been endowed to the fiber by changing the functional nanoparticles for blending before fiber drawing. The advantages of this strategy are simplicity, high productivity, and excellent mechanical strength. However, the functional performance of the as-drown fiber is limited by the interaction between the function of nanoparticles and polymer substrates.

The surface modification strategy usually includes three steps:

1) Functional materials are dissolved or dispersed with adhesive bonding agents into a viscous solution;
2) The fiber substrates should be treated with certain chemical agents or processing technologies to promote further adherence;
3) The viscous solution was coated on the fiber substrates to allow the interfacial reaction or physical transformation.

The surface modification strategy is relatively simple, high efficiency, and scalable. However, the coating of many function materials still requires harsh conditions, such as vacuum and high temperature. Besides, the directional growth of many functional structures on the fiber's bent surface is usually companied by adherence difficulties or compatibility issues.

This part will introduce various fabrication technologies for classical functional fibers.

3.1.1 Wet Spinning

The first method is wet spinning. It usually includes three steps:

1) Nano function materials are dissolved or dispersed in a solvent;
2) The solution is pressed out from a spinneret hole to form a trickle;
3) The nano-functional materials in the trickle are solidified with the fiber-forming process.

The wet spinning process has highly economy and high productivity. However, the homogeneity of the nano function materials constructed inside the fiber usually needs to be improved.

Wet spinning emerged as a means to produce synthetic polymer fibers such as nylon and spandex. With the development of nano function materials, wet spinning has also been widely applied in blending nano function materials into the classical raw materials to form functional fibers. For example, in 2017, Zhou et al. [1] reported a wearable textile-based humidity sensor for the first time using high strength (~750 MPa) and ultra-tough (energy-to-break, 4300 J/g) single-walled carbon nanotube (SWCNT)/poly(vinyl alcohol) (PVA) filaments via a wet-spinning process (Figure 3.1a). In 2018, Wu et al. [2] presented that continuous preparation of graphene oxide (GO) fibers had been realized via a wet spinning of GO aqueous solution (Figure 3.1b). Furthermore, the ammonia added GO fibers resulted in good mechanical properties, the tensile strength of 141 MPa at 5.6% elongation and Young's modulus of 3.7 GPa. At the same time, Tang et al. [3] reported that highly stretchable polymer elastomer-wrapped carbon nanocomposite piezoresistive core-sheath fibers were successfully prepared using a facile and scalable one-step wet-spinning assembly approach (Figure 3.1c). Moreover, the fibers showed excellent electrical conductivity with a low percolation threshold (0.74 vol%).

3.1.2 Electrospinning

The second method is electrospinning. It usually includes three steps:

1) Functional materials are dispersed in a polymer melt or solution;
2) The polymer solution is ejected through a spinneret driven by a high-voltage electric field;
3) The functional fibers are formed with the solidification of the ejecting polymer solution.

Figure 3.2a shows the schematic process of electrospinning [4]. Electrospinning technology is also a low-cost and simple method for the fabrication of various functional fiber materials.

As early as 1934, Formalas et al. [8] had reported a patent field in the high-voltage electrostatic spinning. However, until in 1993, this technology was beginning to be defined as electrospinning. For the application of electrospinning in the fabrication of fiber devices in 2015, Joseph et al. [5] demonstrated that biodegradable polymers and polycaprolactone were electro-spun obtain yarns. Moreover, they integrated conventional textile technology with electrospinning to form intelligent textile (Figure 3.2b). The advantage of this integration was that the textile could be drug/dye loaded and be developed into a core-sheath architecture with greater functionality. In 2016, Huang et al. [6] reported the development of high-performance wearable supercapacitor (SC) fabrics based on flexible metallic fabrics (Ni-cotton), on which nanofiber webs containing multi-walled carbon nanotubes (MWCNTs) were directly electrospinning (Figure 3.2c). The as-prepared fabric devices showed a high areal capacitance of 973.5 mF/cm^2 (2.5 mA/cm^2) and ultrahigh stability in

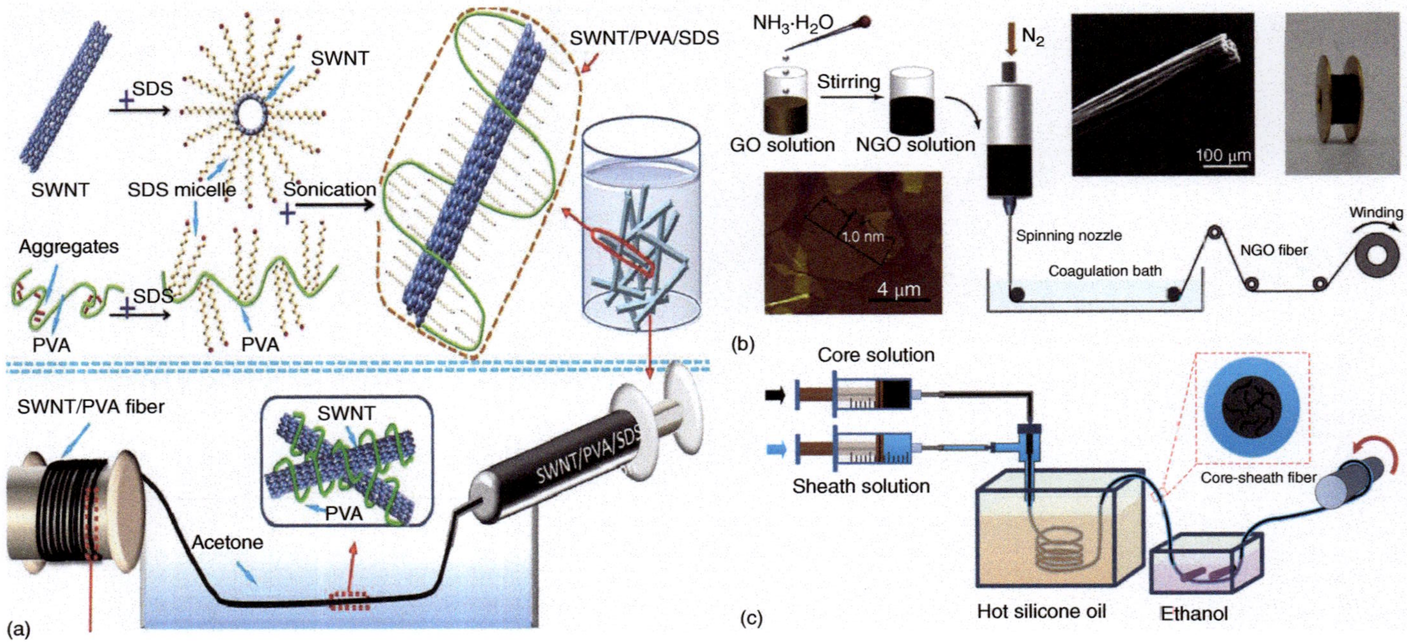

Figure 3.1 The development of wet-spinning process: (a) Schematic process of the wet-spinning used to produce the SWCNT/PVA composite filament. Source: Zhou et al. [1]. Licensed under CC BY 3.0. (b) Schematic illustration of continuous preparation of NGO fibers. Source: Adapted with permission from Wu et al. [2]. © 2018, Elsevier. (c) Schematic illustration of the coaxial spinning process for the highly stretchable fibers. Source: Tang et al. [3]. © 2018, American Chemical Society.

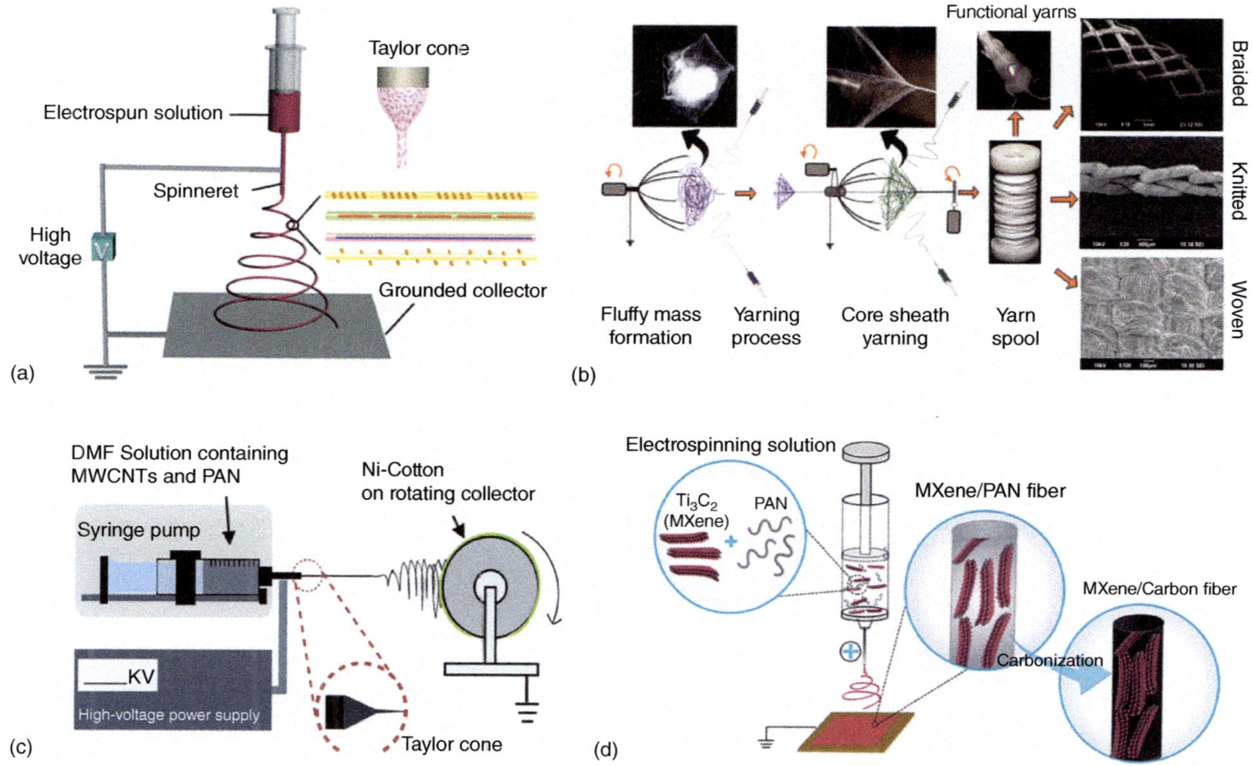

Figure 3.2 The mechanism and development of electrospinning: (a) Schematic representation of the laboratory setup for electrospinning. Source: Yu and Zhang [4]. © 2016, Royal Society of Chemistry. (b) Biodegradable polymers such as PLLA and PCL were electrospun to obtain yarns. Source: Adapted with permission from Joseph et al. [5]. © 2015, American Chemical Society. (c) Schematic illustration of the fabrication procedures of supercapacitor fabrics. Source: Huang et al. [6]. © 2016, Royal Society of Chemistry. (d) Schematic process of the production of $Ti_3C_2T_x$ MXene/carbon nanofibers via electrospinning. Source: Levitt et al. [7]. © 2018, Royal Society of Chemistry.

the bending test at a minimal bending radius (2 mm). Recently, Levitt et al. [7] presented that free-standing $Ti_3C_2T_x$ MXene/carbon nanofiber electrodes were prepared via electrospinning $Ti_3C_2T_x$ MXene flakes with polyacrylonitrile (PAN) and carbonizing the fiber networks (Figure 3.2d). Moreover, the electrodes exhibited an areal capacitance of up to 244 mF/cm^2.

3.1.3 Dry Spinning

The third method is dry spinning, which usually includes three steps. Firstly, functional materials are dispersed into a polymer solution with a low boiling point, good solubility. Secondly, the solution is fed at a constant and controllable rate through jet nozzles into a spinning duct in the heated gas. Thirdly, the solvent in the solution is volatilized, and the polymer strips with nano function materials gradually solidify. Lastly, the functional fibers are obtained through the post-treatment process, such as stretching, shaping, washing, and drying. With a lower boiling point for the solvent and higher temperature than wet spinning and electrospinning, the dry spinning process has continuous production, high spinning speed, and high yield.

In 2013, Kim et al. [9] demonstrated that biscrolling aligned electro-spun fiber sheets and carbon nanotube (CNT) sheets could be fabricated for conductive, functional yarns by a versatile dry composite method (Figure 3.3a). Moreover, the composite yarn's electrical conductivity values were three orders higher than those of other non-conducting polymers/CNT composite fibers electro-spun from polymer solutions containing similar loading of CNTs. In the next year, in 2014, Yue and coworkers [10] regenerated silk fibroin fibers with high strength and extensibility were firstly dry-spun from $CaCl_2$–formic acid solution (Figure 3.3b). The resulting as-spun regenerated silk fibroin (RSF) fibers showed a smooth and uniform surface with a diameter from 14 to 23 μm. The moderate mechanical properties of as-spun RSF fibers could be significantly improved by drawing one to two times in ethanol, which approaching native silk. At the same time, Zhang et al. [11] prepared CNT yarn by dry spinning (Figure 3.3c). The core/sheath yarn composed by dry-spun also showed a high tensile strength (550 MPa), which was about 80% of the pure CNT yarn strength (680 MPa). This was still much higher than the strength of conventional textile yarns.

3.1.4 Thermal Drawing Process

The fourth method is the thermal drawing process. It usually includes five steps. Firstly, functional materials are homogenized in a high-temperature solution and then are added to the macrostructured preform. Secondly, the functional materials-rich phase is cooling for solidification, during which it was drawn into fibers; lastly, the functional fibers are rinsed or evaporated to remove the solvent. Figure 3.4a shows the basic schematic process of the thermal drawing [12]. Besides, the nano function materials and architectures on the surface of fibers can be tunable to show more functions in the thermal drawing process.

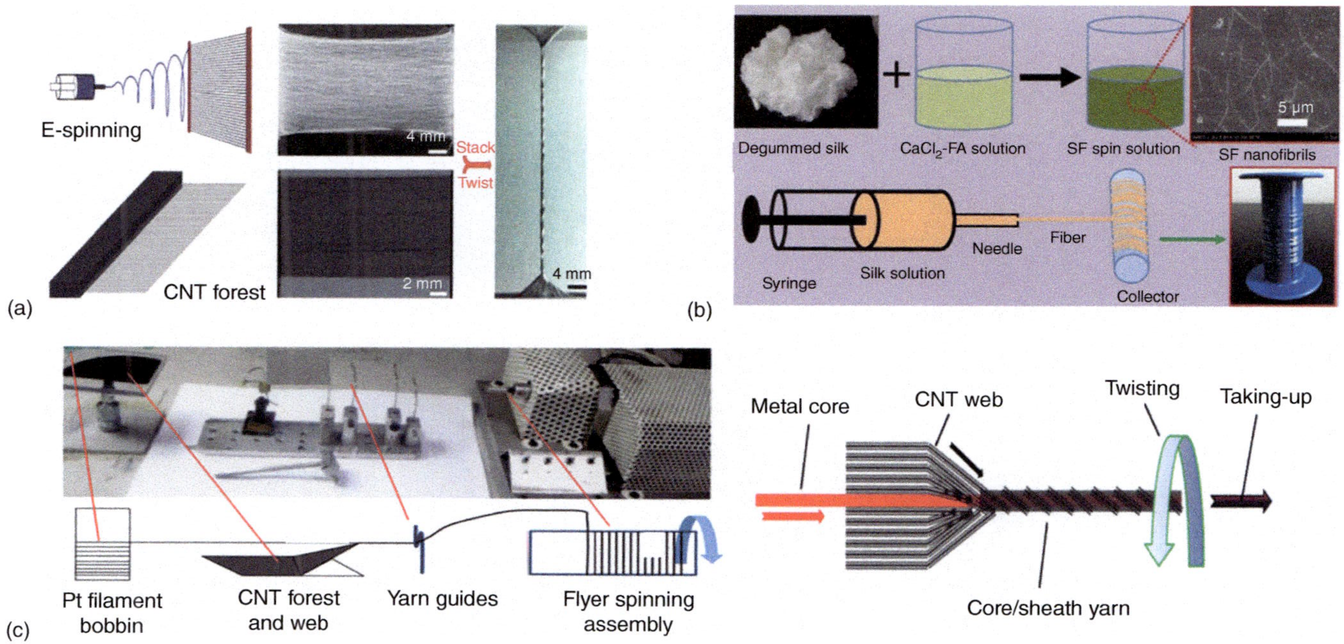

Figure 3.3 The schematic process and development of dry-spinning: (a) Biscrolling process using an aligned electro spun fiber (AEF) sheet and a CNT sheet. Source: Adapted with permission from Kim et al. [9]. © 2013, Royal Society of Chemistry. (b) Schematic process showing formation of the core/sheath yarn structure. Source: Adapted with permission from Yue et al. [10]. © 2014, Elsevier. (c) Continuous spinning of Pt/CNT core/sheath yarns, Photograph of the flyer spinner. Source: Adapted with permission from Zhang et al. [11]. © 2014, American Chemical Society.

The functional fibers fabricated by the thermal drawing are widely used in optics, communications, sensing, imaging, energy harvesting and storage, novel materials, and structure formation. However, it usually needs complicated and specific equipment.

In 2017, Grena et al. [13] demonstrated that spatially selective porous fibers could be produced by separating the polymer solution phase into a solvent-rich phase and a polymer-rich phase during the thermal drawing process (Figure 3.4b). Also, the pore size tunability between 500 and 10 μm was established by regulating the phase separation kinetics. At the same time, Zhang et al. [14] fabricated crystalline thermoelectric micro/nanowires/fibers by thermally drawing. Additionally, a wearable two-dimensional cooling textile was assembled to achieve a maximum cooling of 5 °C (Figure 3.4c). In 2018, Qu et al. [15] demonstrated the superelastic multi-material electronic and photonic fibers and devices via thermal drawing (Figure 3.4d). Furthermore, thermoplastic elastomers can be thermally drawn at high viscosities (above 10^3 Pa s), allowing the encapsulation of various micro-structured, soft, and rigid materials.

3.1.5 Surface Modification Method

In addition to the above common fiber drawing methods (wet spinning, electrospinning, dry spinning, and thermal drawing), people also use surface modification methods to prepare classical functional fiber materials. It usually includes three steps:

1) Functional materials are dispersed in a polymer solution;
2) Fibers to be coated are immersed in the polymer solution and taken from the solution after a short time;
3) The functional fibers are fabricated after dried.

The coating process is simple and highly efficient, applied to any complex shape of the fibers. However, the uneven coating of nano-functional materials on the fiber's bending surface is always a big challenge in this process.

For example, Bashir et al. [16] presented that high-performance fibers such as aramid (Twaron), polyamide (PA6), polyester (PET), and hybrid Twaron/PA6 fibers were transformed into electroactive fibers by coating them with conjugated polymer, poly(3,4-ethylenedioxythiophene) (PEDOT) through vapor phase polymerization (VPP) method. Obtained coated fibers at particular reaction conditions were successfully converted into electroactive fibers (Figure 3.5a). In 2018, Li et al. [17] presented that the reel dip-coating process was developed with deionized water and ethanol, and the eight CNT-aligned fibers were twisted into CNT yarns. As the motor was stably rotated, the MnO_2 was uniformly coated on the surface of the spiral structure CNT yarn (Figure 3.5b). The zinc-ion battery (ZIB) textile delivered a high specific capacity and volumetric energy density (302.1 mAh/g and 53.8 mWh/cm^3, respectively) as well as excellent cycling stability (98.5% capacity retention after 500 cycles).

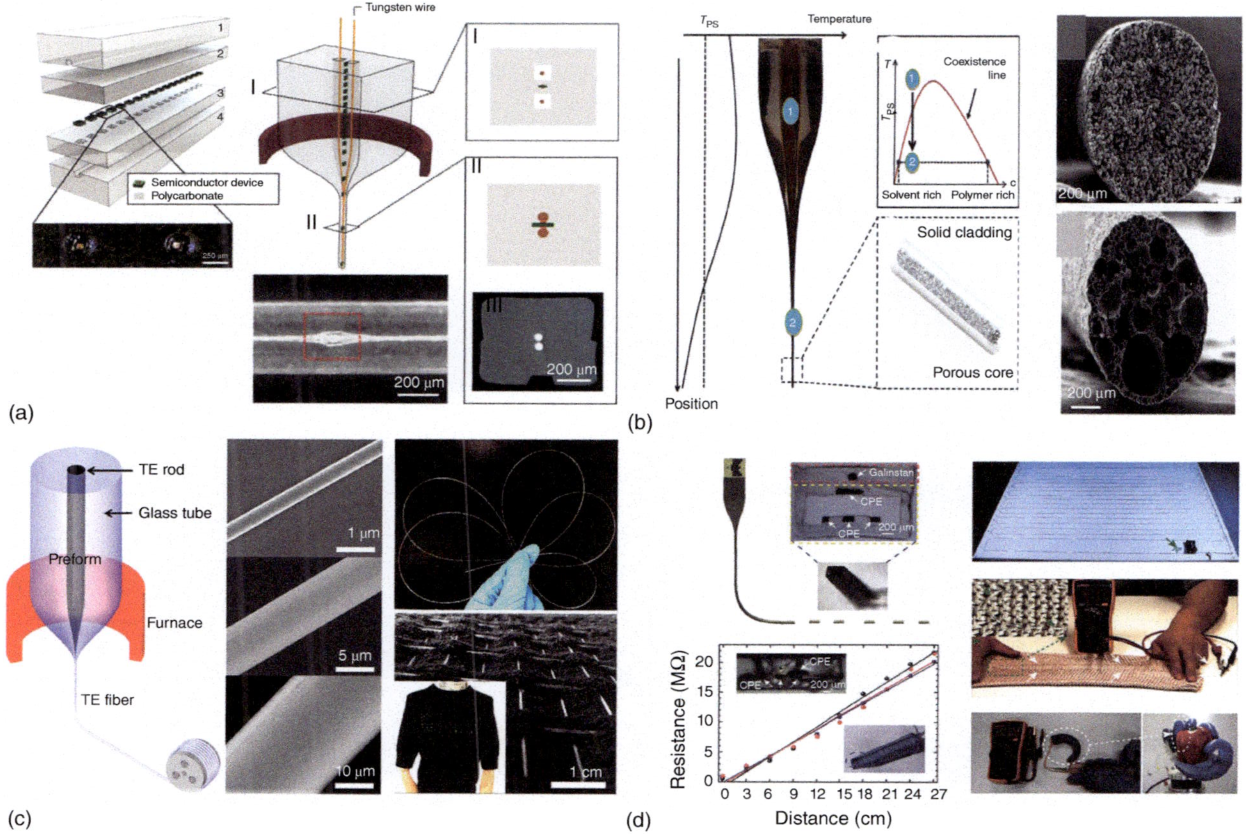

Figure 3.4 The schematic process and development of thermal drawing: (a) Schematic process of the apparatus of the thermal fiber drawing (TFD). Source: Adapted with permission from Michael et al. [12]. © 2018, Springer Nature Limited. (b) Thermal drawing of a fiber with a thermoelectric material in the core. Source: Adapted with permission from Grena et al. [13]. © 2017, Springer Nature Limited. (c) Silicon-core fiber drawn from an aluminum-core–silica-cladding preform. Source: Adapted with permission from Zhang et al. [14]. © 2017, Elsevier. (d) Thermally drawn of a micro-structured multi-material soft electronic fiber. Source: Adapted with permission from Qu et al. [15]. © 2018, WILEY-VCH Verlag GmbH & Co. KGaA, Weinheim.

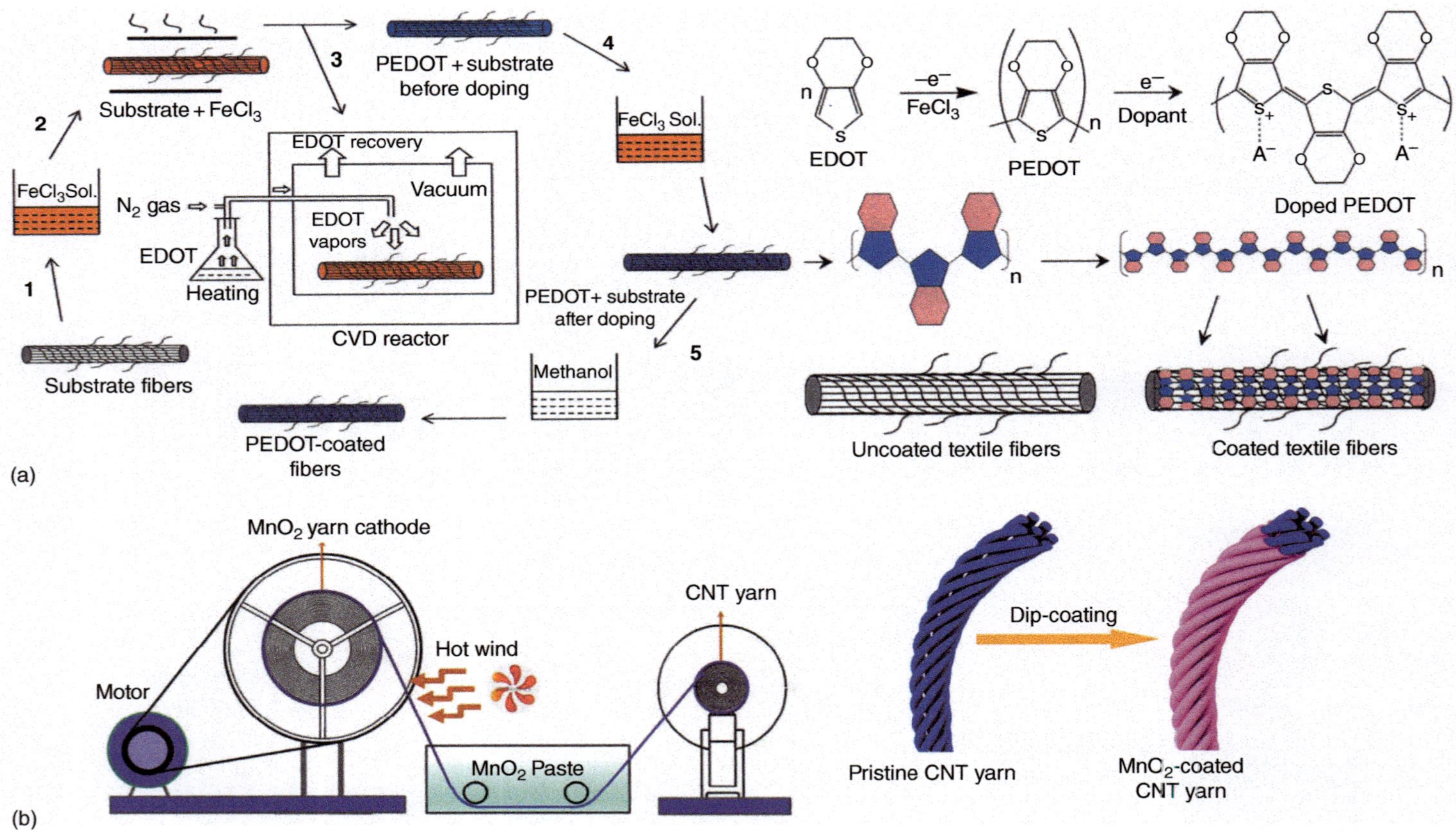

Figure 3.5 The diagrams of coating method: (a) Schematic diagram of poly(3,4-ethylenedioxythiophene) on textile fibers and polymerization and doping mechanism of PEDOT along with schematic process of substrate fibers before and after PEDOT coating. Source: Bashir et al. [16]. © 2017, John Wiley & Sons. (b) Fabrication process of MnO_2 yarn cathode. Source: Li et al. [17]. © 2018, American Chemical Society.

3.2 The Nano/Micro-Assembly on Fiber Materials

Intelligent fibers are fabricated by assembling the nano/micro-functional structures such as nanowires, nano-arrays, nanoballs, nanoflowers, and nanorods on functional fibers. Compared with the functional fibers from simple fiber drawing methods and surface modification methods, the intelligent fibers are fabricated by a more complicated process. One or more layers of ordered nano/micro-functional structures are assembled on fiber to make intelligent fiber electrodes, which can be then twisted or paralleled together to form intelligent fiber devices.

The micro/nano-assembly strategies can be divided into chemical liquid phase deposition (CLPD), physical liquid phase deposition, chemical vapor deposition (CVD), physical vapor deposition (PVD), and liquid phase electro-deposition. These assembly strategies can control nanostructures' size for the mass production of fiber-shaped functional materials, electrodes, and devices. However, they have many challenges in maintaining structural integrity and stability.

This part will introduce the technologies of the nano/micro-assembly on fiber materials.

3.2.1 Chemical Liquid Phase Deposition

In the CLPD, the fiber is immersed in the reaction solution, and then the required nano-functional structure can be in situ constructed on the surface of fiber with the reaction proceeds. The CLPD includes hydrothermal methods [18], and solvothermal synthesis [19], a microwave-assisted hydrothermal method [20], template-assisted synthesis [21], and others. People usually use the CLPD to assemble the nano-functional structures such as nanowires, nanorod, nanotube, nano-arrays, nanoflowers, and hierarchical array-type on the fibers. The CLPD is simple and low cost. However, it needs strict condition-control requirements for parameters, such as reaction time, concentration, and high temperature.

The first CLPD is the hydrothermal method in a sealed pressure vessel at a high temperature and pressure. It usually includes three steps. Firstly, the reaction raw materials are prepared into the solution and put into a Teflon-lined stainless steel autoclave. Secondly, the fibers are put into the autoclave, and then the autoclave is put in a drying oven with a high temperature for the hydrothermal process. Lastly, the nano-functional structures are constructed on the fiber. It is an important technique to prepare nano-materials with controllable structures, sizes, and morphologies. In 2017, Wei and coworkers [22] presented a weavable fiber-shaped one-dimensional SCs with high energy density for smart wearable electronic textiles (SWETs) (Figure 3.6a). The nickel-cobalt oxide hierarchical nano-structured arrays were synthesized via a facile hydrothermal route:

1) The conductive stainless steel yarn (denoted as SSY) substrate was immersed in the nitric acid solution (20 wt%) overnight and then was cleaned with de-ion (DI) water in the ultrasonic bath several times;

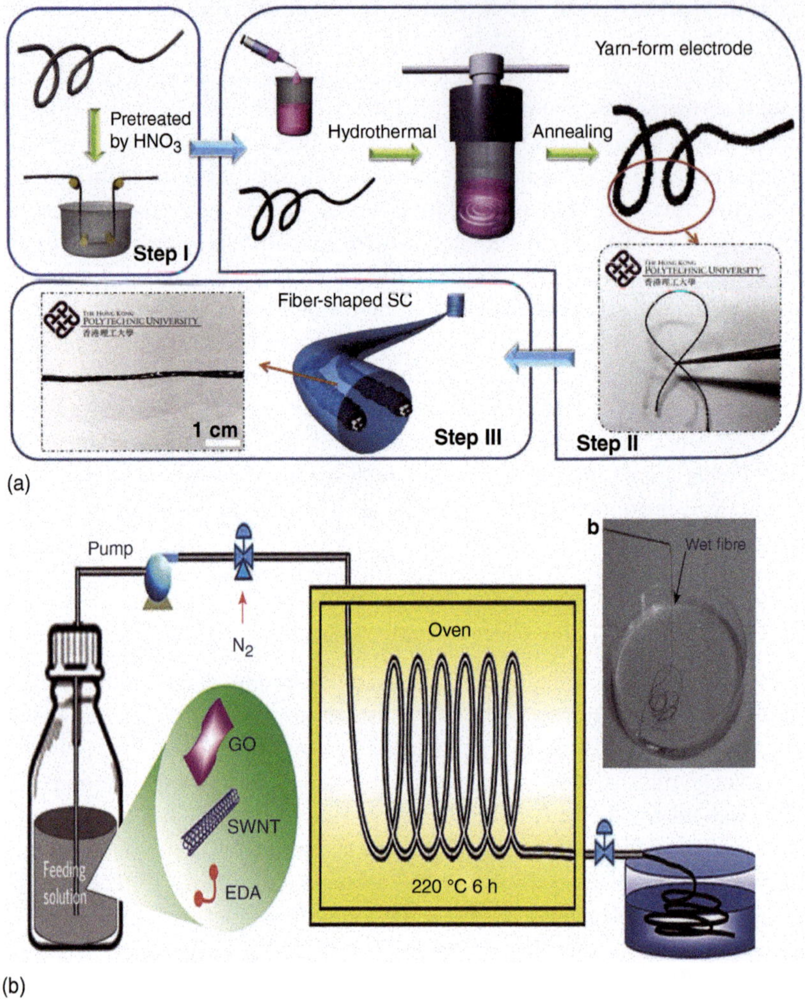

Figure 3.6 The diagrams of hydrothermal method: (a) Flow chart of the protocol for fabricating yarn-form electrodes coated with hierarchical nanostructured nickel-cobalt mixed oxides, and assembling of them into all-solid-state fiber-shaped SCs. Source: Adapted with permission from Wei et al. [22]. © 2018, Elsevier. (b) Schematic illustration for the preparation of CFs@ZnO NWs. Source: Adapted with permission from Xie et al. [23]. © 2018, Elsevier.

2) The as-obtained reaction solution was prepared and transferred into a Teflon-lined stainless steel autoclave, including the SSY substrate, and kept the temperature at 140 °C for 10 hours;
3) SSY coated with nickel-cobalt precursor was washed and dried in an oven overnight at 65 °C;
4) The nickel-cobalt precursor formed the hierarchical nano-structured nickel-cobalt oxide after annealed at 400 °C in the air for 2 hours;

5) The hierarchical nano-structured nickel-cobalt mixed oxides with outstanding electrochemical performance.

The as-fabricated yarn-shaped electrodes exhibit high capacitance, long cycling stability, excellent flexibility, and mechanical stability.

Recently, in 2018, Xie et al. [23] presented that a facile hydrothermal method has been successfully employed to modify CFs with the growth of ZnO nanowires (NWs), and the CFs@ZnO NWs were used to prepare paper-based friction materials based on a wet-forming process (Figure 3.6b).

1) The polydopamine (PDA) coated CFs were pre-prepared;
2) The PDA treated CFs were dipped in the ZnO seed solution, and then the soaked CFs were annealed at 150 °C for 15 minutes for three times to strengthen the adhesion between the ZnO seed and the CFs;
3) The hydrothermal growth of ZnO NWs on the CFs was conducted in the ZnO growth solution obtained by mixing zinc nitrate hexahydrate and hexamethylenetetramine (HMTA) in the deionized water;
4) The tribology properties of paper-based friction materials containing modified CFs were greatly improved compared to those containing pristine CFs.

The dynamic friction coefficient gradually reached a higher and more steady state as a function of time, and the wear rate decreased by 33%.

The second CLPD is the solvothermal method, which is developed based on the hydrothermal method. It usually includes three steps:

1) The precursor is dispersed in an organic or non-aqueous solvent;
2) The fibers are put into the solvents in a closed solvothermal reactor and reacted at a high temperature and pressure;
3) The nano-functional structures are constructed on the fibers.

The solvothermal method can be used to prepare metal oxides with different dimensions and morphologies on the fibers. In 2013, Chen et al. [24] presented that CuOx–Co_3O_4 heterostructure arrays were grown on 3D porous nitrogen-doped carbon nanofiber (PNCNF) substrates by electrospinning and solvothermal methods (Figure 3.7a). When the CuOx–Co_3O_4@PNCNF composite was employed as a free-standing anode for lithium-ion batteries, it exhibited a discharge capacity of 1122 mAh/g at a current density 200 mA/g after 100 cycles. In 2018, Kim et al. [25] presented that the hexagonal pillar-shaped Zn-MOF-74 with a diameter ranging from 300 to 600 nm grow along with the CNFs web by solvothermal method (Figure 3.7b). The C–Zn–MOF-74@CNFs showed improved oxygen reduction reaction (ORR) electrocatalytic activities favor a direct four-electron pathway and better durability than commercial Pt/C.

The third CLPD is the microwave-assisted hydrothermal method [26], where microwave interaction transforms the nano-functional structure. It usually includes three steps. Firstly, functional materials are dispersed in a polymer solution. Secondly, the solution is transferred in the microwave-assisted hydrothermal synthesis system. The solution is reacted at a high temperature for a while. Lastly, the nano-functional structure is constructed on the fibers after washing and drying.

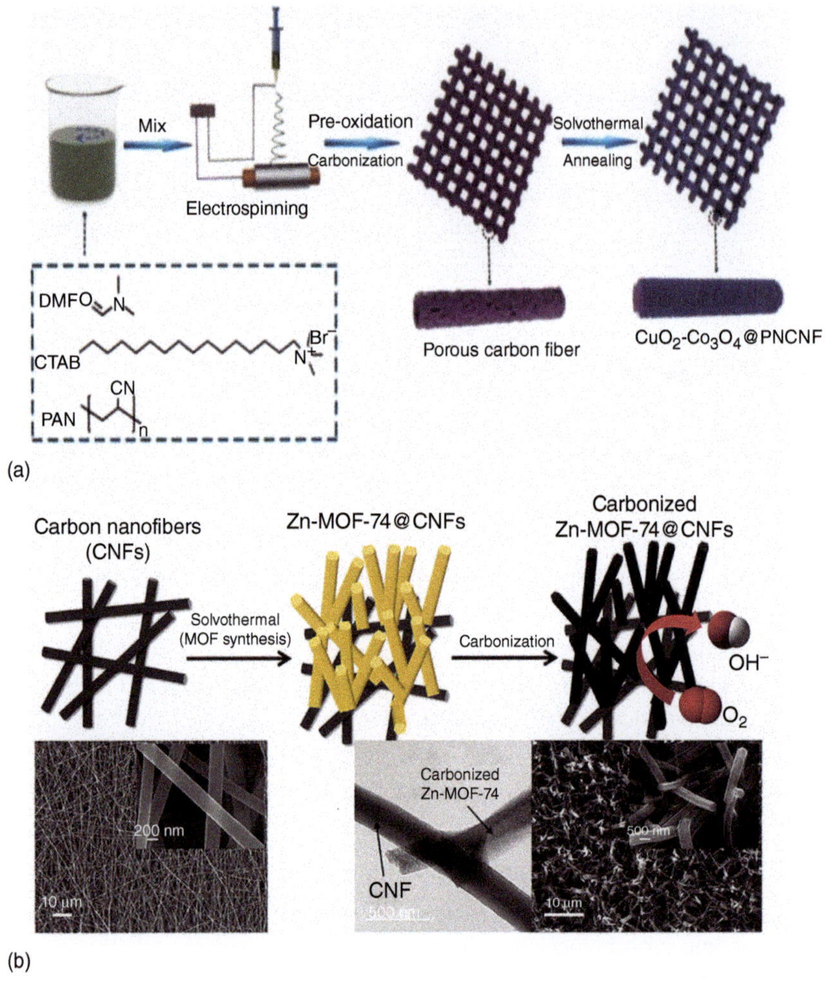

Figure 3.7 The diagrams of solvothermal method: (a) Schematic diagrams for the formation of CuOx–Co_3O_4@PNCNF composites. Source: Adapted with permission from Chen et al. [24]. © 2013, Royal Society of Chemistry. (b) Illustration of the preparation of C–Zn–MOF–74@CNFs. Source: Adapted with permission from Kim et al. [25]. © 2018, Elsevier.

Compared with traditional external heating methods, the microwave-assisted hydrothermal method is rapid volumetric heating without the heat conduction process, which can achieve uniform heating in a short period. It is a rapid, energy-saving, and promising route for the synthesis of functional nano-materials. However, it needs strict control of reaction conditions, and it also needs the precise control synthesis of functional nano-materials.

The microwave-assisted hydrothermal method has been successfully employed in multiple areas such as material sciences, nanotechnology, polymer chemistry, and biochemical processes. Han et al. [27] presented that this method could

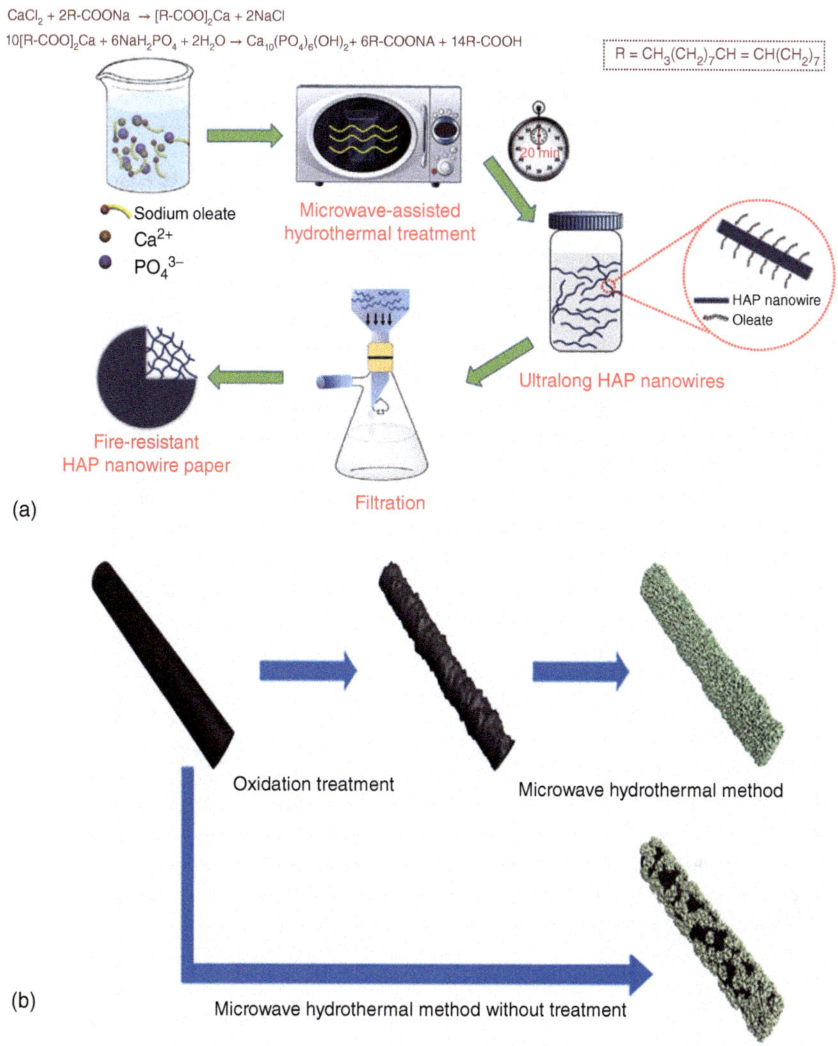

Figure 3.8 The schematic process of various microwave hydrothermal method: (a) Schematic illustration of the microwave-assisted hydrothermal strategy for the rapid synthesis of ultralong HAP nanowires. Source: Han et al. [27]. © 2018, Elsevier. (b) Schematic illustration of microwave hydrothermal synthesis of TiO_2/CF with CF treatment. Source: Meng et al. [28]. © 2014, Elsevier.

synthesize ultralong hydroxyapatite nanowires on the functional fibers with a length of several hundred micrometers and high flexibility (Figure 3.8a). The as-synthesized ultralong hydroxyapatite nanowires were the ideal raw material for fabricating the highly flexible and fire-resistant inorganic paper. Ultralong hydroxyapatite nanowires with diameters of tens of nanometers and lengths of hundreds of micrometers could be prepared in a short period (within 20 minutes). This significantly shortens the synthetic time by about two orders of magnitude compared with the conventional solvothermal method, showing an ultrahigh efficiency and significant energy saving. Recently, Meng et al. [28] represented that a systemic strategy was developed, employing loaded nano-TiO_2 on carbon fiber (CF) utilizing microwave hydrothermal method for the fast and efficient preparation of photocatalysts (Figure 3.8b). The experimental results indicated that TiO_2/CF possesses favorable photocatalytic activity, achieving a 97% degradation rate of Rhodamine B within one hour of UV light irradiation.

The fourth CLPD is template-assisted synthesis. It usually involves the following three steps:

1) The solution with nano function materials combined with templates by immersion method;
2) The solution is acted under a certain condition;
3) Nano-functional structures are constructed on the fibers after template removal.

It is a cost-effective process. However, it needs a complete removal of the template, and the breaking of the fibers' structure easily happens in this process.

Template-assisted synthesis is one of the most effective methods for fabricating nanofibers for diverse materials (metals, carbons, semiconductors, metal oxides, and conductive polymers). In 2009, Liu et al. [29] reported that a dual template method in which both a porous membrane and a small sphere were used as a template, and the diameter and pore diameter were controlled by adjusting the channel size of the membrane and the diameter of the sphere (Figure 3.9a). The nanoporous Pt–Co alloy nanowires exhibited distinctly enhanced electrocatalytic activities toward methanol oxidation compared to the current state-of-the-art Pt/C and PtCo/C catalysts' substantial promise as efficient anode electrocatalysts in direct methanol fuel cells. In 2012, Zhu et al. [30] represented that CNT@Ni_3S_2 one-dimensional hybrid nanostructures were prepared by a multi-step conversion route using CNT as a template (Figure 3.9b). Remarkably, this hybrid 1D hierarchical structure exhibits high capacitance with good cycling stability and significantly enhanced photocatalytic activity in photocatalytic H_2 production from water splitting. In 2017, Sin et al. [31] discovered that the mesoporous carbon nanotubes synthesized using MnO_2 nanorods as templates were reported (Figure 3.9c). They exhibited an excellent reversible capacity (~72.8 mAh/g) at 1 °C, capacity retention of ~98.4% after 100 cycles, and an excellent high-rate capability.

The fifth CLPD is the sol–gel method [32] that includes the following four steps:

1) Solid functional nanoparticles are dispersed in the monomer solution, yielding a colloidal suspension (sol);
2) The solution is polymerized after the aging process, and then an interconnecting network is formed between phases (gel);
3) The polymer nanoparticle 3D network is extended inside the liquid during the hydrolysis process;
4) The special nano-functional structures are constructed on the fibers after dried.

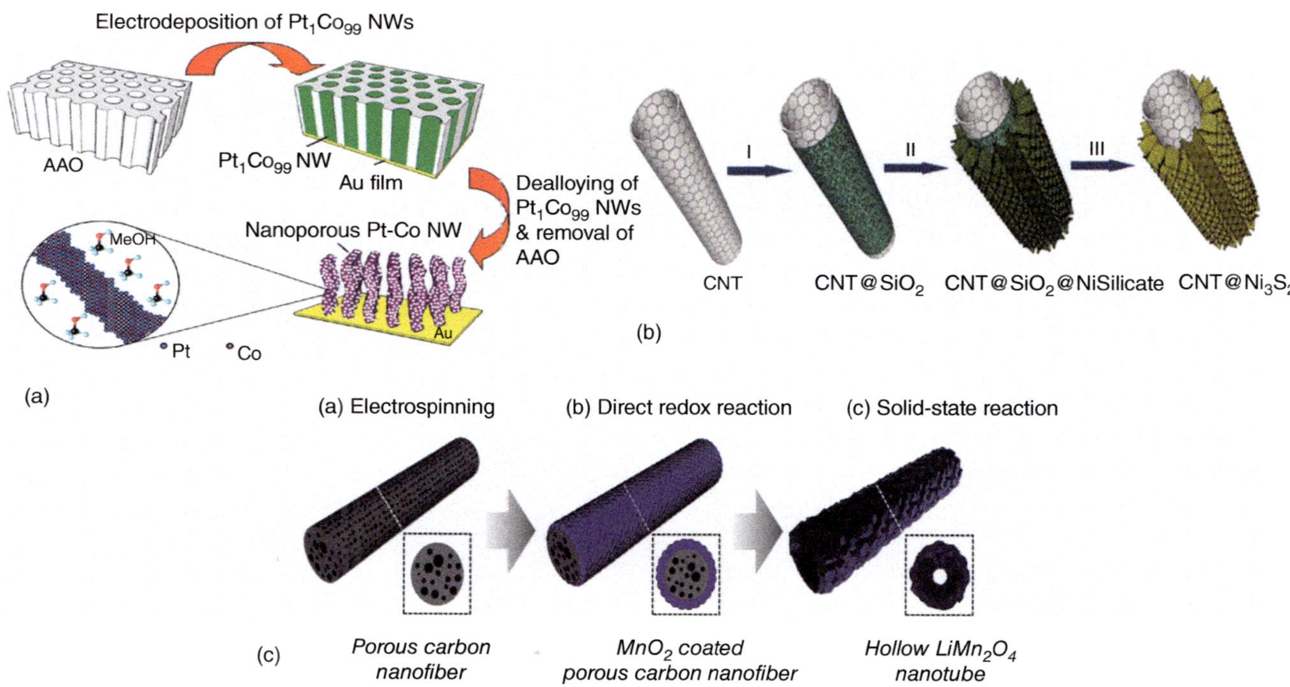

Figure 3.9 The schematic process of template-assisted synthesis: (a) Schematic process for fabricating porous Pt-Co alloy nanowires fiber. Source: Liu et al. [29]. © 2009, American Chemical Society. (b) Schematic process of the formation. Source: Zhu et al. [30]. © 2012, John Wiley & Sons. (c) Schematic illustration of process steps to fabricate the hollow $LiMn_2O_4$ nanotubes fiber. Source: Sin et al. 2017 [31]. © 2017, Elsevier.

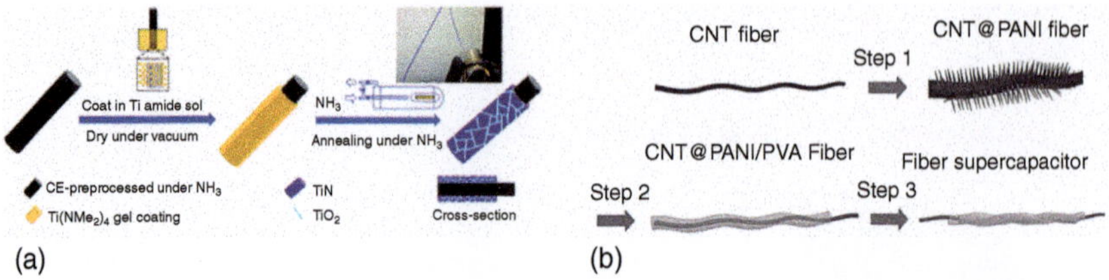

Figure 3.10 The schematic process of sol–gel method: (a) Schematic illustration of the flexible supercapacitors electrodes. Source: Adapted with permission from Zhang et al. [33]. © 2018, The Royal Society of Chemistry. (b) Schematic illustration of the assembly procedure for the fiber supercapacitor using polyaniline–CNT composite fibers. Source: Wang et al. [34]. © 2013, John Wiley & Sons.

The sol–gel process is widely used for synthesizing metal oxides, ceramic, and glass materials, especially 0-D, 1-D, 2-D, and 3D nano/micro-structures. In 2018, Zhang et al. [33] presented that smooth titanium oxide nitride coatings had been deposited on carbon fibers using a sol–gel method (Figure 3.10a). Moreover, the maximum electrical conductivity of the composite fabrics was 6.3 S/cm. In 2019, Wang and coworkers [34] used the sol–gel process to fabricate the conductive polyaniline array of nanowires deposited on the CNT fibers (Figure 3.10b). The resulting SC was highly flexible and could be easily woven or knitted into conventional textile fabrics for wearable electronics. At the current density of 0.01 mA/cm^2, the CNT@PANI fiber-based SC showed capacitance of 38 mF/cm^2.

Electro-depositions refer to the electroplating or electrophoretic deposition in the solution with nano-functional materials. It concludes the following four steps:

1) The functional materials are dispersed in the polymer-solvent, and the functional fibers are immersed in the solution;
2) The surface of functional fibers is activated under the action of high voltage;
3) Polar molecules' positive and negative charge centers are separated and adsorbed on the activated reaction points to become activated molecules;
4) The nano-functional materials are crystallized on the functional fibers.

People use the electro-deposition to assemble the metal hydroxides nano function materials with the morphology of vertically aligned nanosheets on the fibers.

Zhang et al. [35] synthesized highly capacitive hierarchically-structured MnO_2@PEDOT:PSS@oxidized carbon nanotube fibers (MnO_2@PEDOT:PSS@OCNTF) positive fiber-shaped electrode and flower-like MoS_2 nanosheets@CNTF (MoS_2@CNTF) negative fiber-shaped electrode. Moreover, the optimized stretchable fiber-shaped asymmetric supercapacitor (FASC) device exhibited a remarkable specific capacitance of 278.6 mF/cm^2 and a superior energy density of 125.37 μWh/cm^2 (Figure 3.11a). In 2018, Li et al. [17] developed a reel electro-deposition process to manufacture fiber-shaped zinc anodes with nanostructure (Figure 3.11b). The twisted yarn ZIB delivered a high specific capacity and volumetric energy density (302.1 mAh/g and 53.8 mWh/cm^3, respectively) and excellent cycling stability (98.5% capacity retention after 500 cycles).

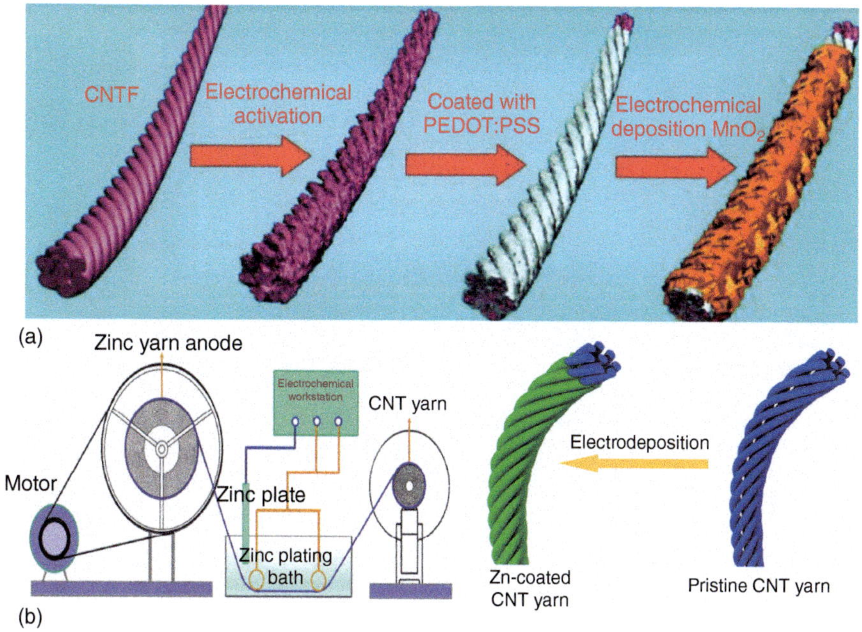

Figure 3.11 Schematic process of the various liquid-phase electrodeposition: (a) Schematic process of the hydrothermal-electrodeposition of hierarchical porous core/shell nanowire arrays. Source: Zhang et al. [35]. © 2017, Elsevier. (b) The roll-electrodeposition processes demonstrate strong potential for a completely scalable system. Source: Li et al. [17]. © 2018, American Chemical Society.

3.2.2 Plasma Spraying Method

The plasma spraying method [36] refers to a purely physical process, during which the molten or heat softened functional material is sprayed onto a surface of the fiber. It usually includes the following three steps:

1) The deposited material-typically as a powder, sometimes as a liquid, suspension, wire, or others -is introduced into the plasma jet;
2) In the jet, where the temperature is 10 000 K, the material is melted and propelled towards a substrate. There, the molten droplets flatten, rapidly solidify and form a deposit;
3) The deposits remain adherent to the substrate as coatings; free-standing parts can also be produced by removing the substrate.

It does not require harsh experimental conditions, and it can be mass-produced in nano/micro-functional structures. However, it has a restriction to the nano/micro-structure thickness and slow deposition rates. People usually use the plasma spraying method to assemble the nano-functional structures such as nanowires, nanorod, nanotube, nano-arrays, nanoflowers, nanoclusters, and hierarchical array-type on the fibers.

In 2018, Bu et al. [37] described that a good quality SiO_2/SiC coating was successfully fabricated on CF by a novel electrolytic plasma spraying method (Figure 3.12a).

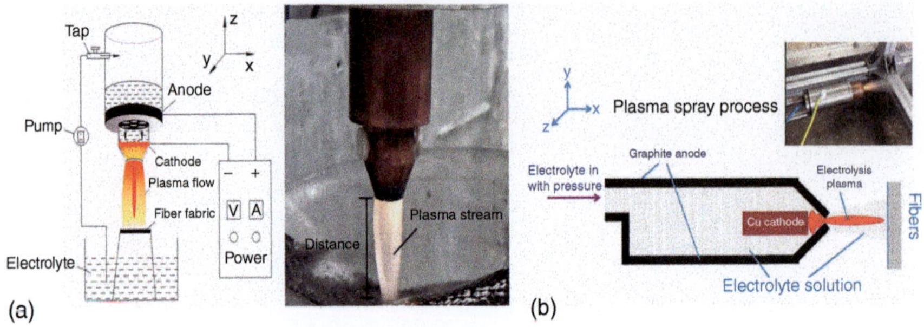

Figure 3.12 Schematic process of the low-pressure plasma spraying method: (a) Schematic process of the low-pressure plasma spray process. Source: Adapted with permission from Bu et al. [37]. © 2018, MDPI. (b) The schematic diagram of the experimental process with the spraying photo shown in the inset. Source: Adapted with permission from Bu et al. [38]. © 2020, MDPI.

Moreover, the tensile strength increased by 73 MPa in their work. Recently, in 2019, Bu and coworkers [38] presented that the synthesis of Al_2O_3 coating onto quartz fiber by plasma electrolysis spray for enhanced thermal conductivity and stability (Figure 3.12b). The nano- and micro-sized clusters were partially observed on the fiber. The deposition of the Al_2O_3 coating increased the tensile strength from 19.2 to 58.1 MPa, while the coated quartz fiber's thermal conductivity was increased by ~45% compared to the bare fiber.

3.2.3 Chemical Vapor Deposition

CVD refers to a process in which gaseous nano-functional substances are chemically reacted on a fiber surface and deposited on the surface using heat energy, plasma discharge, ultraviolet irradiation, individually or comprehensively, to form a nano-functional structure. People usually use the CVD to assemble the nano-functional structures such as nanowires, nanoball, nanotube, and nanoclusters on the fibers. The formation speed of nano-functional structures is fast, the composition of the nano-functional structure is precisely controllable, and the adhesion to the fibers is good in this process. However, it still has a low deposition rate.

The common CVD method has a problem with the controlled growth within a precise lithographically patterned area and usually needs a high substrate temperature for deposition. People have further developed advancing CVD methods by adding pre-patterned catalyst as the floating catalytic chemical vapor deposition (FCCVD) and intensifying the infiltration inside porous substrates as the chemical vapor infiltration (CVI).

For the FCCVD, the catalyst particles are just formed on the substrate with desired patterns. It usually includes four steps:

1) The catalyst is deposited on the substrate, and nucleation of the catalyst is carried via thermal annealing;

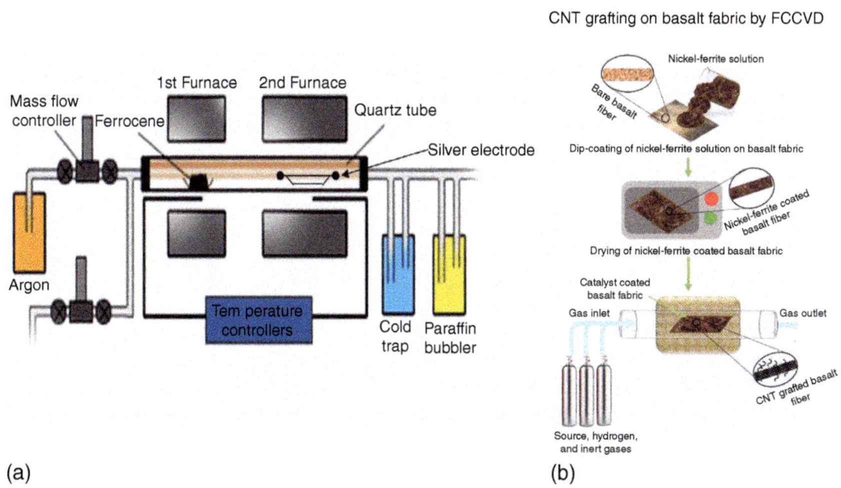

Figure 3.13 Schematic process of the floating catalytic chemical vapor deposition: (a) Schematic diagram of floating catalytic chemical vapor deposition method. Source: Zhou et al. [39]. Licensed under CC BY 4.0. (b) Schematic representation of the fabrication of CNT grafted basalt fabric reinforced epoxy composite. Source: Adapted with permission from Garima and Kyong [40]. © 2020, Elsevier.

2) The catalyst layer is patterned by chemical etching (as ammonia) and then is placed into a tubular reactor for the growth process;
3) After the reactor/furnace is heated to a sufficient reaction temperature, a gaseous mixture of raw materials is made to react in a reaction chamber over the surface of metal catalysts for a given time;
4) When the precursor's decomposition occurs, the function structure grows on the area with the catalyst particle.

Figure 3.13a shows the schematic diagram of a floating catalytic CVD method [39]. The FCCVD is critical for applications such as field emitters, sensors, electronic and photonic devices operating at micro- and nanoscale. It is a high yield and low cost. However, it has a low synthesis rate, and the catalyst is difficult to disperse on the functional fibers uniformly. For example, Garima et al. [40] presented that hierarchical structures of CNT-coated basalt fiber fabric (BF-CNTs) were made by FCCVD (Figure 3.13b). Furthermore, the BF-CNT600 fiber was shown volume resistivity in $4.5 \times 10^{12}\ \Omega$ cm.

The CVI usually includes the following three steps. Firstly, the gaseous precursor enters the reactor from the feed system at the bottom of the reactor. Secondly, the fibrous preform undergoes a chemical reaction at high temperature with the gaseous precursor. Thus the latter infiltrates in the fiber or preform crevices. Lastly, as the reaction between the fiber surface and the gaseous precursor takes place, the nano-functional structures are formed on the fiber surface. Figure 3.14a shows the schematic process of CVI. It can be applied for massive production with good process repeatability. For example, Xie et al. [42] produced porous carbon nanofibers by infiltrating diethylzinc vapor (Figure 3.14b). The nanofibers with porous carbon

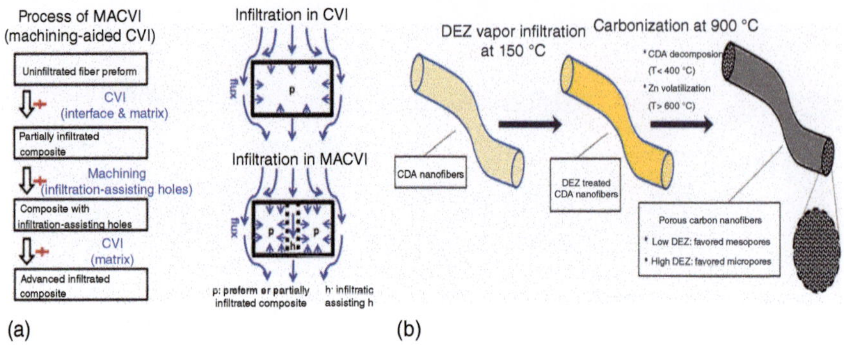

Figure 3.14 Schematic process of the various chemical vapor deposition: (a) Schematic process of the process of CVI. Source: Zhang et al. [41]. © 2017, Elsevier. (b) Schematic process of the DEZ infiltration process applied to prepare carbon nanofibers. Source: Xie et al. [42]. © 2018, American Chemical Society.

were produced with a higher surface area (1700 m^2/g) and pore volume, and they were promising materials for energy-storage applications.

3.2.4 Physical Vapor Deposition

PVD uses evaporation or sputtering strategies to make nano-functional materials into vapor and condenses and deposits on the functional fibers' surface. It is a purely physical process and is also the most important way to prepare nano-functional structures. PVD includes thermal evaporation and magnetron sputtering. People have employed the PVD to assemble the nanowires, nano-arrays, nanoflowers, nanoclusters, and hierarchical array-type on the fibers. The PVD has a high structural purity and fast nano-functional structure forming rate. However, the adhesion between the fibers and the substrate remains to be a challenge.

For the thermal evaporation method, the source material is evaporated by a resistive heat source in a vacuum–the vacuum facilities the vapor particles to travel directly to the target object (fibers). Figure 3.15a shows the schematic illustration of the thermal evaporation [43]. The thermal evaporation has a high deposition rate. The nanostructures of functional oxides, including nanobelts, nanowires, nanosheets, and nanodiskettes, can be synthesized in the thermal evaporation. For example, Yang et al. [43] synthesized a high-yield, high-purity SiC/SiO_2 core–shell nanocable on a CF felt in the absence of a catalyst by a simple thermal evaporation method (Figure 3.15b). The as-prepared nanocables with a 10–40 nm diameter have a crystalline SiC core and an amorphous shell of SiO_2. This composite nanostructure will provide opportunities for technological applications in electronic nanodevices.

For magnetron sputtering, it includes four steps:

1) The target cathode for nano function materials is put in a chamber filled with inert gas;
2) The electrons collide with the argon atom to form ionized Ar positive ions and new electrons under the action of the electric field;

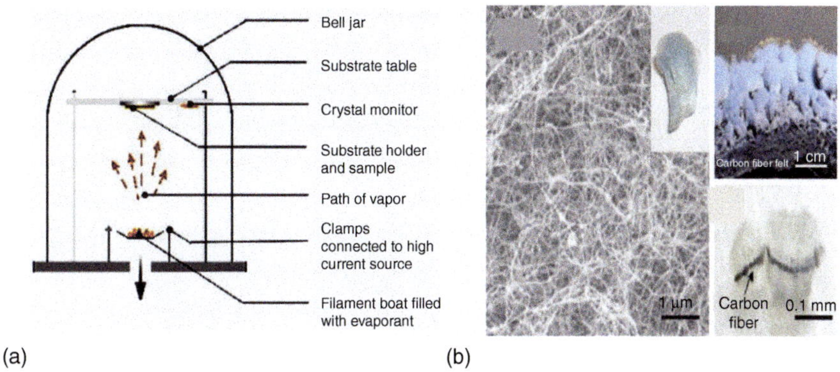

Figure 3.15 Schematic process of the various physical vapor depositions: (a) schematic illustration of the thermal evaporation. (b) The SEM image of large quantities of nanocables in the inner cross section of the lump formed on the carbon fiber felt (inset). Source: (a) Yang et al. [43]. © 2006, Elsevier, (b) Adapted with permission from Yang et al. [43]. © 2006, Elsevier.

3) The Ar positive ions accelerate to the target cathode under the electric field's action and bombard the target cathode's surface with high energy, causing the target to sputter;
4) The neutral target atom or molecule in the sputtering process will ionize many Ar ions to bombard the target cathode affected by the electric and magnetic field;
5) Neutral target atoms with function structures are efficiently deposited on the substrate like fibers.

Figure 3.16a shows the schematic process of magnetron sputtering [44]. It can be used for depositing metals, alloys, or inorganic oxide on fibers and textiles with a thickness of up to 5 μm. For example, in 2016, Liu et al. [45] successfully prepared CFs/ZnO core–shell hybrids (Figure 3.16b). the CFs/ZnO core–shell hybrids were used to fabricate a microfiber-shaped electrode to determine ascorbic acid (AA) and UA. Furthermore, the AA and UA's linear range was 600–2000 μm and 20–200 μm, respectively.

3.3 Device Assembly from Fibers to Textiles

Clothing is a feature of all human societies. Textile-like materials date back to 34 000 before the common era, even possibly to prehistoric times. Textiles' functionalities have been limited to skin protection, body temperature maintenance, decoration, and aestheticism. Nowadays, merging electronics with textiles is becoming increasingly desirable, as this can lead to an excellent electronic function platform for maintaining features of textiles. People have assembled multi-layered functional materials with specific nanostructures on fibers, like fabricating an electronic device to fabricate an intelligent fiber and textile. During which, the functional fiber materials usually act as the electrode for intelligent fiber device. The assembly strategies of

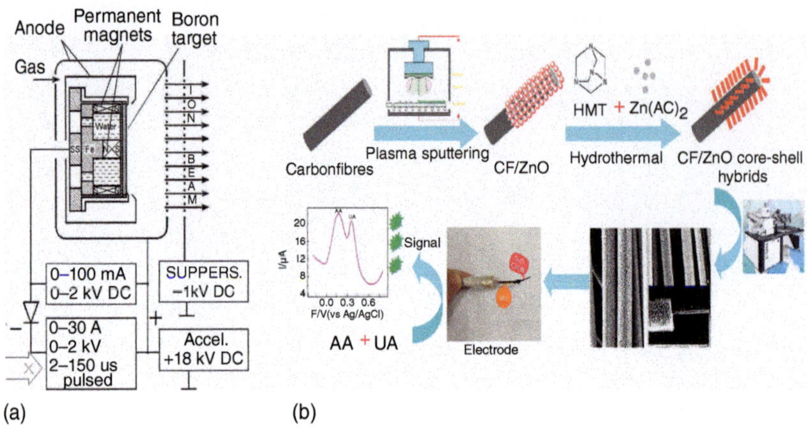

Figure 3.16 Schematic process of the various physical vapor deposition: (a) Schematic process of the magnetron sputtering process. Source: Mohammad [44]. © 2018, Elsevier. (b) Schematic illustration for preparation of CF/ZnO CS hybrids and biosensor. Source: Adapted with permission from Liu et al. [45]. © 2016, Elsevier.

intelligent textile usually include the direct coating based on fabric, the layer stacking of fabric electrodes, the interweaving of fiber electrodes, the weaving of fiber devices, and others. The following sections present a broad picture of intelligent fibers' assembly strategies into intelligent textile.

3.3.1 Direct Coating Based on Fabric

The first method is a direct coating based on fabric. It usually includes three steps. Firstly, various conductor and semiconductor materials were successively coated on a textile substrate layer by layer. Secondly, a layer of conductor, as the counter electrode, was coated on the semiconductor layer. Lastly, intelligent textile was fabricated after the encapsulation process.

Liu et al. [46] presented proposed a novel fabrication process for a monolithic-structured solid-state dye-sensitized solar cell (ssDSSC) on textile using all solution-based processes (Figure 3.17). The fabrication process for monolithic-structured ssDSSC on a single textile consists of five main steps. All functional layers were directly deposited onto the textile substrates using screen printing and spray-coating processes:

1) The silver bottom electrode was screen printed onto the dried polyimide to further reduce the textile's surface roughness;
2) The spray pyrolysis deposition of TiO_2 CL used a pre-designed shadow mask aligned with the silver bottom electrode on the fabric substrate;
3) The mesoporous TiO_2 layer was deposited on the TiO_2 CL and sintered to form photo-anodes, followed by the dye staining process;
4) The freshly prepared solid-state electrolyte solution was deposited on the dye stained devices and left for 16 hours to allow the solid-state electrolyte to infiltrate into the TiO_2 mesoporous film;

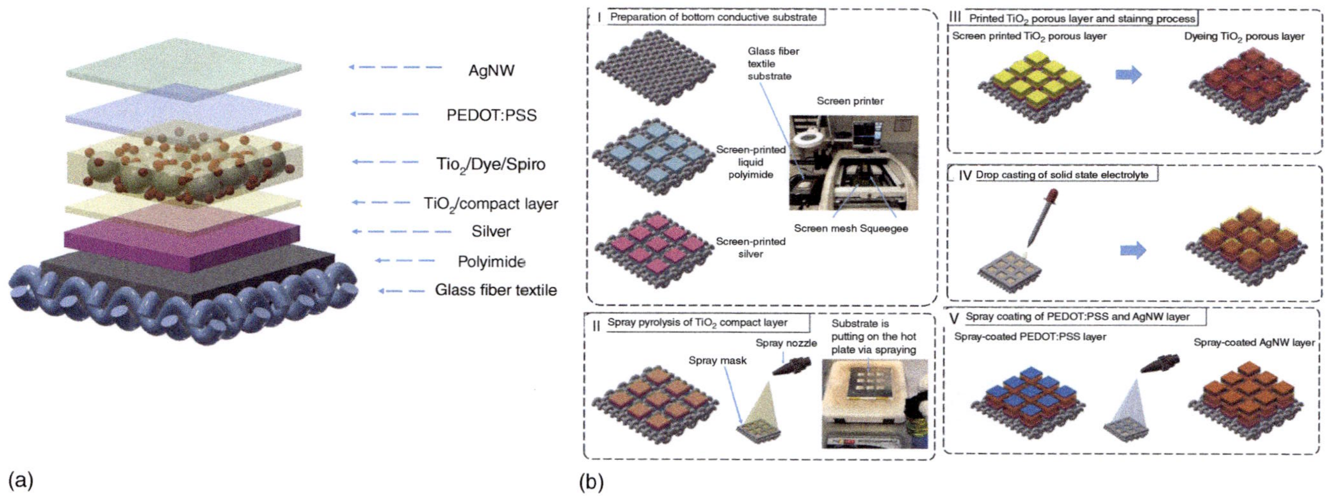

Figure 3.17 The examples of the assembly from normal fabr cs to functional fabrics: (a) Textile based ssDSSC device structure. (b) Fabrication process diagram of textile based ssDSSC. Source: Adapted with permission from Liu et al. [46]. © 2019, Springer Nature.

5) The fabric devices were spray-coated with PEDOT: PSS and AgNW top electrodes to complete the fabrication.

The textile device showed a maximum PCE of 0.4% with a V_{OC} of 0.31 V, while the FF and J_{SC} were 0.25 and 5.2 mA/cm^2, respectively.

3.3.2 Layer Stacking of Fabric Electrodes

The second method is the stacking of two or more textile-type electrodes, like a sandwich. The fabric-type electrode was fabricated by a direct coating of conductor and semiconductor materials on a textile substrate, like the method to fabricate a conventional plate-type device. The coating process can be accomplished by dip coating, spray coating, and in situ chemical growth. After that, the working fabric electrode and counter fabric electrode were stacked together to obtain a double-layered textile, easily integrated into various fabrics or other flexible structures. Unlike the first method, the counter electrode was stacked as another fabric electrode, instead of coating a conductor layer as the counter electrode on the semiconductor layer.

In 2014, Pan and coworkers [47] had developed a new universal method to produce flexible, wearable dye-sensitized solar cell textiles by stacking two textile electrodes (Figure 3.18a). For working electrodes, the titanium wires with a diameter of 130 μm were firstly woven into textiles. Secondly, aligned TiO_2 nanotubes were then perpendicularly grown on the titanium wire surface by electrochemical anodizing. Lastly, after the absorption of the dye N719, the Ti wire textile was used as the working electrode. For the counter electrode, spinnable CNT arrays were first synthesized by CVD. Secondly, the CNTs were continually pulled out of the array and then stacked into layered CNT films. Thirdly, the layered CNT films were further twisted into a CNT fiber. Finally, the CNT fibers were woven into textile electrodes as the counter electrode. After fabricating the two fabric electrodes. The dye-sensitized solar cell (DSSC) textile was fabricated by stacking the CNT fiber textile onto the modified Ti textile, followed by incorporation of electrolyte. The wire-shaped DSSC showed an energy conversion efficiency of 4.6 %.

Yun and coworkers [48] presented a flexible, efficient DSSC textile fabricated by sewing textile-structured electrodes onto casual fabric such as cotton, silk, felt, or paper (Figure 3.18b). One electrode of the 3-D textile fulfills the photo-anode's role, and the other acts as the counter electrode. For the photo-anode, TiO_2 paste was printed on electrode I using the floating printing method, which is suitable for the textile substrate, and then a sintering process was conducted. For the counter electrode, a platinum (Pt) particle coated with activated carbon paste was printed on electrode II. The printing process after fabrication of the 3-D textile prevents cracking or peeling of the deposited electrode. A non-volatile succinonitrile (SN)-based electrolyte was infiltrated onto the 3-D textile after dye-loading on the photo-anode. Then the fabrication process was completed by sealing with PET film using the pouching method. Through this process, 3-D textile DSSCs were successfully fabricated, and there is a possibility of a continuous roll-to-roll manufacture. The fabricated 3-D textile DSSCs has a short-circuit current density of 4.63 mA/cm^2 and an open-circuit voltage of 0.64 V.

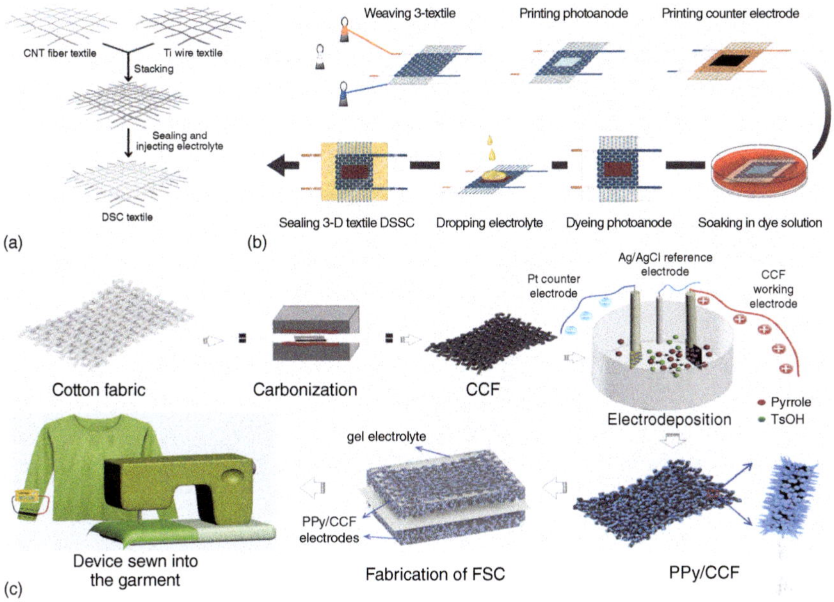

Figure 3.18 The examples of the assembly from fabric electrodes to fabric devices: (a) Fabrication of the dye-sensitized solar cell textile. Source: Pan et al. [47]. © 2014, John Wiley & Sons. (b) The process used to create the textile-based flexible solar cells and schematic process of fabrication process. Source: Yun et al. [48]. Licensed under CC BY 4.0. (c) Flow diagram for the fabrication of electrode materials and all-solid-state supercapacitor. Source: Sun et al. [49]. © 2018, Elsevier.

Recently, in 2018, Sun et al. [49] presented the first time applying the synergistic effect of polypyrrole (PPy) and carbonized cotton fabric to prepare high capacitance electrodes for weaving, making flexible and wearable SCs. In this study, the pristine cotton fabric was carbonized at high temperatures to obtain carbonized cotton fabric (CCF) material. Then PPy was coated on the CCF through in situ electro-deposition, and then the PPy/CCF fabric electrodes were fabricated. The flexible supercapacitor (FSC) was assembled with two identical PPy/CCF electrodes, immersed in the gel electrolyte for 1 hour. After being taken out, the two fabric electrodes were parallel located on a cellulose film. When the LiCl/PVA gel solidified at room temperature for two hours, the FSC was obtained (Figure 3.18c). Also, the FSC obtained the maximum value of 500.06 mF/cm^2 (5.9 F/cm^3) at the current density of 2 mA/cm^2.

3.3.3 Interweaving of Fiber Electrodes

Although the intelligent textile based on layer stacking has a convenient and straightforward fabrication process, stacking functional layers with required spacers between them increases device thickness. It undermines its inherent wearing comfort, such as flexibility, breathability, and sensitivity to biomechanical deformation induced by body movement. These concerns can be eliminated by developing woven or knitted intelligent textile based on the fiber electrodes' interweaving.

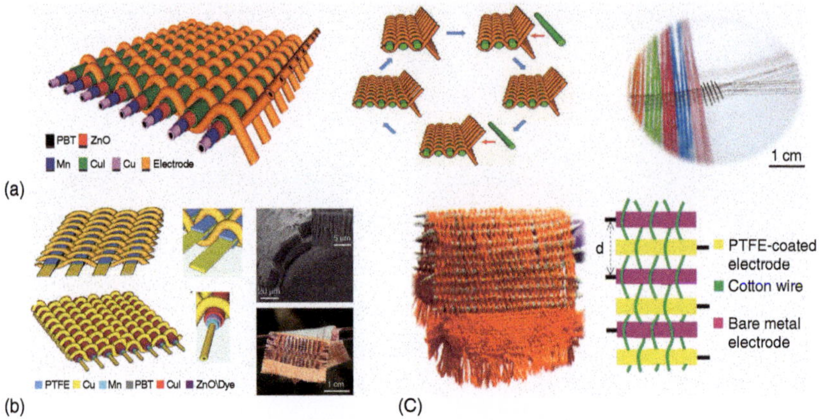

Figure 3.19 The examples of the assembly from fiber electrodes to fabric devices: (a) Structure design of the photovoltaic textile. Source: Adapted with permission from Zhang et al. [50]. © 2015, WILEY-VCH Verlag GmbH & Co. KGaA, Weinheim. (b) Structural design of the hybrid power textile. Source: Adapted with permission from Chen et al. [51]. © 2016, Springer Nature Limited. (c) Photograph of a flexible sensor textile. Source: Adapted with permission from Zhang et al. [52]. © 2018, American Chemical Society.

Different fiber electrodes were fabricated and then fixed on a weaving machine as the weft and warp. After interwoven together by a so-called shuttle-flying process (Figure 3.19a), the intelligent textiles were formed.

In 2015, for the first time, Fan and coworkers [50] presented a solution to power portable electronics in a wearable manner by fabricating an all-solid photovoltaic textile. The photovoltaic textile had a single-layered interlaced structure, composed of wire-type photo-anodes and counter electrodes (CE). The wire-type photo-anode was fabricated by growing a layer of ZnO nano-arrays on a lightweight and low-cost Mn-plated polymer wire as the substrate. After being dye-sensitized, a layer of CuI was deposited onto the ZnO nano-arrays as the hole-transfer material. The counter electrode is made of Cu-coated polymer wires. Subsequently, strings of wire-shaped photo-anodes and counter electrode were woven in a staggered way on an industrial knitting machine to form the photovoltaic textile based on a shuttle-flying process (Figure 3.19a). Under a standard light intensity of 100 mW/cm^2, an open-circuit voltage of 4.6 V and a short-circuit current density of 7.8 mA/cm^2 were achieved, corresponding to an efficiency of 1.3% for a single cell unit.

In 2016, Fan and coworkers [51] reported a micro-cable power textile for simultaneously harvesting energy from ambient sunshine and mechanical movement. Solar cells fabricated from lightweight polymer fibers into micro-cables are then woven via a shuttle-flying process with fiber-based triboelectric nano-generators to create a smart fabric (Figure 3.19b). This hybrid power textile, fabricated with a size of 4 cm by 5 cm, was demonstrated to charge a 2 mF commercial capacitor up to 2 V in 1 minute under ambient sunlight in the presence of mechanical excitation, such as human motion and the wind blowing.

In 2018, Fan and coworkers [52] also presented a flexible self-powered all-in-one fluid sensor textile for simultaneously monitoring the velocity, acceleration, and chemical composition based on an enhanced liquid–solid triboelectric effect (Figure 3.19c). The process of assembling the textile includes two steps. Firstly, Ni dendrites were directly deposited on metal fibers through a nontemplate electro-deposition process at a suitable range of current density using self-made all-round reactors, and a thin layer of polytetrafluoroethylene (PTFE) nanofibers were deposited on dendrite electrode by the spray-coating method. Secondly, all of the electrodes were employed as the weft, and cotton wire was employed as the warp. Both were woven to form a textile via a shuttle-flying process. Furthermore, a textile of 6 cm^2 could charge a commercialized capacitor to 1 V within 80 seconds by harvesting flow energy on the liquid/solid interface, showing a potential use as the power supply of a signal-processing circuit.

3.3.4 Weaving of Fiber Devices

The textile device assembly method, via the weaving of fiber devices, usually includes three steps:

1) The fiber devices were assembled from the fiber electrode;
2) The fiber devices, as a whole, were fed into the weaving machine as the weft or warp;
3) By weaving together with cotton or other polymer wires, intelligent textiles can be obtained.

For example, in 2020, Fan and coworkers [53] presented a photo-rechargeable fabric as a high-performance, sustainable, and stable power source for the on-body sensor network. The fabric consists of a photovoltaic energy-harvesting component and a rechargeable fabric battery to form a self-charging power unit. For the weaving of the energy-storage device, firstly, MnO_2 and Zn electrodes were assembled on conductive MWCNTs/polyethylene terephthalate (PET) wires, which were formed by embedding a strand of PET wires inside a net of MWCNTs. Secondly, the MnO_2 electrodes were prepared by brush coating a layer of MnO_2 nanofibers, while the Zn electrodes were prepared by electro-depositing a layer of metal Zn. Thirdly, a MnO_2 electrode and a Zn electrode were aligned in parallel as a pair and buried inside a neutral aqueous gel-type electrolyte. The battery wire was encapsulated in a polymethyl methacrylate (PMMA) layer to avoid possible contact of the electrolyte with human skin after solidifying gel electrolyte. Thirdly, entwined the battery wire inside thin colored cotton wires instead of the commonly used plastic pipe, which protects the battery core and endows the battery wire with cotton-like softness. Lastly, the energy-storage textile was directly woven via a scalable shuttle-flying process with various functional fibers, including polymer warp and battery wires (Figure 3.20a, b). The fabric could constantly deliver electric power for 10 minutes at 0.1 mA after being charged for 1 minute under the standard 1-sun condition.

3.3.5 Other Assembly Methods

In addition to the above four device assembly methods, other assembly methods, such as couching embroidery, have been proposed as an industrially scalable

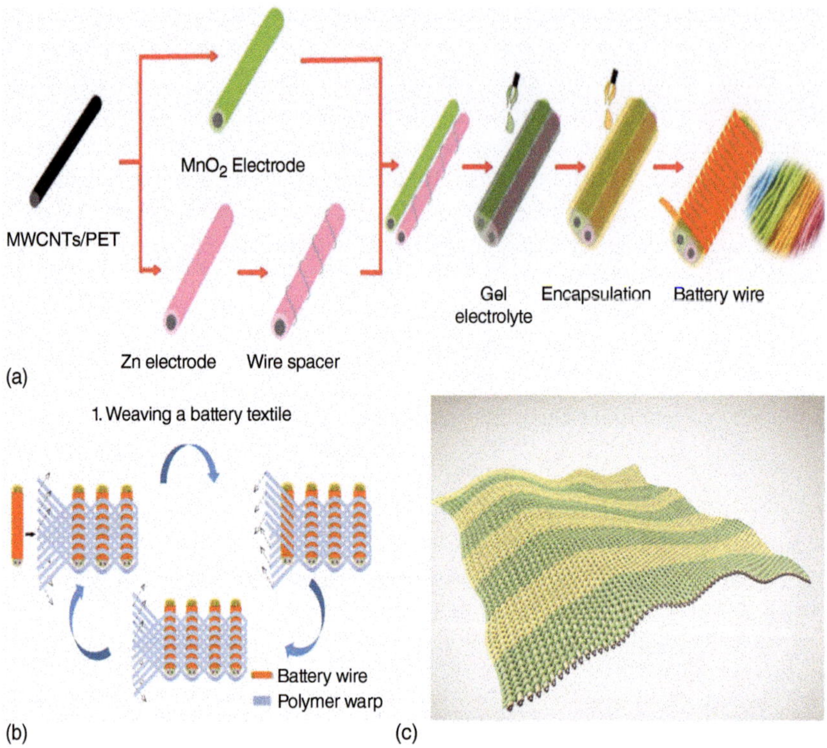

Figure 3.20 The example of the assembly from fiber devices to fabric devices: (a) Fabrication process of the battery wire. (b) Fabrication process of the energy-storage textile. (c) Schematic illustration. Source: Zhang et al. [53]. © 2020, Elsevier.

electrode-to-device assembling strategy to provide a light and breathable intelligent textile adaptable for various weather conditions and body movements. It usually includes the following four steps. First, a conductive thread pierced down through a bottom tulle following the lead of a needle. Second, the end of the conductive thread rotated in a circle with the assisting of a shuttle hook, during which the conductive thread would also circle the bottom thread. Third, the needle would lift above the bottom tulle again, and the conductive thread would be tightened with the bottom thread. Fourth, a function electrode cable fed in-between the conductive thread and the bottom tulle would move a small step together with the bottom tulle following a specified trace. When the needle pierced down again, the conductive thread would stitch the function electrode cable on the bottom tulle and function as a counter electrode to form a device unit. Unlike the current device assembling strategies of layer-by-layer electrode stacking or shuttle-flying weaving, both long and short cable electrodes can be conveniently and reliably stitched together via the couching embroidery strategy as one device with required energy output. This technology was more conducive to integrating more wearable electronic devices into daily human life, making wearable technology more portable, fast, and comfortable.

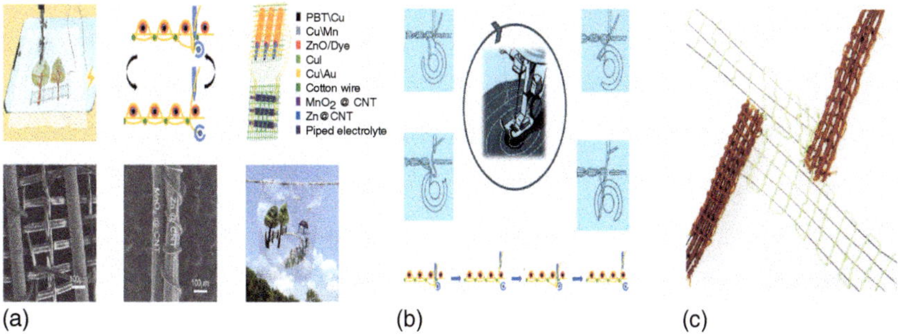

Figure 3.21 Construction of the photo rechargeable energy fabric: (a) construction of the photo-rechargeable energy fabric. (b) Schematic illustrations of the embroidery process of the energy fabric. (c) The couching embroidery of the photo-anode and battery. Source: Huang et al. [54]. © 2019, American Chemical Society.

Fan and coworkers [54] proposed a photo-rechargeable energy fabric consisting of both the photovoltaic unit and Zn–MnO_2 rechargeable battery unit was fabricated in the couching embroidery way (Figure 3.21). For the photovoltaic unit, conductive threads, such as Cu-coated polymer wires or Cu wires, could be directly employed as the top threads, which then served as the counter electrodes assembling or the connecting wires for circuit integration. The lightweight photo-anode wires stitched on the bottom tulle were assembled based on Mn-coated polymer wires and coaxially coated with layers of ZnO nano-arrays and CuI in sequence (Figure 3.21a). For the battery unit, lightweight and rechargeable fiber Zn–MnO_2 batteries of any length could be directly and reliably stitched on bottom tulle by either conductive or insulated top threads. In each battery, MnO_2 and Zn electrodes were arranged in pair and then were sealed in a tube together with the neutral aqueous gel-type electrolyte. The MnO_2 electrodes or Zn electrodes were fabricated on very light conductive wires formed by coating a layer of MWCNT on a strand of PET wires. The MnO_2 electrode was assembled by coating a layer of MnO_2 nanofibers, while the Zn electrode was prepared by electro-depositing a layer of Zn metal. The Zn electrode was twisted with a thin cotton wire to prevent short-circuiting. Well-patterned battery cables were twined with colorful threads to endow the device fabric with almost the same softness, wear resistance, and esthetics as common cotton fabric. Besides, a gauze glove of such a device could drive commercialized electric watches under dark conditions for over 1 hour after 10 minutes of standard solar illumination.

In this chapter, we have discussed the fabrication process of intelligent fibers and fabrics. Inside the intelligent fabric, functional fibers serve as the fundamental materials of intelligent fibers and textiles. Moreover, we need to assemble functional materials with a nano function structure layer by layer on a fiber to form an electrode, similar to assembling an electronic device. By weaving or twisting different fiber electrodes, intelligent fibers or textile devices can be assembled. A scalable fabrication technology can be regarded as a critical challenge for the real application of intelligent fiber and textiles, especially for integrating the intelligent fibers into commercialized textile devices.

References

1 Zhou, G., Byun, J.H., Oh, Y. et al. (2017). Highly sensitive wearable textile-based humidity sensor made of high-strength, single-walled carbon nanotube/poly (vinyl alcohol) filaments. *ACS Applied Materials & Interfaces* 9 (5): 4788.

2 Wu, C., Wang, X., Zhuo, Q. et al. (2018). A facile continuous wet-spinning of graphene oxide fibers from aqueous solutions at high pH with the introduction of ammonia. *Carbon* 138: 292–299.

3 Tang, Z., Jia, S., Wang, F. et al. (2018). Highly stretchable core-sheath fibers via wet-spinning for wearable strain sensors. *ACS Applied Materials & Interfaces* 10: 6624–6635.

4 Yu, S. and Zhang, C. (2016). Spraying functional fibres by electrospinning. *Materials Horizons* 3: 266–269.

5 Joseph, J., Nair, S.V., and Menon, D. (2015). Integrating substrateless electrospinning with textile technology for creating biodegradable three-dimensional structures. *Nano Letters* 15 (8): 5420–5426.

6 Huang, Q., Liu, L., Wang, D. et al. (2016). One-step electrospinning of carbon nanowebs on metallic textiles for high-capacitance supercapacitor fabrics. *Journal of Materials Chemistry A* 4: 6802–6808.

7 Levitt, A.S., Alhabeb, M., Hatter, C.B. et al. (2018). Electrospun mxene/carbon nanofibers as supercapacitor electrodes. *Journal of Materials Chemistry A* 7 (1): 269–277.

8 Formalas, A. (1934). US patent 1, 975, 504.

9 Kim, S., Sim, H., Shin, M. et al. (2013). Conductive functional biscrolled polymer and carbon nanotube yarns. *RSC Advances* 3 (46): 24028–24033.

10 Yue, X., Zhang, F., Wu, H. et al. (2014). A novel route to prepare dry-spun silk fibers from $CaCl_2$-formic acid solution. *Materials Letters* 128: 175–178.

11 Zhang, D., Miao, M., Niu, H., and Wei, Z. (2014). Core-spun carbon nanotube yarn supercapacitors for wearable electronic textiles. *American Chemical Society Nano* 8: 4571–4579.

12 Michael, R., Valentine, D.F., Chong, H. et al. (2018). Diode fibres for fabric-based optical communications. *Nature* 560: 2166.

13 Grena, B., Alayrac, J.B., and Levy, E. (2017). Thermally-drawn fibers with spatially-selective porous domains. *Nature Communications* 8: 364.

14 Zhang, T., Li, K., Zhang, J. et al. (2017). High-performance, flexible, and ultralong crystalline thermoelectric fibers. *Nano Energy* 41: 35.

15 Qu, Y., Nguyên, D.T., Page, A.G. et al. (2018). Superelastic multimaterial electronic and photonic fibers and devices via thermal drawing. *Advanced Materials* 30: 1707251.

16 Bashir, T., Skrifvars, M., and Persson, N.K. (2017). High-strength electrically conductive fibers: functionalization of polyamide, aramid, and polyester fibers with PEDOT polymer. *Polymers for Advanved Technologies* 29: 310–318.

17 Li, H., Liu, Z., Liang, G. et al. (2018). Waterproof and tailorable elastic rechargeable yarn zinc ion batteries by a cross-linked polyacrylamide electrolyte. *American Chemical Society Nano* 12: 3140.

18 Seon, H.K., Tae, U.K., Hyun, C.K. et al. (2013). Piezoelectricity of ZnO nano-rods array prepared by hydrothermal method. *Proceedings of SPIE-The International Society for Optical Engineering* 8816 (22): 4476–4485.

19 Zhang, L. and Zhu, Y. (2009). ZnO micro- and nano-structures: microwave-assisted solvothermal synthesis, morphology control and photocatalytic properties. *Applied Physics A* 97 (4): 847.

20 Xu, L., Ding, Y., Chen, C. et al. (2008). 3D flowerlike alpha-nickel hydroxide with enhanced electrochemical activity synthesized by microwave-assisted hydrothermal method. *Chemistry of Materials* 20 (1): 308–316.

21 Hsieh, C.T. and Chen, W. (2007). Gaseous adsorption of carbon tetrachloride onto carbon nanofiber arrays prepared by template-assisted synthesis. *Diamond and Related Materials* 16 (11): 1945–1949.

22 Wei, H., Hu, H., Feng, J. et al. (2018). Yarn-form electrodes with high capacitance and cycling stability based on hierarchical nanostructured nickel-cobalt mixed oxides for weavable fiber-shaped supercapacitors. *Journal of Power Sources* 400: 157–166.

23 Xie, F., Hu, W., Ning, D. et al. (2017). ZnO nanowires decoration on carbon fiber via hydrothermal synthesis for paper-based friction materials with improved friction and wear properties. *Ceramics International* 44 (4): 4204–4210.

24 Chen, H., He, J., Li, Y. et al. (2019). Hierarchical CuOx–Co_3O_4 heterostructure nanowires decorated on 3D porous nitrogen-doped carbon nanofibers as flexible and free-standing anodes for high-performance lithium-ion batteries. *Journal of Materials Chemistry A* 7: 7691–7700.

25 Kim, T., Shin, S., and Shin, M.W. (2018). Development of 3D interconnected carbon materials derived from ZnMOF-74@carbon nanofiber web as an efficient metal-free electrocatalyst for oxygen reduction. *Carbon* 135: 35–43.

26 Phuruangrat, A., Ham, D.J., Hong, S.J. et al. (2010). Synthesis of hexagonal WO_3 nanowires by microwave-assisted hydrothermal method and their electrocatalytic activities for hydrogen evolution reaction. *Journal of Materials Chemistry* 20: 1683–1690.

27 Han, P., Zhu, Y., and Lu, B. (2018). Highly efficient and environmentally friendly microwave-assisted hydrothermal rapid synthesis of ultralong hydroxyapatite nanowires. *Ceramics International* 44: 12352–12356.

28 Meng, H., Hou, W., Xu, X. et al. (2014). TiO_2-loaded carbon fiber: microwave hydrothermal synthesis and photocatalytic activity under UV light irradiation. *Particuology* 14: 38–43.

29 Liu, L., Pippel, E., Scholz, R., and Gösele, U. (2009). Pt–Co alloy nanowires: fabrication, characterization, and electrocatalytic properties sol-del method electrodepositions. *Nano Letters* 9 (12): 4352–4358.

30 Zhu, T., Wu, H., Wang, B. et al. (2012). Formation of 1D hierarchical structures composed of Ni_3S_2 nanosheets on CNTs backbone for supercapacitors and photocatalytic H_2 production. *Advanced Energy Materials* 2 (12): 1497–1502.

31 Sin, D., Koo, B.R., and Ahn, H.J. (2017). Hollow lithium manganese oxide nanotubes using MnO_2-carbon nanofiber composites as cathode materials for hybrid capacitors. *Journal of Alloys and Compounds* 696: 290–294.

32 Mahani, R.M., Battisha, I.K., Aly, M., and Abou-Hamad, A.B. (2010). Structure and dielectric behavior of nano-structure ferroelectric $BaxSr_1{-}xTiO_3$ prepared by sol–gel method. *Journal of Alloys and Compounds* 508 (2): 354–358.

33 Zhang, J., Andrew, L.H., Samantha, S. et al. (2018). Effects of ammonolysis and of sol-gel titanium oxide nitride coating on carbon fibres for use in flexible supercapacitors. *Journal of Materials Chemistry A* 6: 5208–5216.

34 Wang, K., Qing, H., Zhang, Y. et al. (2013). High-performance two-ply yarn supercapacitors based on carbon nanotubes and polyaniline nanowire arrays. *Advanced Materials* 25 (10): 1494–1498.

35 Zhang, Q., Sun, J., Pan, Z. et al. (2017). Stretchable fiber-shaped asymmetric supercapacitors with ultrahigh energy density. *Nano Energy* 39: 219–228.

36 Konaka, T., Sankawa, I., Matsuura, T. et al. (1988). Preparation of Y–Ba–Cu–O superconducting tape by atmospheric plasma spraying. *Japanese Journal of Applied Physics* 27: 1092–1093.

37 Bu, A., Zhang, Y., Zhang, Y. et al. (2018). A novel electrolytic plasma spraying preparation SiO_2/SiC coating on carbon fiber fabric. *Coatings* 8 (10): 344.

38 Bu, A., Zhang, Y., Xiang, Y. et al. (2020). Plasma electrolysis spraying Al_2O_3 coating onto quartz fiber fabric for enhanced thermal conductivity and stability. *Applied Sciences* 10 (2): 702.

39 Zhou, F., Tien, H., Xu, W. et al. (2017). Ultrathin graphene oxide-based hollow fiber membranes with brush-like CO_2-philic agent for highly efficient CO_2 capture. *Nature Communications* 8: 2107.

40 Garima, M. and Kyong, Y.R. (2018). Hierarchical structures of CNT@basalt fabric for tribological and electrical applications: impact of growth temperature and time during synthesis. *Composites Part A: Applied Science and Manufacturing* 115: 8–21.

41 Zhang, J., Cheng, L., Liu, Y. et al. (2017). Fabricating thick-section carbon fiber/silicon carbide composites by machining-aided chemical vapor infiltration. *Ceramics International* 43 (16): 13276–13281.

42 Xie, W., Saad, K., Orlando, J.R., and Gregory, N.P. (2018). Derived from cellulose diacetate via vapor phase infiltration of diethyl zinc. *ACS Sustainable Chemistry & Engineering* 6 (11): 13844–13853.

43 Yang, Z., Zhou, W., and Zhu, F. (2006). SiC/SiO_2 core-shell nanocables formed on the carbon fiber felt. *Materials Chemistry and Physics* 96 (2–3): 439–441.

44 Mohammad, M.H. (2018). Antimicrobial coatings for textiles. In: *Handbook of Antimicrobial Coatings* (ed. A. Tiwari), 321–355. Elsevier.

45 Liu, H., Gu, C., Hou, C. et al. (2016). Plasma-assisted synthesis of carbon fibers/ZnO core-shell hybrids on carbon fiber templates for detection of ascorbic acid and uric acid. *Sensors and Actuators, B. Chemical* 224: 857–862.

46 Liu, J., Li, Y., Yong, S. et al. (2019). Flexible printed monolithic-structured solid-state dye sensitized solar cells on woven glass fibre textile for wearable energy harvesting applications. *Scientific Reports* 9 (1): 1362.

47 Pan, S., Yang, Z., Chen, P. et al. (2014). Wearable solar cells by stacking textile electrodes. *Angewandte Chemie International Edition* 53 (24): 6110–6114.

48 Yun, M., Sim, Y.H., Cha, S. et al. (2019). Three-dimensional textile platform for electrochemical devices and its application to dye-sensitized solar cells. *Scientific Reports* 9 (1): 2322.

49 Sun, C., Li, X., Cai, Z., and Ge, F. (2018). Carbonized cotton fabric in-situ electrodeposition polypyrrole as high-performance flexible electrode for wearable supercapacitor. *Electrochimica Acta* 296: 617–626.

50 Zhang, N., Chen, J., Huang, Y. et al. (2016). A wearable all-solid photovoltaic textile. *Advanced Materials* 28 (2): 263–269.

51 Chen, J., Huang, Y., Zhang, N. et al. (2016). Micro-cable structured textile for simultaneously harvesting solar and mechanical energy. *Nature Energy* 1: 16138.

52 Zhang, L., Zhang, N., Yang, Y. et al. (2018). Self-powered all-in-one fluid sensor textile with enhanced triboelectric effect on all-immersed dendritic liquid–solid interface. *American Chemical Society Applied Materials & Interfaces* 10: 30819–30826.

53 Zhang, N., Huang, F., Zhao, S. et al. (2020). Photo-rechargeable fabrics as sustainable and robust power sources for wearable bioelectronics. *Matter* 2: 1260–1269.

54 Huang, F., Yu, H., Xiang, S. et al. (2019). Embroidering a filmsy photo-rechargeable energy fabric with wide weather adaptability. *American Chemical Society Applied Materials & Interfaces* 12: 3654–3660.

4

Energy Harvesting Fibers

Next-generation wearable electronics demands systems directly worn on soft and curved human body. In order to adapt to the working environment of rough and high curvature surface, people put forward higher requirements for the flexibility and stretchability of electronic equipment while the development of wearable equipment is severely restricted by the traditional planar rigid base energy supply system. Therefore, people are increasingly demanding light, cheap, flexible, and portable sustainable power supply system and expecting to continuously supply power for intelligent wearable equipment by collecting and transforming energies in the environment, such as solar energy, wind energy, sound energy, mechanical energy, etc.

The rapid development of nanoscience and nanotechnology in the recent decade has accelerated the miniaturization process of electronic devices. In particular, flexible power sources with efficient energy conversion and storage can automatically generate electricity and contribute to self-driven energy systems, which are very promising candidates for wearable electronics. Fiber-based cloth has been made and used by humans for thousands of years as they are soft and deformable, breathable, durable, and washable, thus being ideal to be explored as the platforms for future wearable electronics. Among the energy technology innovations, flexible fiber-shaped energy harvesting devices, including photovoltaic fibers, piezoelectric fibers, triboelectric fibers, thermoelectric fibers, etc., have boomed in recent years and are attracting increasing attention. These fiber assemblies possess many unique characteristics appropriate to wearable systems.

This chapter mainly introduces the structure, working principle, key materials, and device performance of various fiber-shaped devices, including photovoltaic fibers, piezoelectric fibers, triboelectric fibers, and thermoelectric fibers, which are expected to power various electronic devices.

4.1 Photovoltaic Fibers

As an inexhaustible source of energy, solar energy has long been recognized as an important source of energy for solving the human energy crisis. Photovoltaic devices, also called solar cells, which convert solar energy to electric energy, represent a promising technology to use renewable energies. With the rapid development of

Textile-Based Energy Harvesting and Storage Devices for Wearable Electronics, First Edition.
Xing Fan, Nannan Zhang, and Yi Wang.

traditional industry and the increasing demand of modern electronics, integrated electronic products are taking the form of small, portable, lightweight, and smart merchandise and gradually expanding in the direction of wearable. However, solar cells with a flat-panel structure cannot be worn well. In order to be compatible with body wear, people apply traditional photovoltaic materials to the fiber substrate to achieve high fit and high comfort of the wearable. Emerging fiber-shaped solar cells can also demonstrate the conversion of solar energy to electricity, including a special photovoltaic structure made on a fiber substrate, which gives the device special performance and allows innovation.

The research on fiber-shaped solar cells has mainly gone through two stages, including the coaxial-structure photovoltaic fibers and the twisted-structure photovoltaic fibers. On the photovoltaic fibers with coaxial structure, photoactive materials were coated onto a metal wire layer by layer to yield one-dimensional (1D) coaxial dye-sensitized solar cells (DSSCs), which appear early since 2002 [1]. In 2007, fiber organic/organic polymer solar cells were manufactured by coaxially coating photosensitive materials on optical fibers [2]. And later in 2014, to further enhance the power conversion efficiency (PCE), coaxial fiber perovskite solar cells (PSCs) were produced to form all-solid-state candidates [3].

On the photovoltaic fibers with twisted structure, two fiber-shaped electrodes fabricated by coating different types of photoactive materials on metal wires, respectively, were twisted together, which appear early since 2008 [4]. In 2009, Lee et al. [5] displayed a twisted polymer-based solar cell; a phase-separated, photovoltaic layer is coated onto a thin metal wire. A second wire, coated with a silver film, serving as the counter electrode, is wrapped around the first wire, and efficiency values of the wires range from 2.79% to 3.27%. Later in 2015, Deng et al. [6] designed stretchable nanostructured fiber and spring-like modified titanium wire as two electrodes. First, after twisting the carbon nanotube (CNT) sheets on elastic rubber fibers, they are applied to twisted modified titanium wires and elastic conductive fibers. The two fiber electrodes are stably tied together and coated with a perovskite material, thereby realizing the realization of an elastic PSC. By weaving these photovoltaic fibers through advanced textile technology, flexible materials for wearable applications can be obtained.

4.1.1 Fiber-Shaped Inorganic Solar Cell

The main fabrication process included four steps: (i) select a metal wire as the conductive fiber, (ii) deposit the p-type inorganic semiconductor layer on the conductive fiber, (iii) deposit the n-type inorganic semiconductor layer, and (iv) assemble the counter electrode.

The working principle of inorganic solar cell is based on the PN junction of the photovoltaic effect: n-type and p-type semiconductors with certain structures form a PN junction at their interface, forming a space charge area due to diffusion of majority carriers and an ever-growing built-in electric field directing from the n-type region to the p-type region, leading to reverse drifting of the majority carriers and finally reaching equilibrium. When sunlight or other light irradiates PN junction of

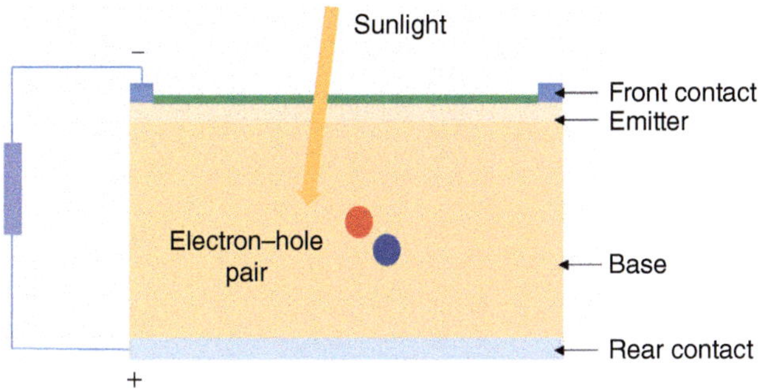

Figure 4.1 Working principle of inorganic solar cell.

semiconductor, voltage will appear on both sides of PN junction, which is called photogenerated voltage. When light irradiates the PN junction, the electron-hole pair is produced. The carrier generated near the PN junction in the semiconductor is not compounded and reaches the space charge area. Attracted by the internal electric field, the electron flows into the n area, and the hole flows into the p area. As a result, the n area stores excess electrons, and the p area has excess holes. They form a photogenerated electric field near the PN junction in the opposite direction to the barrier. In addition to partially counteracting the barrier electric field, the photoelectric field also makes the p-region positively charged and the n-region negatively charged. The thin layer between the n-region and the p-region produces the electromotive force, which is the photovoltaic effect (Figure 4.1) [7].

The material of fiber-shaped inorganic solar cells is mainly based on silicon, because of its excellent reliability and acceptable cost. Silicon solar cells, depended on the highly naturally abundant elements, have experienced a long research history, and silicon technology based on vapor deposition could be adapted to different substrates, which is a good choice to construct solar cells and detectors on flexible fibers. Early explorations since 2006 have claimed the flexible fiber-shaped poly-Si solar cells on glass fiber, but the device performance was very low. Firstly, in 2010, uniform deposition of poly-Si on the cylindrical-shaped carbon fiber was achieved by low-pressure chemical vapor deposition (CVD) with the silicon layer thickness of 0.85-10 μm [8]. Then in 2012, a flexible fiber-shaped $CuInSe_2$ (CIS) solar cell was prepared, and the CIS layer was electrodeposited on a flexible Mo wire as a substrate. Subsequently, CdS, ZnO, and indium tin oxide (ITO) layers are deposited on the Mo/CIS wires in an orderly manner, and each upper layer ensures complete contact with the lower layer, resulting in excellent structural uniformity along the circumference of the flexible supercapacitor (FSC) (Figure 4.2A) [9]. The single-wire solar cell achieved energy conversion efficiencies of up to 2.3% and exhibited excellent long-term stability (stored at 60 °C for 0-600 h), as shown in Figure 4.2B. Later on in 2013, Badding and coworkers [10] fabricated flexible coaxial silicon fiber pores with a p-i-n photodiode junction (Figure 4.2C) through high-pressure CVD. Flexible silicon fibers with length of up to 10 m were deposed, which demonstrated a potential for scalable preparation. The low light absorption

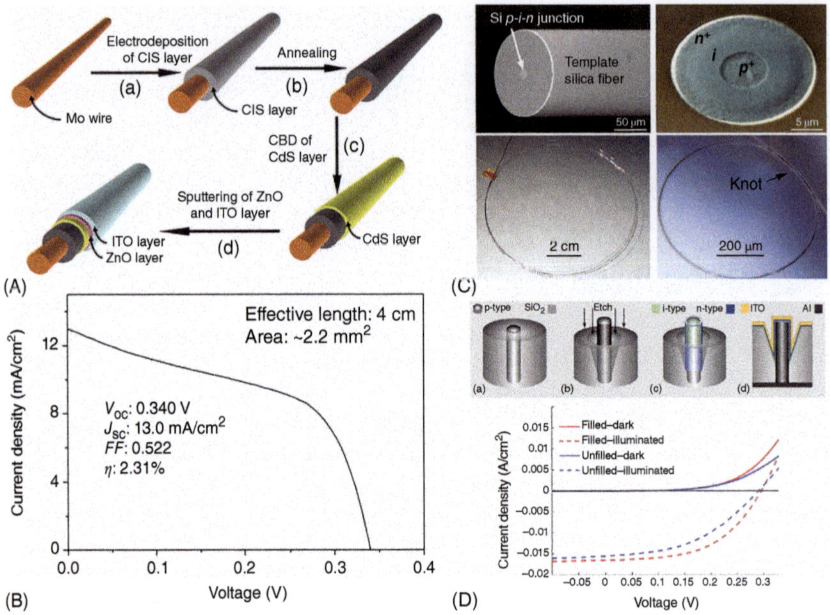

Figure 4.2 Structure and performance of fiber-shaped silicon solar cells. (A) Fabrication of Mo/CIS/CdS/ZnO/ITO single-wire flexible solar cells. Source: Zhang et al. [9]. © 2012, Elsevier. (B) Photocurrent-voltage characteristics of single-wire flexible solar cells. Source: Zhang et al. [9]. © 2012, Elsevier. (C) Silicon p-i-n junction fibers fabricated in capillary templates via high-pressure chemical vapor deposition. Source: Adapted with permission from He et al. [10]. Copyright (2013) by WILEY-VCH Verlag GmbH & Co. KGaA, Weinheim. (D) Stepwise solar cell fabrication procedure and solar cell performance of p-i-n junction solar cells. Source: Martinsen et al. [11]. © 2014, Springer Nature.

of the active layers resulted in an overall conversion efficiency of 0.5%. Later on, Gibson and coworkers [11] reported silica sheath fibers by using bulk glass draw techniques with low purity p-type silicon polycrystalline as the starting material (Figure 4.2D). Finally, they used vapor deposition techniques to create p-i-n junction solar cells. The silicon-core glass fiber achieved efficiency up to 3.6%, leaving room for improvements over the presented prototype. This fabrication method has the potential to reduce the energy cost and the silicon volume required for solar cell production. Significant improvement and great chances for applications are available considering the performance of traditional inorganic photovoltaic cells and the solid foundations of the silicon industry.

4.1.2 Fiber-Shaped Organic Polymer Solar Cell

Distinguished from the conventional planar organic polymer solar cell, the fiber-shaped organic polymer solar cell exhibits a unique structure based on the use of fiber electrode. The fiber-shaped organic solar cell could be divided into twisted type and coaxial type. For the twisted type, the main fabrication process included three steps. First, a metal wire or polymer fiber (with a conductive layer coated on the outside) is used as the conductive fiber. Second, the organic p- and n-semiconductor layers are deposited on the fiber conductive substrate respectively, and then polymer layer is coated as a photosensitive material. Third, the counter electrode was twisted around the above electrode.

For the coaxial type, the main fabrication process also included three steps. First, a metal wire or polymer fiber (with a conductive layer coated on the outside) is used as the fiber conductive substrate. Second, the polymer layer is coated as a photoactive material. Third, the counter electrode (usually metal) is deposited on the upper layer.

As discussed above, in manufacturing processes for both twisted organic polymer solar cell and coaxial organic polymer solar cell, researchers usually deposit different type of organic semiconductor layer (optical functional layer) on the conductive fiber (metal wire) and then assemble the counter electrode.

The working principle of organic polymer solar cell can be described as follows: the optical functional layer is phthalocyanine materials, thiophene oligomer, triphenylamine, and other organic small molecules or polymer materials and their hybrid materials with inorganic nano-semiconductors. When they are illuminated, free-charge carriers are generated by a photoinduced electron transfer from donor (p-type semiconductor) to acceptor (n-type semiconductor). A donor-to-acceptor heterojunction is the core part in solar cell that dominates the photoexcitation and charge separation that is similar with the PN junction in traditional silicon-based solar cells. The working principle can be described as follows: When a semiconductor layer is illuminated, the donor material absorbing light is excited and generates excitons, associated electron-hole pairs. The excitons diffuse to the donor-acceptor interface and dissociate into free-charge carriers, which then transfer to respective electrodes (Figure 4.3) [12].

In the development of fiber-shaped organic polymer solar cell, research also follows two directions: one is the twisted-structure organic polymer solar cell, and the other is the coaxial-structure organic polymer solar cell.

On the aspect of the twisted-structure organic polymer solar cell, the first one was reported in 2009, for which the twisted structure is made by twisting two fiber

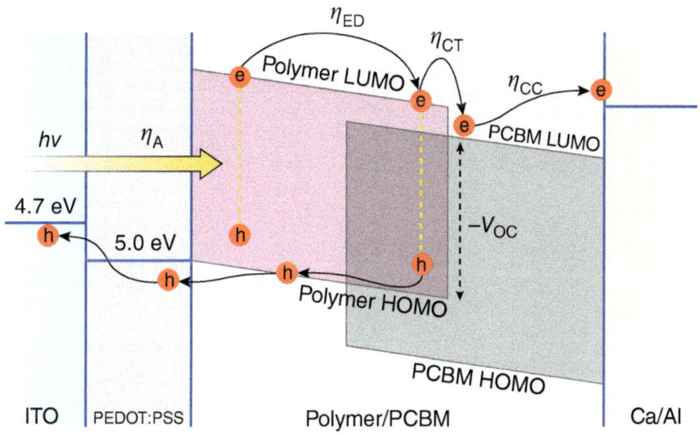

Figure 4.3 The basic working principle of polymer photovoltaic. Source: Li et al. [12]. © 2012, Springer Nature.

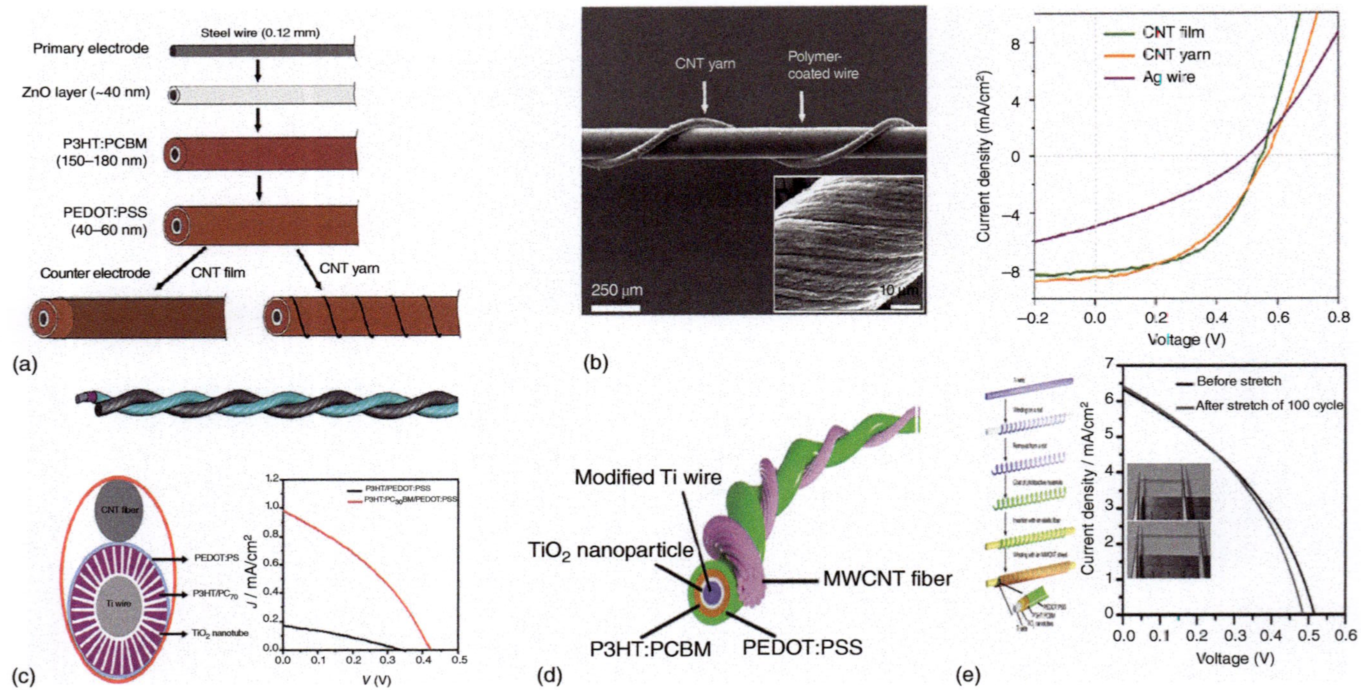

Figure 4.4 Twisted structure and performance of fiber-shaped organic polymer solar cell. (a) Fabrication of wire-shaped organic polymer solar cells with two architectures. Source: Liu et al. [13]. © 2012, American Chemical Society. (b) *J-V* curves of polymer FSCs with different counter electrode. Source: Adapted with permission from Liu et al. [13]. Copyright (2012) by American Chemical Society. (c) Schematic illustration of a polymer photovoltaic wire [14]. Source: Chen et al. [14]. © 2012, Royal Society of Chemistry. (d) Schematic illustration to the wire-shaped organic polymer solar cell. Source: Zhang et al. [15]. © 2014, John Wiley & Sons. (e) Schematic illustration of the fabrication of a stretchable fiber-shaped PSC. Source: Adapted with permission from Zhang et al. [16]. Copyright (2014) by WILEY-VCH Verlag GmbH & Co. KGaA, Weinheim.

electrodes based on metal wires. The main parts of the organic polymer solar cell, electron transport layer (TiO_x), photoactive layer (P_3HT:PCBM), and hole transport layer (PEDOT:PSS), were constructed on a metal wire by dip coating. The other electrode, a silver coated wire, was twisted with the primary electrode to conduct electrons [5]. The twisted structure ensured the electrical contact between the two electrodes, and efficiency values of the wires range from 2.79% to 3.27% [5]. A poly(3,4-ethylenedioxythiphene):poly(styrenesulfonate) layer with thicknesses of 40-60 nm was then coated to improve the hole transport and electrical contact to the transparent counter electrode (Figure 4.4a) [13]. A typical FSC with a yarn electrode showed a V_{oc} of 0.56 V, a J_{sc} of 8.49 mA/cm^2, and η of 2.11% (Figure 4.4b), comparable with the efficiency obtained in the FSC using a film electrode (2.31%). Later on, the TiO_2 nanotube-modified titanium wire was immersed in P_3HT:PC70BM blend followed by dip-coating PEDOT:PSS to work as primary electrode. At last, the CNT fiber was twined around the primary electrode (Figure 4.4c) [14]. The two electrodes could be closely contacted, but the resulting PSC wire showed an efficiency of 0.15%.

For another example, Figure 4.4d schematically shows the twisted structure of the wire-shaped PSC with a Ti wire and an aligned multiwalled carbon nanotube (MWCNT) fiber as cathode and anode, respectively [15]. The resulting Ti wire was finally wound with an aligned MWCNT fiber to produce the wire-shaped PSC. The PCE without the treatment has the V_{oc} of 0.50 V, J_{sc} of 6.48 mA/cm^2, and fill factor (FF) of 41%, and the conversion efficiency was lower than 1.31%. As shown in Figure 4.4e, a Ti wire was first wound onto a fiber substrate such as steel wire and then removed from the steel wire to produce a spring-like structure. Aligned TiO nanotubes were then grown in a direction perpendicular to the outer surface of the Ti spring using an electrochemical process. Then, the photoactive materials, P_3HT:PCBM, and the hole-transfer layer consisting of PEDOT:PSS were sequentially coated onto the modified Ti spring. The results show the comparison of a 1D PSC before and after stretching (stretched by 30% for 100 times). Before stretching, the PSC showed a PCE of 1.23% with an open-circuit voltage of 0.51 V, short-circuit current density of 6.33 mA/cm^2, and FF of 37.9%. After stretching, the PCE decreased slightly to 1.19%. Note that although the highest PCE of 1.23% was achieved, these fiber-shaped perovskite solar cells (FPSCs) still generally exhibited efficiencies around 0.7% [16].

The aspect of the coaxial-structure organic polymer solar cell is shown in Figure 4.5a [17]. The conductive ITO layer and photoactive layer of P_3HT or PCBM were sequentially coated on the optical fiber by a dip-coating process, and the aluminum layer was then deposited on the outer surface by thermal evaporation method. Compared with the planar counterpart, the reflective and transmissive losses can be minimized, and an efficiency of 1.1% was obtained (Figure 4.5b). The fiber-based photovoltaic (PV) cells in this work were fabricated by vacuum thermal evaporation of concentric thin films onto 0.48 mm thick, rotating, polyimide-coated silica fibers (Figure 4.5c) [17]. Figure 4.5d compares the current density-voltage (J-V) characteristics of the fiber device to the analogous planar devices, in which the fiber PV shows the $\eta_p = 0.50\%$. Besides the optical fiber, modified plastic fibers

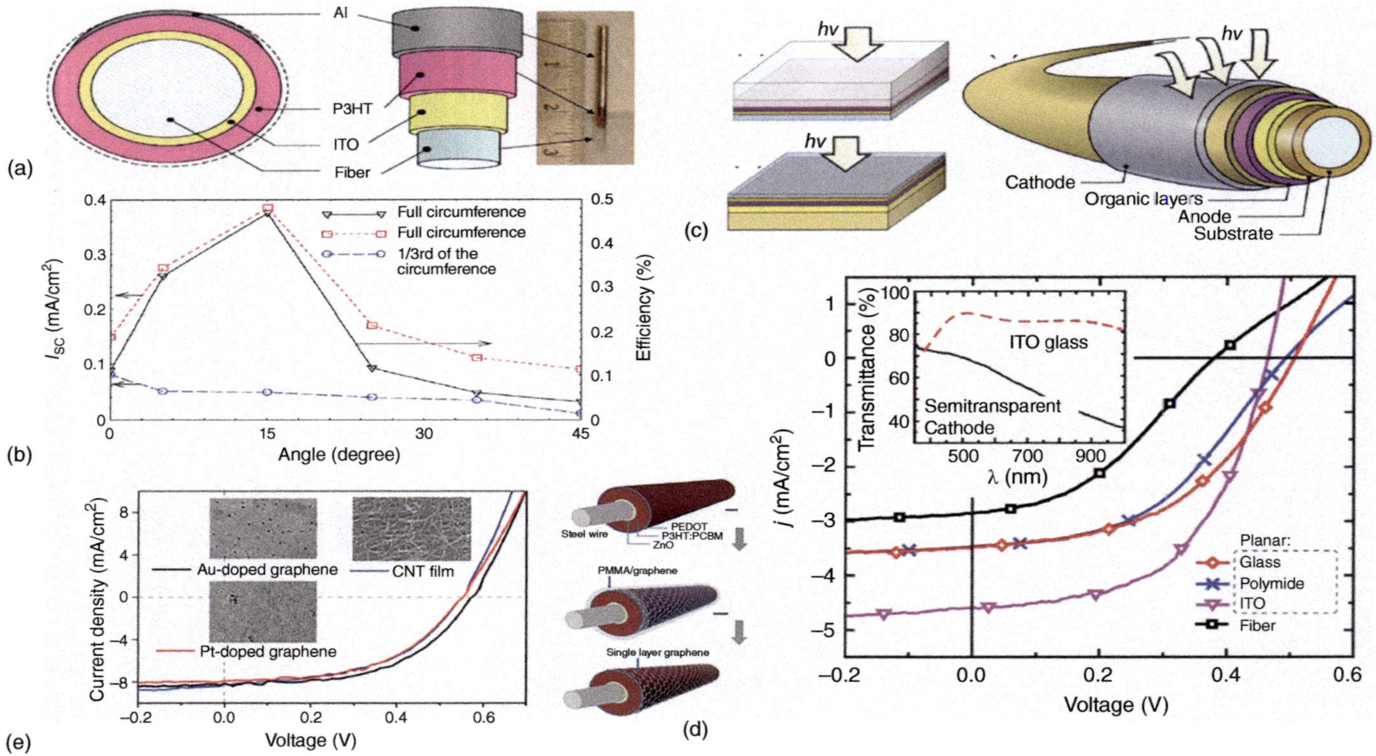

Figure 4.5 Coaxial structure and performance of fiber-shaped organic polymer solar cell. (a) A wire-shaped PSC based on an optical fiber. Source: Adapted with permission from Liu et al. [2]. Copyright (2007) by American Institute of Physics. (b) Dependence of the current density and efficiency on incident angles with respect to the axis of the optical fiber with diameter of 1.5 mm. Source: Liu et al. [2]. © 2007, AIP Publishing. (c) An archetypal organic solar cell deposited onto ITO-coated glass. Source: O'Connor et al. [17]. © 2008, AIP Publishing. (d) The current density-voltage (*J-V*) relationship for the fiber and planar devices. O'Connor et al. [17]. © 2008, AIP Publishing. (e) Illustration of the FSC structure based on a polymer-coated steel wire (primary electrode) and wrapped by a SLG sheet as counter electrode and comparison of three FSCs with different types of electrodes including an Au-doped graphene layer, a Pt-doped graphene layer, and a carbon nanotube film, respectively. Source: Adapted with permission from Liu et al. [18]. Copyright (2013) by The Royal Society of Chemistry.

such as polyimide-coated silica fiber and polypropylene fiber had been also used as the fiber substrate in PSCs. Furthermore, another example shows the photovoltaic fibers that were prepared using the PEDOT:PSS layer in polypropylene (PP) fiber, the photoactive layer, and a metal-based electrode, and the result shows that the highest values of the photovoltaic parameters are obtained with an open-circuit voltage of 360 mV, a short-circuit current density of 0.11 mA/cm^2, a FF of 24.5%, and a PCE of 0.010% [19]. As presented in Figure 4.5e, a single-layer graphene sheet was prepared from CVD on a copper substrate. Then the graphene sheet was introduced with gold nanoparticles to enhance its conductivity and then transferred to wrap around the primary electrode. An efficiency of 2.53% was obtained with V_{oc} of 0.570 V, J_{sc} of 8.14 mA/cm^2, and FF of 54.5%.

4.1.3 Fiber-Shaped Dye-Sensitized Solar Cell

The fiber-shaped DSSCs could be divided into twisted type, coaxial type, and parallel type.

For the twisted DSSCs, the main fabrication process included three steps. First, the metal wire is used as the fiber conductive substrate. Second, the semiconductor layer is deposited on the fiber substrate, dye is soaked, and electrolyte is loaded, that is, a fiber-shaped DSSC photoanode is prepared. Third, another fiber electrode (counter electrode) is twisted on the photoanode at a certain torsion angle. When the winding angle is 0, it is DSSCs with parallel structure.

For the coaxial DSSCs, the main fabrication process included three steps. First, a metal wire is used as a fiber conductive substrate. Second, a semiconductor layer is deposited on the fiber substrate, dye is soaked, and electrolyte is loaded, that is, a fiber-based DSSC photoanode is prepared. Third, another fiber electrode (counter electrode) is deposited directly on the photoanode.

The working principle of DSSC mimics that of the process of photosynthesis, as shown in Figure 4.6. When illuminated, the dye molecules harvest photons and get excited. The excited dye molecules then inject electrons into the conduction band of the nearby semiconductor. Electrons transport through the semiconductor layer and travel throughout the external circuit. The triiodide ions in electrolyte obtain electrons at the counter electrode and are reduced into iodide ions that release electrons to the oxidized dye molecules to regenerate the dyes. Meanwhile, the charge injection and transportation are accompanied by the recombination, which diminishes the output performances of the DSSC. In the process of converting light energy into electrical energy, the performance and stability of the battery depend on the regenerability of the dye and the electrolyte. During the cycle of the reaction cycle, the overall state of the whole system remains unchanged.

The electrolyte used in the earliest dye-sensitized solar cell was liquid electrolyte, and the DSSC with liquid electrolyte has the highest photoelectric conversion efficiency [20]. But unfortunately, the liquid electrolyte presents several technological problems such as dye desorption, solvent evaporation, dye degradation, and leakage of the volatile solvent and the redox couple. These problems limit the long-term stability of these wet solar cells. Therefore, many efforts were made

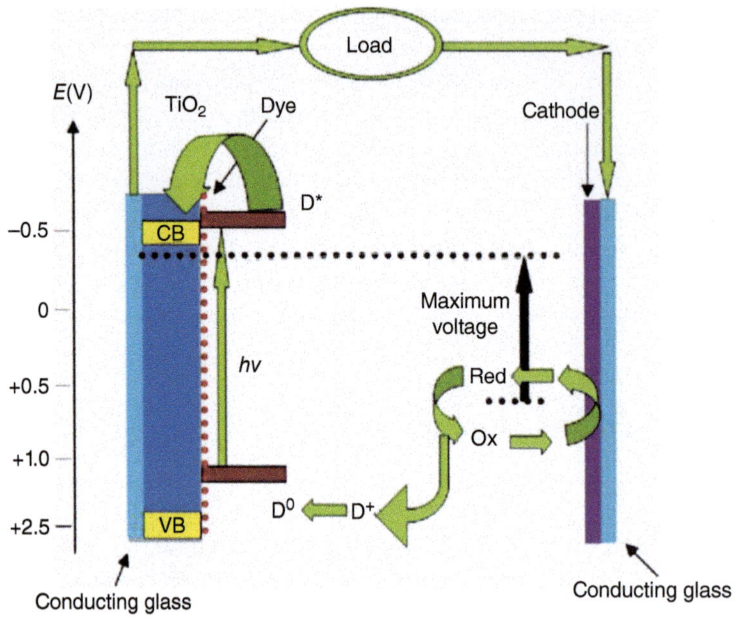

Figure 4.6 The basic working principle of dye sensitized solar cell. Source: Zou et al. [7]. © 2010, Elsevier.

to replace the liquid electrolyte with solid- or quasi-solid-type hole transport materials. For example, room-temperature molten salts, p-type semiconductors, conducting organic polymers, polymer gel electrolyte, ionic conductive polymer electrolytes, and hole transport medium were used to fabricate DSSCs. We know that in fabricating solid-state solar cells, there are two ways to fill solid-type electrolyte into the pores of the dyed TiO_2 film. One is a general "solution method," in which a p-type semiconductor is usually dissolved in a suitable solvent and filled into dyed TiO_2 porous film to substitute the electrolyte while heating the film to evaporate the solvent. Several p-type semiconductors [21], such as CuI, $CuBr_3S(C_4H_9)$, and CuSCN, have been successfully used to fabricate a complete solid-state DSSC. This method is very simple and easy to handle. The development of solid-state electrolyte provides the foundation for all-solid-state fiber-shaped DSSC and promotes the application of wearable photovoltaic energy collection fabric. Another is electrochemical deposition. Grätzel and coworkers fabricated a completely solid-state ZnO/dye/CuSCN solar cell by electrochemical deposition of CuSCN [22]. This method was regarded as a perfect way to fill the holes with a p-type semiconductor, but it is limited by complexities such as dye degradation, so it is not widely applicable.

In the development of fiber-shaped DSSCs, research also follow three directions: twisted structure, coaxial structure, and parallel structure.

For twisted one, in 2008, Zou and coworkers [4] firstly reported flexible fiber-shaped dye-sensitized solar cells (FDSSCs) based on double twisted metal fibers (Figure 4.7a), which are just like deoxyribonucleic acid (DNA) double-helix structure. The open voltage (V_{oc}), current density (I_{sc}), and FF were 0.61 V, 1.2 mA/cm^2, and 0.38, respectively. Besides the direct twist of two fiber electrodes, the other modified structure had also been explored. Later on, singly twisted fiber-shaped DSSCs were developed, where the counter electrode (Pt wire) was wound around the Ti wire-based fiber photoanode (Figure 4.7b) were prepared [23], and the device shows the current-voltage curves of the fiber solar cell. The open voltage (V_{oc}), current density (I_{sc}) and FF were 0.728 V, 12.28 mA/cm^2, and 0.786, respectively (Figure 4.7c). Fiber-like DSSCs with capillary package achieved efficiency up to 7.02%. Later in 2014, Fan and coworkers [24] had successfully developed different type of Ni-based composite fiber electrode, which was prepared by simple aqueous electroplating and was applied for the first time in a ZnO-based fiber-shaped DSSCs (Figure 4.7d). For an all-solid-state fiber cell with a thin layer of semiconductive oxide (2.5 μm), the J_{sc} reaches about 2.2 A/cm^2, the V_{oc} reaches about 0.4 V, and the efficiency reaches about 0.48%. Furthermore, all-solid fiber-shaped DSSCs based on Cu/Mn, Ti/Mn, and Fe/Mn have been developed (Figure 4.7e). The Mn-based fiber-shaped DSSCs show the results that the short-circuit photocurrent (J_{sc}) of 4 mA/cm^2 and open-circuit photovoltage (V_{oc}) over 0.4 V were achieved at an optimal Mn electroplating voltage of 3.1 V [25].

A lot of efforts have been devoted to the design of twisted fiber-shaped DSSCs with different fiber counter electrodes. Graphene/Pt composite fibers for FDSSCs (Figure 4.8a) contributed to an energy conversion efficiency of 8.5% [26]. The catalytic fiber was surrounded by a helical photoanode (Figure 4.8b), and an efficiency of 7.2% was achieved in the FDSSCs with a high J_{sc} of 19.35 mA/cm^2 (Figure 4.8c) [27]. The intrinsic electrochemical properties, electrical conductivity of the TiN nanoplates, and their specific nanostructure (e.g. active sites, large macropores, tight fusion connectivity, and vertically aligned configuration) may function jointly for the photovoltaic performance. $CoNiS_2$ nanoribbon (NR) and NR on carbon fibers with good catalytic activity to I^{3-}/I^- redox couple were applied to FDSSCs (Figure 4.8d). The former achieved higher efficiency (7.0%) than the latter (4.1%). The distinct morphologies and exposed crystal facet are responsible for the significant distinction in performance [28]. With a spiral wire photoanode, a novel stretchable photovoltaic device was fabricated on elastic conductive rubber fibers twisted with CNT fiber (Figure 4.8e). The photovoltaic devices achieved an energy conversion efficiency of 7.1%, and the efficiency was well maintained under bending and stretching, which are important in portable electronic devices and facilities [29]. Additionally, CdS, CdSe, and N719 were incorporated in the fiber photoanode (Figure 4.8f) to realize both multiple exciton generation effects and multiple electron transmission paths. An efficiency of up to 6.4% was achieved, which may have resulted from the special anode design [30].

For the coaxial one, the coaxial fiber-shaped DSSC was emerged considering that the twisted two electrodes are not reliable and readily to slip out especially when deformed. A coaxial structure that integrated optical fibers and nanowire arrays as three-dimensional (3D) DSSCs was shown in Figure 4.9a [31]. Aligned zinc oxide (ZnO) nanowires are grown from a ZnO layer deposited on the ITO layer as a seed layer, followed by sensitizing with dye, and the internal space of the device is

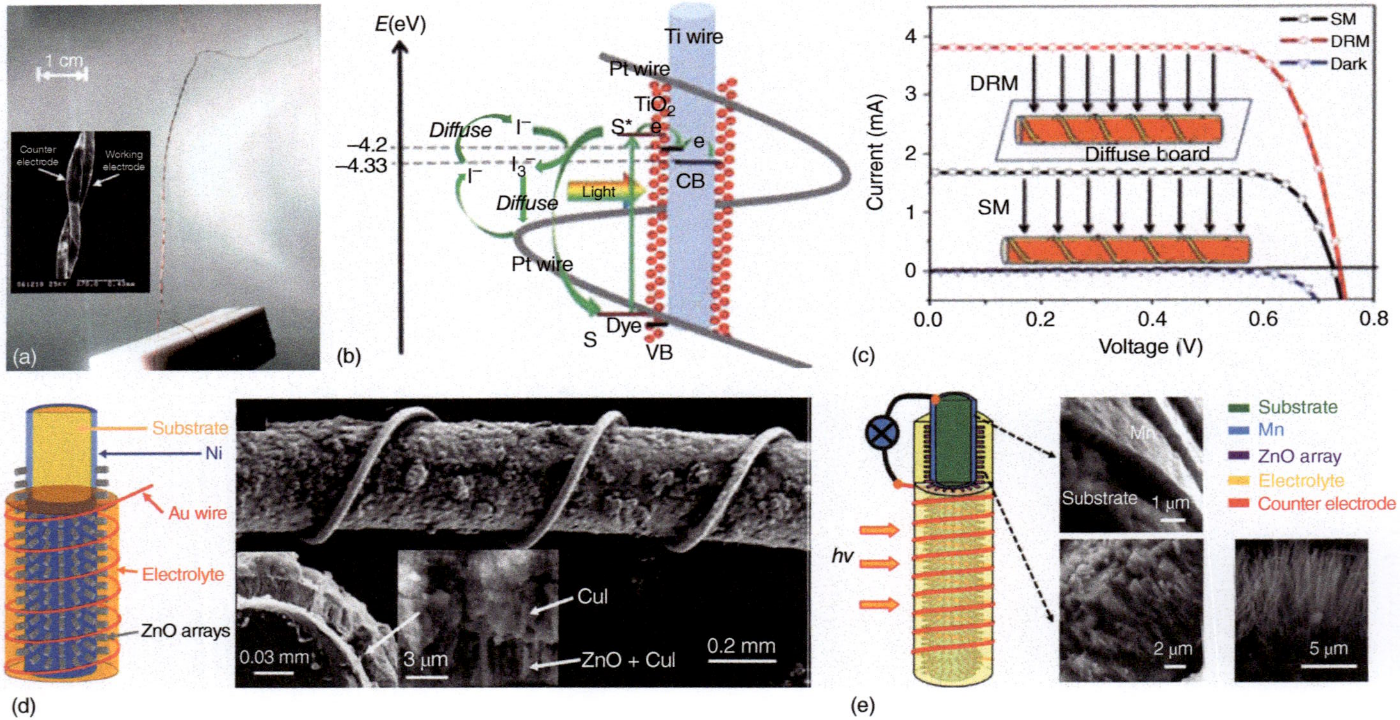

Figure 4.7 Twisted structure and performance of fiber-shaped dye sensitized solar cell. (a) Flexible wire-shaped dye-sensitized solar cell based on twisted metal electrodes. Source: Adapted with permission from Fan et al. [4]. Copyright (2008) by WILEY-VCH Verlag GmbH & Co. KGaA, Weinheim. (b) Schematic of singly twisted FDSSCs where the counter electrode (Pt wire) is wound around the fiber photoanode. Source: Fu et al. [23]. © 2011, Royal Society of Chemistry. (c) Current-voltage curves of a fiber-shaped solar cell under different irradiation modes. Source: Fu et al. [23]. © 2011, Royal Society of Chemistry. (d) All solid fiber-shaped DSSCs based on Ni-based substrate. Source: Adapted with permission from Feng et al. [24]. Copyright (2014) by The Royal Society of Chemistry. (e) An all solid fiber-shaped DSSCs based on Mn-based substrate. Source: Adapted with permission from Fan et al. [25]. Copyright (2014) by Elsevier.

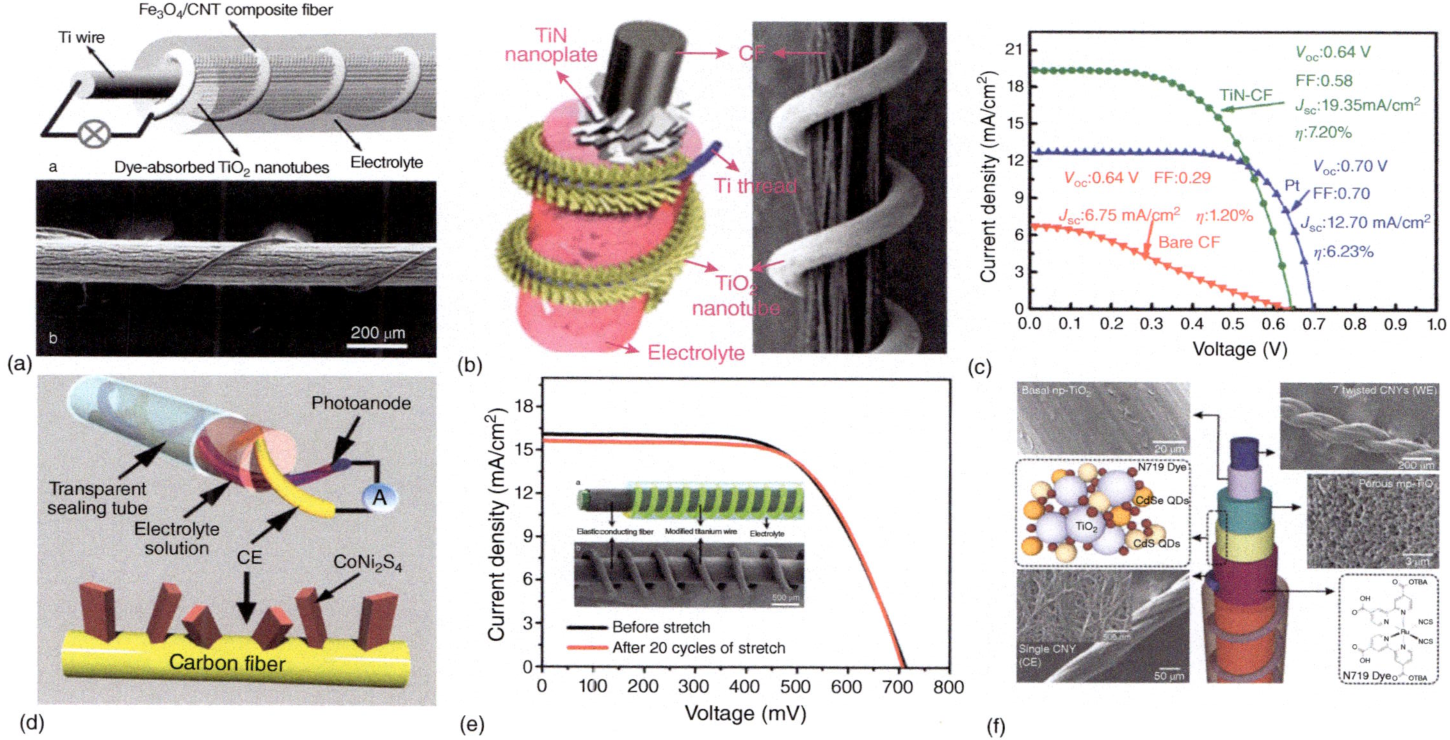

Figure 4.8 Structure and performance fiber-shaped DSSC. (a) Spirally wound graphene/Pt composite fiber as the counter electrode. Source: Adapted with permission from Yang et al. [26]. Copyright (2013) by WILEY-VCH Verlag GmbH & Co. KGaA, Weinheim. (b) TiN nanoplate-functionalized CFs were twisted with the TiO_2 nanotube array-coated Ti thread for FDSCs. Source: Adapted with permission from Chen et al. [27]. Copyright (2014) by The Royal Society of Chemistry. (c) *I*-*V* curves of FDSSCs based on TiN-CF, Pt wire and bare CF as CEs. Source: Chen et al. [27]. © 2014, Royal Society of Chemistry. (d) FDSCs based on $CoNi_2S_4$ nanoribbon- and NR-modified carbon fibers. Source: Chen et al. [28]. © 2015, Elsevier. (e) Stretchable photovoltaic device based on spiral wire photoanode and elastic rubber fibers twisted with CNT fibers. Source: Adapted with permission from Yang et al. [29]. Copyright (2014) by WILEY-VCH Verlag GmbH & Co. KGaA, Weinheim. (f) The wire-shaped CdS, CdSe, and N719 co-sensitized solar cell are prepared by using a platinized CNY as the counter electrode and a braid of seven-twisted carbon nanotube yarns with hybrid coating as the working electrode. Source: Adapted with permission from Yan et al. [30]. Copyright (2014) by WILEY-VCH Verlag GmbH & Co. KGaA, Weinheim.

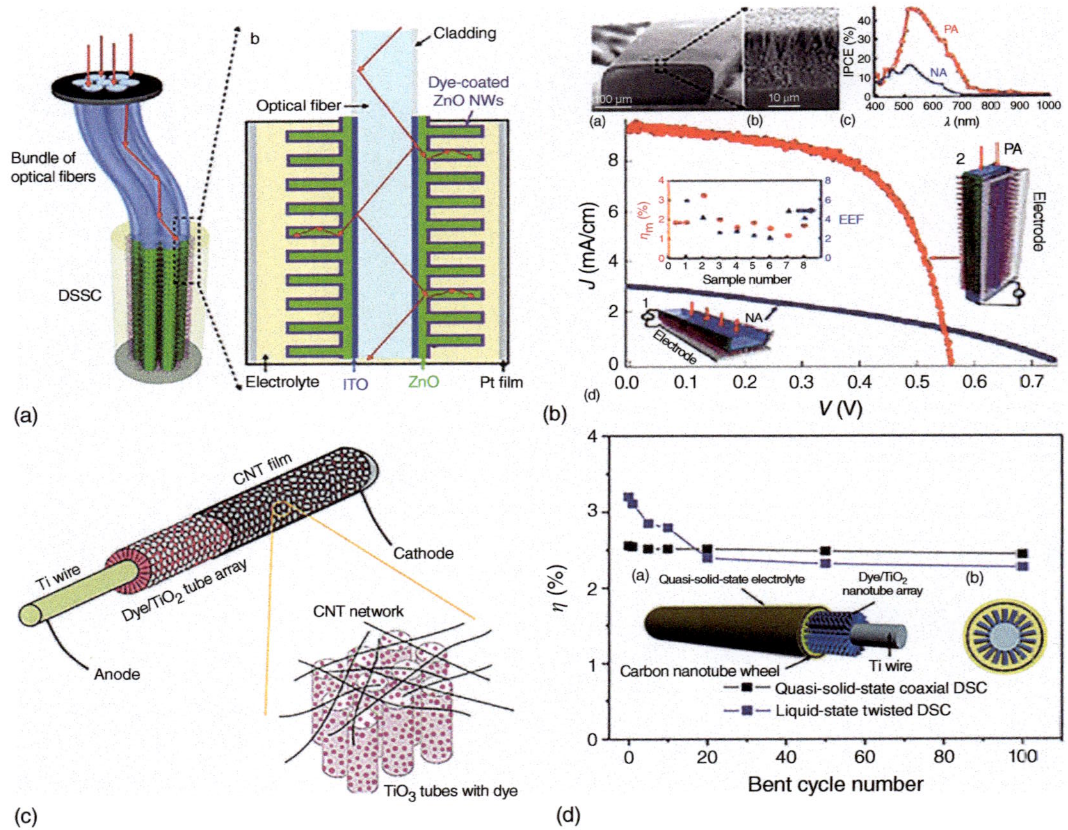

Figure 4.9 Core-sheath structures of FDSSCs based on different materials. (a) A 3D DSSC was constructed based on ZnO NWs vertically grown on an optical fiber. The internal light reflection along the fiber was guided to the photoanode. Source: Weintraub et al. [31]. © 2009, John Wiley & Sons. (b) *J*-*V* curves of a DSSC under one full-sun illumination oriented. Source: Adapted with permission from Weintraub et al. [31]. Copyright (2009) by WILEY-VCH Verlag GmbH & Co. KGaA, Weinheim. (c) Flexible single-wire DSSCs by wrapping a carbon nanotube film around Ti wire-supported TiO_2 tube arrays as the transparent electrode. Source: Zhang et al. [32]. © 2011, American Chemical Society. (d) The dependence of the energy conversion efficiency on the bent cycle number. The black and blue dots correspond to the quasi-solid-state coaxial DSSC fiber and liquid-state twisted DSC wire, respectively. The liquid-state twisted DSSC used the same amount of CNT as the quasi solid-state coaxial DSSC. Source: Sun et al. [33]. ©2014, Royal Society of Chemistry.

injected with a liquid electrolyte. These FDSCs achieved an efficiency of 3.3%, and a corresponding efficiency enhancement factor (EEF) of 4.3 has been achieved by converting the two-dimensional (2D) DSSC to the 3D DSSC (Figure 4.9b). In the early attempts, commercial pen ink based on well-dispersed carbon nanoparticles was proved to be a good catalytic material for I^{3-}/I^- and T_2/T-based electrolytes. Considering the difficulty in preparing spiral dye-sensitized photoanodes and fixing electrodes, the above 3D architectures have attracted relatively less attention. A freestanding CNT film can be easily wrapped around a TiO_2-modified Ti wire to form a coaxial cell (Figure 4.9c) [32]. In the fabricated coaxial DSSC, an energy conversion efficiency of 1.65% was recorded. An improvement was made by replacing the CNT film with aligned CNT sheet (Figure 4.9d) [33]. The aligned CNT sheets are also flexible and transparent while with higher electrical conductivities. As for different thicknesses of CNT sheets, the open-circuit voltage and FF were slightly changed while the short-circuit current density intensively relied on the thickness of the CNT sheet. Since the thicker sheet diminishes the illumination intensity, dye molecules are not fully excited, leading to a lower photoelectron current. Despite the low efficiency, these works may provide a promising avenue toward lightweight and flexible solar cells.

For the parallel one, Liang et al. [34] fabricated a flexible 3D solar device with multi-working electrodes and one counter fiber electrode, which further extended the 3D light harvesting property of a single-fiber photoanode. The energy conversion efficiency increased gradually from 2.8% to 6.6% as the number of photoanodes increased from 1 to 6 (Figure 4.10a,b). Interface modification with a Nb_2O_5 energy barrier refreshed the efficiency to 9.1%, which was retained at about 93% under bending with different angles. Core-sheath CNT/reduced graphene oxide (rGO) NR, CNT/rGO composite, and rGO fibers benefit from the convenience of physical and/or chemical modifications, which bring more concepts for materials science and device designs. Wang and coworkers [35] firstly demonstrated a CF/TiO_2 NR array-based FDSC with efficiency of 0.76%, and a bunched-NR-based device achieved a higher efficiency of 1.28% because of the higher J_{sc}, attributed to the larger surface area (Figure 4.10c,d).

4.1.4 Fiber-Shaped Perovskite Solar Cell

This section refers to the recent breakthrough of PSC as the pioneer in the next-generation photovoltaics. The FPSCs could be also divided into twisted type and coaxial type.

For the twisted PSC, the main fabrication process included five steps:

(1) Select metal wires or fibers coated with conductive materials as the conductive substrate.
(2) Deposit an electron transport layer (such as TiO_2).
(3) Deposit or coat a perovskite layer.
(4) Deposit a hole transport layer.
(5) Twist and assemble the electrode.

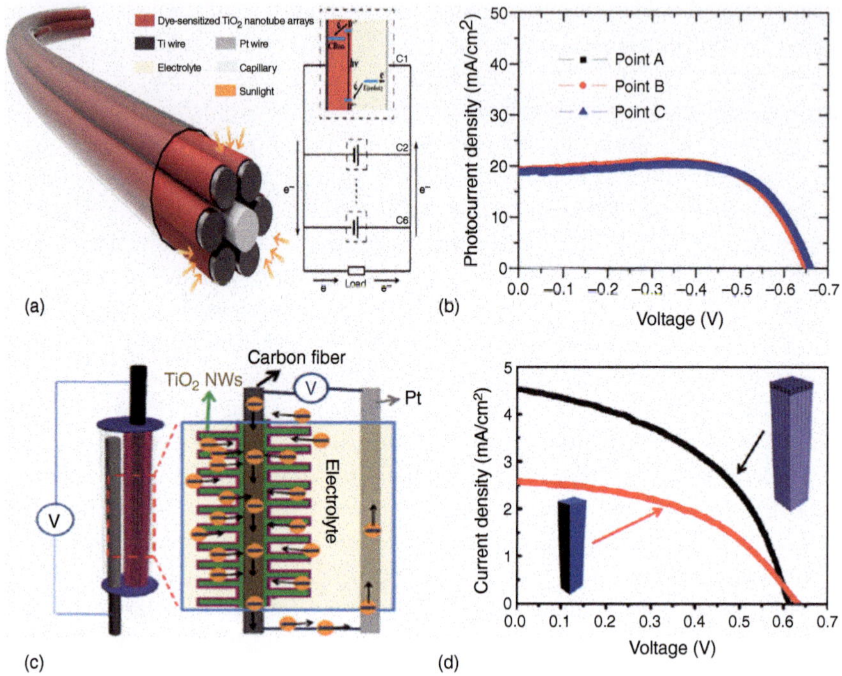

Figure 4.10 Parallel structures of FDSSCs. (a) Schematic of configuration of the CF/TiO_2 NR array-based tube-shaped DSSCs. Source: Liang et al. [34]. © 2015, Elsevier. (b) The three *J-V* curves acquired with the cell respectively illuminated from the three vertical incident directions. Source: Liang et al. [34]. © 2015, Elsevier. (c) Flexible 3D solar device with multi-working electrodes and one counter fiber electrode. Source: Guo et al. [35]. © 2012, American Chemical Society. (d) *J-V* curves of DSSCs based on TiO_2 NR arrays and bunched TiO_2 NR arrays on CFs. Source: Guo et al. [35]. © 2012. American Chemical Society.

For the coaxial PSC, the main fabrication process is different from the twisted PSC on the last step, where a conductive layer is deposited on the counter electrode, rather than twisted around.

The working principle of PSC is shown in Figure 4.11 [36]. To put it simply, we need perovskite as a light-absorbing layer. When a photon with sufficient energy illuminates the solar cell, the photon will excite electrons in the semiconductor valence band into the conduction band. A hole is left at the position of the valence band, and the pair of electrons and holes that were originally together pass through the electron transporting layer (ETL) and hole transporting layer (HTL), respectively. When they meet in an external circuit (External Circuit), a current is generated, and this is the basic principle of photovoltaic power conversion from solar cells. The striking advantage is high open-circuit voltage (V_{oc}) generated under illumination. The open-circuit voltage indicates the maximum energy available from absorbed light that is delimited by the bandgap of the perovskite. Normally, the obtained V_{oc} is lower than the bandgap potential. For PSCs, their V_{oc} ranged from 0.9 to 1.15 V, giving loss in potentials of about 0.4-0.65 eV, which is better than ~0.75 eV loss in potential of the best DSSC and ~0.8 eV of PSC [37].

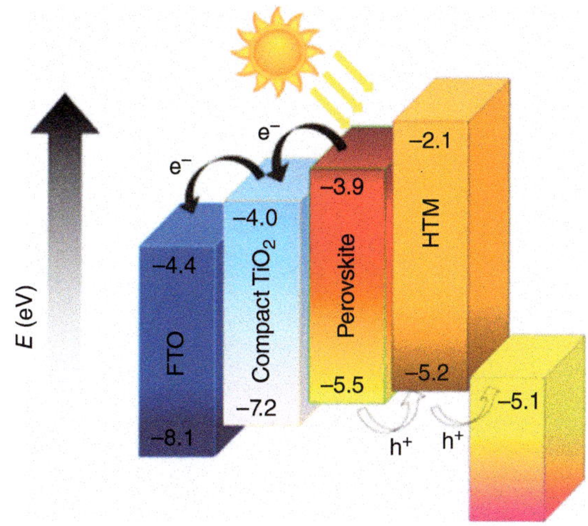

Figure 4.11 The basic working principle of PSC. Source: Zhu et al. [36]. © 2016, Elsevier.

Since 2012, PSCs have been a hot topic, and perovskite materials can be easily prepared by a simple process, such as one-step solution process, anti-solvent engineering, and two-step solution process. Peng and coworkers [3] demonstrated perovskite fiber solar cells by winding an aligned MWCNT sheet onto a TiO_2-modified steel wire and incorporating a perovskite layer between them via a one-step solution process (Figure 4.12a). The coaxial fiber devices (Figure 4.12b) achieved an efficiency of 3.3%, and the photovoltaic parameters (V_{oc} = 0.664 V, J_{sc} = 10.2 mA/cm^2, and FF = 0.487) remained almost constant under different incident light angles. No damage was observed on the sheet electrode, and the energy efficiencies remained at 95% after bending for 50 cycles.

For another approach, Lee et al. [38] improved the surface morphology of the perovskite film by anti-solvent engineering and achieved a PCE of 3.85% (Figure 4.12c). As shown in Figure 4.12d, Zou and coworkers integrated the structure of Ti/C-TiO_2/meso-TiO_2/perovskite/2,2′,7,7′-Tetrakis[N,N-di(4-methoxyphenyl) amino]-9,9′-spiro-bifluorene (spiro-OMeTAD)/Au into the fiber format [39]. The FPSCs achieved a 5.3% PCE under AM 1.5 illumination and an apparent 8.4% efficiency in the diffuse model. The device design required no transparent conductive oxides, and a thin film of gold nanoparticles was employed as a semitransparent top electrode. All the processes for device fabrication were easy to handle and energy saving. The fiber devices exhibited high reproducibility, which could open new doors toward efficient solid fiber-shaped photovoltaic cells.

With a two-step solution process, the corresponding perovskite fiber-shaped solar cells based on a ZnO array achieved 3.8% efficiency (Figure 4.13a) [40]. PbI_2 was firstly dissolved in dimethyl sulfoxide (DMSO) and then dip-coated onto the ZnO nano-obelisk array to form a thin and dense layer. The as prepared $CH_3NH_3PbI_3$ crystals typically had a cubic structure, and their sizes were decreased when CH_3NH_3I concentrations were increased from 4 to 10 mg/ml. Then, the PbI_2 crystal plate reacted with CH_3NH_3I to form perovskite $CH_3NH_3PbI_3$ nanocrystals. When the modified Ti wire and transparent aligned CNT sheet are used as two electrodes

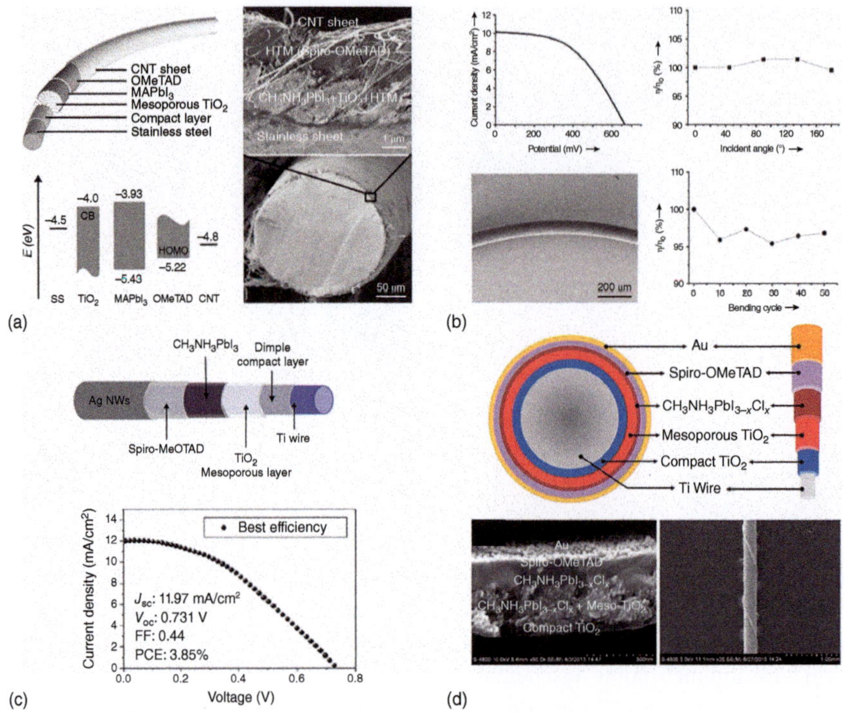

Figure 4.12 Structure and performance of different perovskite fiber solar cells. (a) Structure and energy level diagram of flexible perovskite fiber-shaped solar cells based on a carbon nanotube sheet. Source: Adapted with permission from Qiu et al. [3]. Copyright (2014) by WILEY-VCH Verlag GmbH & Co. KGaA, Weinheim. (b) *J-V* curve of a coaxial fiber-shaped perovskite solar cell. Source: Adapted with permission from Qiu et al. [3]. Copyright (2014) by WILEY-VCH Verlag GmbH & Co. KGaA, Weinheim. (c) A fiber-shaped PSC using silver nanowires as the top electrode. Photograph of a fiber-shaped PSC. Source: Lee et al. [38]. © 2015, Royal Society of Chemistry. (d) Schematic of the fiber-shaped perovskite solar cell (FPSC) structure and SEM cross-sectional image of the FPSC. Source: Adapted with permission from Hu et al. [39]. Copyright (2015) by The Royal Society of Chemistry.

to produce a coaxial perovskite solar cell fiber (PSCF), a high PCE of 7.1% also with a high open-circuit voltage of 0.85 V has been generated (Figure 4.13b) [41]. In addition, a wide-noodle-shaped PSC has also been reported, which had a PCE of 9.49% with a piece-shaped active area of 1 mm × 5 mm (Figure 4.13c) [42]. These studies show that the coverage and grain size of perovskite layers on curved surfaces are critical for the high PCE of FPSCs. Zou and coworkers [43] determined that the fast perovskite crystal growth kinetics by lead acetate can more easily form high-quality perovskite films on highly curved fiber substrates than traditional lead sources. Finally, the maximum PCE and V_{oc} of the FPSC achieved 7.53% and 0.96 V, respectively, which are the best reported values for solid fiber-shaped solar cells (Figure 4.13d). This study will promote the development of efficient and large-sized devices of solid fiber-shaped solar cells. The research may also be further developed into wearable devices as power sources.

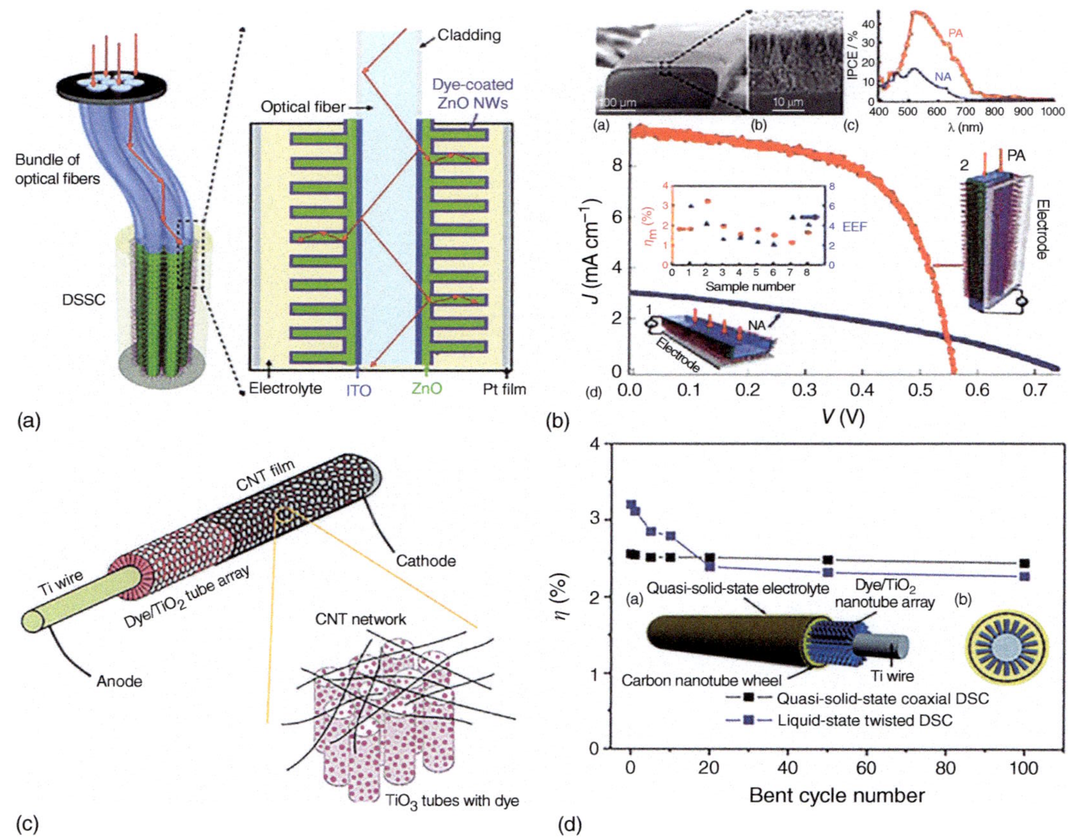

Figure 4.13 Structure and performance of different perovskite fiber solar cells. (a) Schematic illustration showing the fabrication process, *J-V* curve, and image of the optimized fiber-shaped PSC. Source: Adapted with permission from He et al. [40]. Copyright (2015) by The Royal Society of Chemistry. (b) Schematic illustration to the fabrication and performance of the perovskite solar cell fiber (PSCF). Source: Qiu et al. [41]. © 2016, John Wiley & Sons. (c) Structure of the fiber-shaped perovskite solar cell (FPSC) and scanning electron microscopy (SEM) image of the compact perovskite layer. Source: Adapted with permission from Qiu et al. [42]. Copyright (2016) by The Royal Society of Chemistry. (d) Schematic and *J-V* curve of the best-performing FPSC in this work. Source: Hu et al. [43]. © 2017, Royal Society of Chemistry.

Advanced planar PSCs have achieved efficiency of 22.1% [44], and the performance of perovskite photovoltaic fibers is catching up. For practical applications, more research are needed to explore repeatable and scalable technology. But the perovskite layer is sensitive to moisture and hole transport material spiro-OMeTAD, suffering from de-doping in air, so the stability and lifetime of the fiber device need further improvement. Another drawback is the toxic elements; environmentally friendly alternatives are welcome. From this point of view, opportunities are there for novel photovoltaic types and fiber photovoltaics. In the ideal situation, reliable packages could help with the stability and pollution leakage problem at the same time. But some systemic work remains to be done before the scheme is acceptable.

4.2 Piezoelectric Fibers

In recent years, the environmental mechanical energy harvesting has provided important and abundant power sources for driving electronic components in microelectronic systems. Among them, piezoelectric devices prepared based on the piezoelectric effect of materials such as zinc oxide (ZnO) and gallium nitride (GaN) have received widespread attention, which can generate electricity at external pressure. Piezoelectric fiber refers to the assembly of piezoelectric materials on a fiber substrate. When these piezoelectric fibers are subjected to external forces, the polarized charges generated by ion polarization in the crystal can effectively change the metal-semiconductor interface barrier and the transmission characteristics of the PN junction, and the piezoelectric polarization charge generated by mechanical strain can be used to adjust the transportation process of the carrier. Since planar piezoelectric devices can hardly meet this demand, fiber-shaped piezoelectric devices have gradually become a research focus.

4.2.1 Working Principle of Piezoelectricity

Piezoelectricity refers to stress-induced electricity based on the piezoelectric effect [45]. The piezoelectric effect is that when certain dielectrics are deformed by external forces in a certain direction, polarization will occur inside them, and at the same time, positive and negative charges will appear on its two opposite surfaces. When the external force is removed, it will return to an uncharged state. Based on the above principles, people have studied an energy harvesting piezoelectric devices (piezoelectric nanogenerators [PENGs]), which can convert nanoscale mechanical energy into electrical energy through the piezoelectric effect [46].

Take a typical metal-insulator-metal structure as an example, consisting of an insulating piezoelectric layer sandwiched between two metal electrodes, which was considered as a readily developed technique for piezoelectric devices [47]. Specifically, the complete electrical generation process of piezoelectric is as follows: at the beginning and undisturbed state, the charge centers of the cations and anions coincide with one another. No polarization can be observed inside the piezoelectric material (Figure 4.14i). When the pressing force is applied, deformation of the

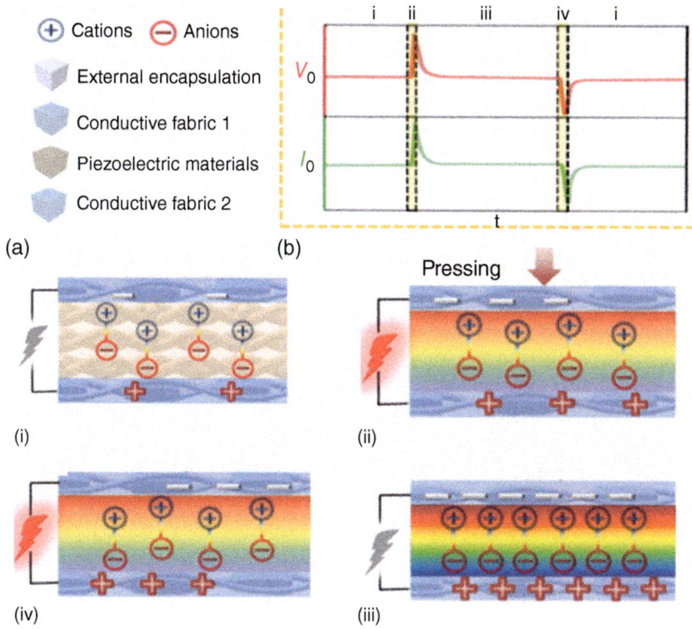

Figure 4.14 Complete electrical generation process of piezoelectric. (a) Schematic illustration of piezoelectric charge generation process. (b) Typical open-circuit voltage (V_{oc}) and short-circuit current (I_{sc}) vs. time curves, in which four parts corresponded to the four movement processes in (a). Source: Dong et al. [48]. © 2019, John Wiley & Sons.

piezo-fabric produces a negative strain and decreased volume. The charge centers separate to form electrical dipoles, and electric dipole moments change, which will result in the formation of a piezopotential between the electrodes. If the electrodes are connected with an external load, the piezopotential will drive electrons to flow through the external circuit in order to partially screen the piezopotential and achieve a new equilibrium state (Figure 4.14ii). As a result, mechanical energy is converted into the electricity. As the full contact between the two conductive fabric electrodes is made, the maximized pressed state is achieved with the highest polarization density (Figure 4.14iii). When the external force is released, electrons flow back to rebalance the charge induced by the strain releasing in the short-circuit condition (Figure 4.14iv). In addition to its special working principle, the progress of devices is always accompanied by the upgrading of materials [48].

4.2.2 Piezoelectric Materials

In piezoelectric fiber-shaped devices, the quality of piezoelectric materials is one key to determine the performance of piezoelectric devices. Since the piezoelectric effect was discovered in 1880, the most widely investigated traditional piezoelectric materials mainly include a mass of inorganic perovskite ceramic crystals and organic polymers [49]. Piezoceramics, or inorganic piezoelectric materials, mainly include semiconducting nanomaterials (e.g. ZnO, GaN, CdS, InN, and ZnS), lead

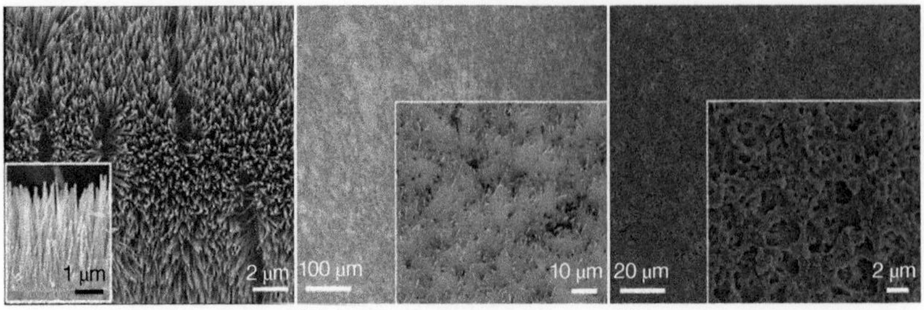

Figure 4.15 The SEM of some piezoelectric materials. (a) An SEM image of a Kevlar fiber covered with ZnO nanowire arrays. Source: Adapted with permission from Qin et al. [56]. Copyright (2008) by Macmillan Publishers Limited. (b) SEM image of ZnO NW array grown on gold film/Kapton substrate. The inset shows a magnified SEM image. Source: Adapted with permission from Lee et al. (2012) [57]. Copyright (2012) by WILEY-VCH Verlag GmbH & Co. KGaA, Weinheim. (c) SEM image of the top surface of PVDF layer on ZnO NW grown on Au/Kapton substrate. The inset shows a magnified SEM image. Source: Adapted with permission from Lee et al. [57]. Copyright (2012) by WILEY-VCH Verlag GmbH & Co. KGaA, Weinheim.

based (e.g. lead zirconate titanate piezoelectric ceramics (PZT), $PbTiO_3$, and $PbZrO_3$), and lead-free ceramics (e.g. $BaTiO_3$, KNN, $KNbO_3$, $LiNbO_3$, $LiTaO_3$, and Na_2WO_3). Moreover, piezopolymers mean piezoelectric class of organic polymers [50, 51], such as polyamide 11 (PA-11), PP, poly(vinylidene fluoride) (PVDF), and its copolymer poly(vinylidene fluoride-*co*-tri-fluoroethylene) (PVDF-TrFE). Specially, some researchers have also demonstrated that incorporating piezoceramics into piezopolymers to create a hybrid material can achieve higher piezoelectric constant and excellent electromechanical coupling factor.

Owing to the intrinsic insulation nature of these materials, the coupling research between piezoelectric-induced polarization and electronic/optoelectronic processes had consequently long been ignored until the discovery of PENGs using ZnO nanowire arrays by Wang and Song in 2006 [52]. By taking advantage of the multiple functionalities offered by piezoelectric semiconductors, Wang proposed the groundbreaking device concepts of piezotronics and piezophototronics in 2007 and 2010, respectively [53, 54]. These piezoelectric semiconductors possess piezoelectric properties but are not as extensively utilized in piezoelectric sensors and actuators as PZT due to their relatively small piezoelectric coefficients [55]. Because the merits of nanostructures include enhanced piezoelectric effect, superior mechanical properties, higher flexibility, and high sensitivity to small forces, so both piezoceramics and piezopolymers can also be further optimized with nanostructures, such as nanowire, nanofiber, nanotube, and nanosheet (Figure 4.15) [58].

4.2.3 Fiber-Shaped Piezoelectric Devices Based on Piezoceramics

At the early stage of the fiber-shaped piezoelectric devices, flexible piezoelectric materials were simply wrapped around fiber substrates with high strength and high modulus. The first fiber-shaped PENG was fabricated by entangling a Kevlar fiber covered by ZnO NWs and the fiber that is further coated with Au electrode together (Figure 4.16a) [56]. The small output current (5 pA) is attributed mainly to the large loss in the fiber due to an extremely large inner resistance. The amplitude of the voltage signal in this case was 1-3 mV. However, its power generation depends on the specific relative brushing between the two fibers, which is inconvenient for daily wearable usage. Later researchers attempted to design fiber-shaped PENGs within a single fiber. The most common method is to sandwich piezoelectric materials between the inner and outer electrodes. For example, ZnO thin film was sandwiched between an inner carbon fiber and an outer silver paste (Figure 4.16b) [59]. Using such a structure, an output peak voltage of 3.2 V and an average current density of 0.15 $\mu A/cm^2$ have been demonstrated. Alternatively, ZnO NWs grown on carbon fibers was further covered by an ITO electrode layer (Figure 4.16c) [61]. However, a potential issue of these fiber-shaped PENGs is that piezoelectrical materials and external covered electrodes are vulnerable to be destroyed under complex mechanical stimuli, which will lead to reducing their energy harvesting capability or even making them useless.

In order to improve their long-term working stability, several approaches have been adopted, such as fixing positions with polymer, wrapping with outer protective layer, fabricating with hybrid materials, and adopting with melt-spun method. In terms of fixing positions with polymer, the roots of ZnO NWs were bonded by PDMS to meet safety requirements and simultaneously guarantee high flexibility (Figure 4.16d) [60]. The protection of ZnO NWs grown on fibers can also be obtained by further wrapping an outer encapsulating layer. For instance, a carbon fiber grown with ZnO NWs was further wrapped by an outer folded Au-coated ZnO NWs paper tube [62]. For a ZnO NW-coated fiber with a diameter of 10 μm and a length of 15 mm, the maximum output voltage reaches 17 mV, and the output current is 400-520 pA. The corresponding current density of the designed nanogenerator (NG) is 0.09 $\mu A/cm^2$. The output power per unit contact area is about 51 $\mu W/m^2$. The established core-shell nanostructure provides more polarization possibilities between inner and outer ZnO NWs. If there are more than two piezoelectric materials existing in one PENG, the hybrid PENGs can be obtained that can improve the mechanical endurance and piezoelectric output performance [63, 64].

4.2.4 Fiber-Shaped Piezoelectric Devices Based on Piezopolymers

The material of piezoelectric device can be also based on piezoelectric polymer. In the process of preparing fiber-shaped piezoelectric polymer, melt spinning has been widely used with the advantages of one-time forming, high forming quality, and fast production speed. As shown in Figure 4.17a, PVDF film was conformally applied on the surface of electrospinning PVDF nanofibers to fabricate a high-performance fiber-based PENG [65]. The resultant unique yarn produced an average peak voltage of 0.52 V, peak current of 18.76 nA, and power density of 5.54 $\mu W/cm^3$ under a cyclic compression of 0.02 MPa at 1.85 Hz, which is significantly higher than the energy output of a yarn with only a PVDF electrospun nanofibrous layer. As shown in Figure 4.17b, a simple melt-spun piezoelectric fiber was developed by wrapping

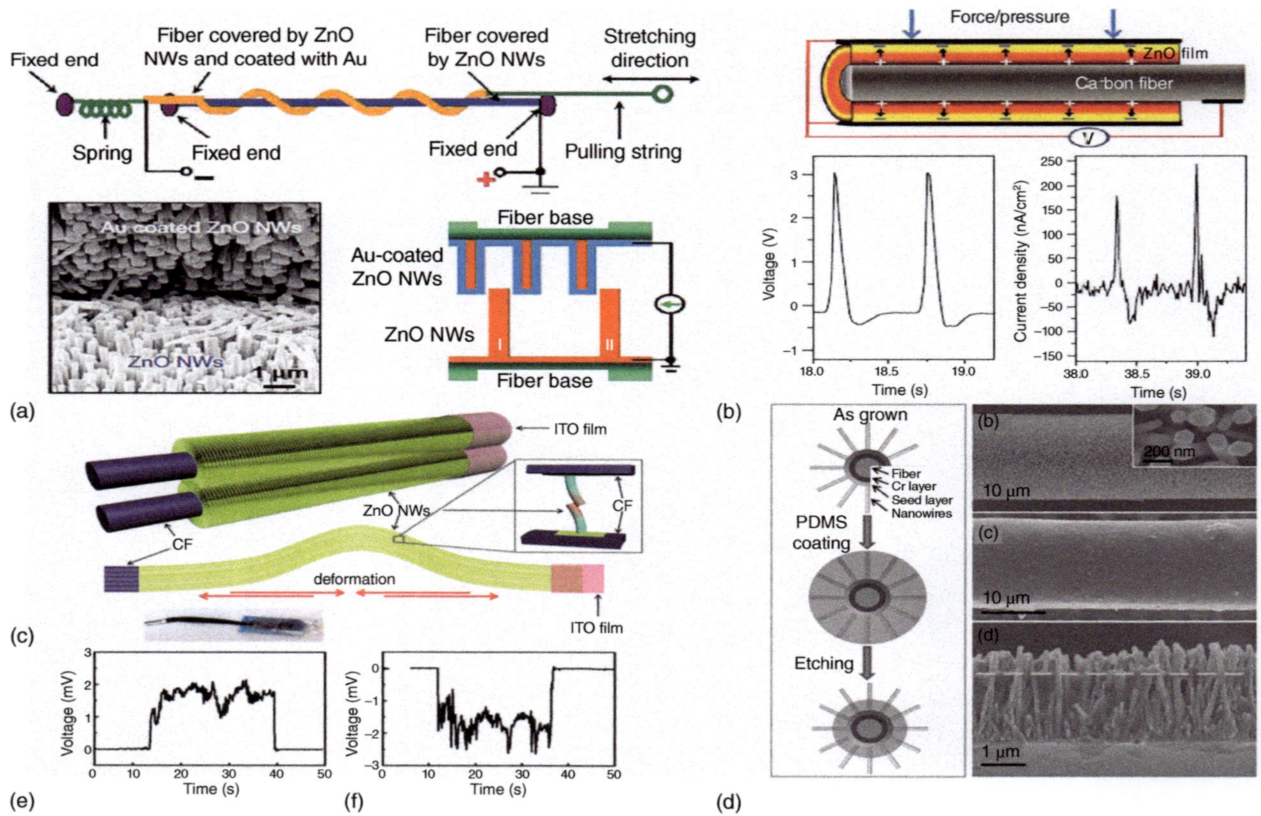

Figure 4.16 Typical single fiber-based PENGs. (a) The first fiber-shaped PENG developed by entangling a Kevlar fiber covered by ZnO NWs and the fiber that is further coated with Au together. Source: Adapted with permission from Qin et al. [56]. Copyright (2008) by Macmillan Publishers Limited. (b) An air/liquid-pressure and heartbeat-driven flexible fiber-shaped PENG designed by covering textured ZnO thin film on the surface of cylindrically carbon fibers. Source: Li and Wang [59]. © 2010, John Wiley & Sons. (c) A fiber-shaped PENG composed of carbon fibers and ZnO NWs based on an insulating interface barrier. Source: Li and Wang [59]. © 2010, John Wiley & Sons. (d) A high-reliability Kevlar microfiber-ZnO NW hybrid PENG. Source: Adapted with permission from Zhang et al. [60]. Copyright (2015) by WILEY-VCH Verlag GmbH & Co. GaA, Weinheim.

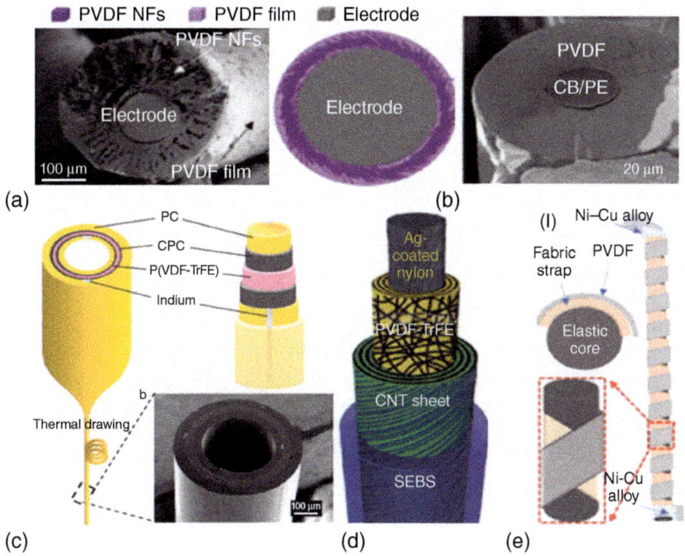

Figure 4.17 PENGs of different materials prepared by melt spinning method. (a) A flexible piezoelectric yarn consisting of a layer of electrospun PVDF nanofibers integrated with PVDF film around an Ag-deposited nylon filament. Source: Adapted with permission from Lund et al. [65]. Copyright (2018) by Springer Nature Limited. (b) A melt-spun piezoelectric microfiber consisting of a conducting core surrounded by β-phase PVDF. Source: Adapted with permission from Lee et al. (2012) [57]. Copyright (2012) by WILEY-VCH Verlag GmbH & Co. KGaA, Weinheim. (c) A multimaterial piezoelectric fiber comprising middle melt-pressed PVDF-TrFE shell and carbon-loaded polycarbonate with indium filaments and a polycarbonate cladding. Source: Adapted with permission from Egusa et al. [66]. Copyright (2010) by Springer Nature Limited. (d) A highly flexible piezoelectric fiber fabricated by sandwiching electrospun PVDF-TrFE mat between inner Ag-coated nylon fiber and outer CNT sheet. The whole system is covered by elastomeric styrene-ethylene-butylene-styrene. Source: Sim et al. [67]. © 2015, John Wiley & Sons. (e) A helical structural fiber-shaped PENG composed of an elastic core and a polymer piezoelectric strap twining the core. Source: Kim and Yun [68]. Licensed under CC BY 4.0.

a fiber core electrode around β-phase PVDF microfibers [57]. With the prepared piezoelectric fibers, a fiber-shaped PENG is fabricated by further spirally winding an outer fiber electrode around it, which can effectively harvest human motion energies and even in rainy days. Instead of twining the outer fiber electrode, other directly deposited the outer electrode on the surface of melt spinning piezoelectric fiber. As a case, a fiber-shaped PENG is obtained by depositing Au on the surface of the melt spinning PTFE-TrFE/Cu coaxial filament via vacuum deposition method [69]. The entire structure of fiber-shaped PENG can be accomplished through one-time thermal drawing in a straightforward manner.

As shown in Figure 4.17c, a multimaterial fiber-shaped PENG containing PVDF-TrFE piezoelectric shell sandwiched by the inner and outer carbon-loaded polycarbonate (CPC)/indium electrodes was fabricated through one-time thermal drawing processes [66]. The piezoelectric Fabry-Perot fiber is electrically driven by a sine wave at frequencies stepped from 1.5 to 8.5 kHz with the amplitude of 10 V. As illustrated in Figure 4.17d, a highly flexible and stretchable fiber-based PENG was fabricated by a simple rolling and coating method [67]. The PVDF-TrFE electrospinning mats were manually wrapped around Ag-coated nylon multifilament yarn and then covered by highly conductive CNTs. The prepared fiber-shaped PENG was finally dip-coated in an elastomeric styrene-ethylene-butylene-styrene (SEBS) to provide mechanical protection and electrical insulation. The fiber-shaped PENG shows piezoelectric voltage outputs to 2.6 V in lateral compression and up to 1.4 V during longitudinal extension. As exhibited in Figure 4.17e, a PVDF piezoelectric strap and a fabric band were helically twined around an elastic core in counter directions [68]. The helix could produce an alternating potential during continuous stretching and contraction motions. The stretchability of piezoelectric fibers can also be directly achieved by twisting electrospun PVDF-TrFE nanofibers into stretchable ribbons [70]. The melt-spun PVDF filaments were first braided around Ag-coated nylon yarns (inner electrode), and then the whole structure was braided a second time with Ag-coated nylon yarns (outer electrode). The developed triaxial braided fiber-shaped PENG exhibited improved mechanical and piezoelectric properties. Whatever the materials and structures are selected, one common feature of single-fiber-shaped PENGs is that all of them present coaxial and core-shell structure. Flexible piezoelectric materials are usually existed between the inner and outer fiber electrodes.

4.2.5 Fiber-Shaped Piezoelectric Devices Based on Piezocomposites

In order to enhance the piezoelectric response of piezoelectric device, piezoceramics are usually incorporated into piezopolymers to form piezocomposite during their thermal drawing processes. For example, $BaTiO_3$, PZT, or CNT-enhanced PVDF acting as piezoelectric layers is alternatively wrapped around CPC to fabricate an all-polymer flexible piezoelectric fiber with a spiral multilayer cladding structure (Figure 4.18a) [71]. The designed spiral multilayer cladding structure can greatly increase the active area of piezoelectric composites, resulting in higher power generation efficiency. A similar example that PVDF polymer is infiltrated into ZnO NW forests is displayed in Figure 4.18b [57]. The ZnO NW arrays not only serve as the piezoelectric material but also assist the formation of PVDF polymer on the outermost surface of the device. By attaching hybrid-fiber device, with a length of ~2 cm, on the human arm, the output voltage (Figure 4.18c), current density, and power density reached 0.1 V, 10 nA/cm^2, and 16 $\mu W/cm^3$, respectively, under folding-releasing of an elbow for ~90°. Recently, some new piezoelectric materials and fabrication techniques have been adopted to fabricate fiber-shaped PENGs. As illustrated in Figure 4.18d, a flexible fiber-shaped PENG based on perovskite $PbTiO_3$ nanotube arrays (PTO NTs) was fabricated by embedding two PTO NTs/Ti core-shell fibers in PDMS [72]. The PTO NTs were grown by using hydrothermal method with anodized TiO_2 NT as a positive template. The perovskite-based fiber-shaped PENG allowed arbitrary-direction bending and continuous harvesting mechanical energy from our natural environment. In terms of fabrication techniques, a novel triaxial braided structure has been developed to design fiber-based PEGNs based on melt-spun PVDF fibers and Ag-coated nylon yarns [73].

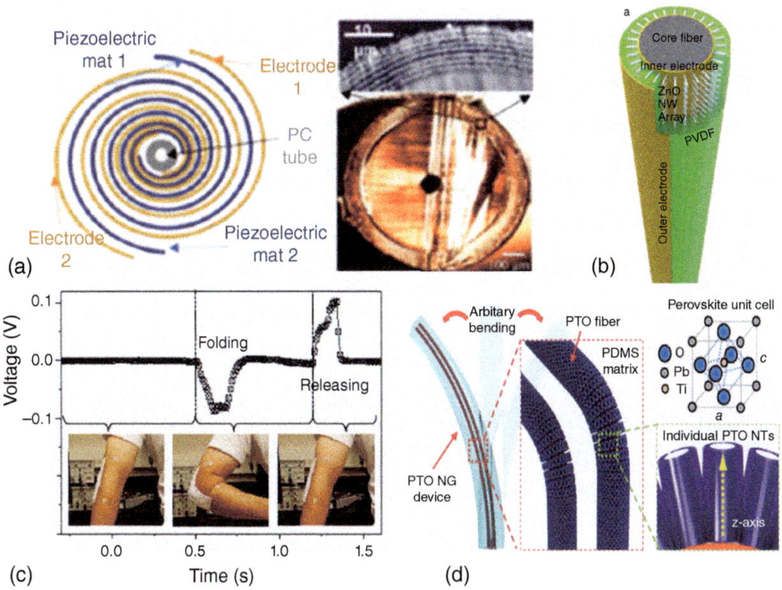

Figure 4.18 Fiber-shaped piezoelectric polymer PENGs. (a) An all-polymer flexible piezoelectric fiber featuring a soft hollow polycarbonate core surrounded by a spiral multilayer cladding consisting of alternating layers of piezoelectric nanocomposites and conductive polymer. Source: Adapted with permission from Lu et al. [71]. Copyright (2017) by The Electrochemical Society. (b) A hybrid-fiber PENG with ZnO NWs and a PVDF infiltrating polymer as the hybrid piezoelectric layer. Source: Lee et al. [57]. © 2012, John Wiley & Sons. (c) Open-circuit voltage output of a fiber device depending on the folding and releasing of the elbow. A voltage output appeared on the event of the folding and releasing action of the elbow. The elbow was field for an angle of ~90°. Source: Adapted with permission from Lee et al. (2012) [57]. Copyright (2012) by WILEY-VCH Verlag GmbH & Co. KGaA, Weinheim. (d) A fiber-shaped PENG based on perovskite $PbTiO_3$ nanotubes synthesized by anodic oxidation and hydrothermal method using TiO_2 NT as a positive template. Source: Lee et al. [72]. © 2016, John Wiley & Sons.

As a summary of the piezoelectric fibers, there is a genuine need for energy harvesting for on-person wearable devices. Recent developments in piezo-properties of piezo-materials, production method of piezo-fibers, biological friendliness performance, fabric architectures, and others present a promising future for powering electronics.

On the aspect of enhancing piezoelectric properties of piezo-materials, metal powders, nanowires, nanotubes, and copolymers have already been used. Furthermore, there are still possibilities of exploring other organic and inorganic materials producing hybrid piezoelectric structures as new piezo-materials to improve piezo-properties.

On the aspect of producing piezo-fibers, several research has been conducted on using various processes for energy harvesting, such as melt spinning, electrospinning, and others. The focus currently is on faster, cheaper, and more reliable methods for the production of piezo-fibers and the corresponding piezo-fabrics.

On the aspect of the biological friendliness performance, reducing toxicity is another focus for wearable applications, which has made materials more biofriendly than PVDF as an alternative option, such as GaN, AlN, ZnO, $BaTiO_3$, etc.

Besides, further work will continue to improvements in fabric architecture and the easy integration of these piezo-fibers with other flexible conductors, as well as sensing and actuating fibers. These developments can lead to a whole new generation of smart fabrics that could ultimately reduce the need to carry parasitic batteries, especially for on-person, low-power devices.

4.3 Triboelectric Fibers

Except for piezoelectric devices, triboelectric nanogenerator (TENG) is another type of device to convert the mechanical energy easily and effectively, such as muscle stretching and deformation, running energy, etc., in the living/working environment into electrical energy. In order to achieve better wearability, triboelectric fiber devices began to attract researchers' attention, which refer to the assembly of materials with different triboelectric properties on the fiber. Fiber-shaped or fabric-type TENG has been proven to collect all kinds of mechanical energy that is ubiquitous but wasted in our daily lives, such as human body motion [74], mechanical triggering [75], vibration [76], wind [77], flowing water [78], etc. The power output of fiber-shaped or fabric-type TENG can be used not only as wearable power supply but also as self-powered sensors.

4.3.1 Working Principle of Triboelectric Nanogenerator

Triboelectric effect is a kind of charged effect caused by contact, which is that on one material will be charged when it rubs against another material. It is the source of all kinds of static charges in our life. The positive and negative charges of materials depend on the relative polarity of contact materials. In recent years, a triboelectric nanogenerator based on the combination of triboelectrification effect and electrostatic precipitator has been developed. This new-generation technology can convert mechanical energy into electrical energy. The greater the polarity difference between them, the greater the difference of the ability of gaining and losing electrons, and the higher the output performance of triboelectric nanogenerator.

The basic working modes of triboelectric nanogenerator are divided into four types: vertical contact separation (CS), lateral sliding mode, single-electrode (SE) mode, and freestanding triboelectric layer mode [79], which can be selected for specific applications to prepare suitable energy devices. These structures are the basic components that provide micro and nano energy for small electronic devices. Large-scale power generation can be achieved by integrating several such basic components.

Vertical Contact Separation Mode

As shown in Figure 4.19a, two different dielectric films having back-coated electrodes are opposed to each other. The physical contact between a pair of dielectric films will create an opposite triboelectric charge on their surface. Once a physical gap is created between the two surfaces by external mechanical motion, a potential difference is created. The periodic contact and separation between the two charged surfaces will cause a periodic potential difference between the paired electrodes, which will drive the electrons to flow back and forth in the external circuit.

Lateral Sliding Mode

The structure of this mode is the same as that of the vertical CS mode. When the two dielectric films are in contact with each other, the relative in-plane sliding with each other also generates electric charges on the contact surface (Figure 4.19b). Therefore, lateral polarization occurs in the sliding direction, thereby driving the electron flow on the back electrode to completely balance the potential difference generated by the triboelectric charge, so that periodically sliding back and forth produces an alternating current (AC) output. In in-plane sliding-mode TENG, the sliding motion can be extended to cylindrical rotation or disk rotation.

Single-Electrode Mode

The first two modes of operation require pairs of electrodes to be connected to an external load, which will greatly limit the movement of the triboelectric layer. In some cases, one of the triboelectric layers must be a movable object, such as a person

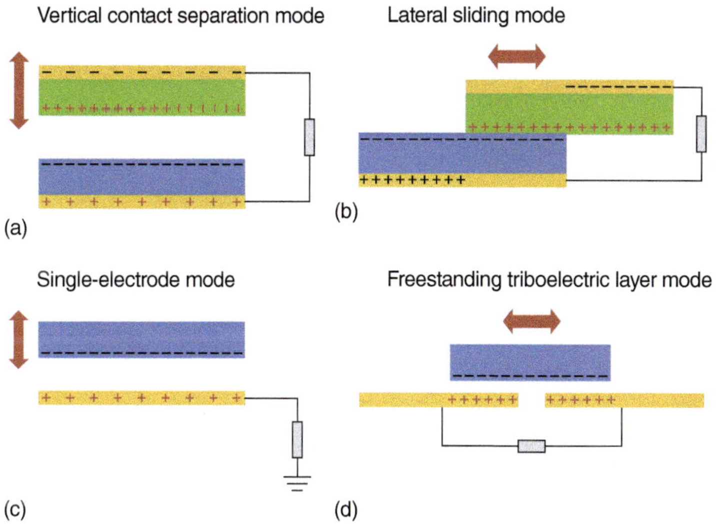

Figure 4.19 Working modes of triboelectric nanogenerator. Source: Luo and Wang [79]. © 2019, Elsevier.

walking on the floor. Therefore, an SE-mode TENG was developed to solve this problem in which the only electrode in TENG is grounded (Figure 4.19c). As the top object approaches or leaves the triboelectric layer, it will also change the distribution of the local electric field, which will cause the flow of electrons to move back and forth between the electrode and the ground. In this mode of operation, the relative motion of the triboelectric layers can also be classified into vertical CS and in-plane sliding motion.

Freestanding Triboelectric Layer Mode

Nonmoving triboelectric charges can be maintained on the contact surface for a long period of time. When there is a symmetrical electrode under the dielectric layer, objects near and away from the electrode will cause electrons to flow between the two electrodes to balance the local potential distribution (Figure 4.19d). When mechanical motion is converted to electricity, direct physical contact between the two triboelectric layers is not required, thereby effectively preventing material wear during TENG operation. This is a nice design that can improve the durability of TENG.

4.3.2 Triboelectrification Materials

The characteristics of triboelectrification materials are one of the factors that determine the performance of triboelectric devices. Many materials have triboelectric electrification effects, including metals, polymers, silk, wood, and so on.[1,2] These materials can be used to prepare triboelectric NGs, so the material selection of triboelectric NGs is very wide. But the material's electronic ability depends on its polarity. There is a series of triboelectrics on electrostatic charges as shown in Figure 4.20.

Positive ↑		Negative ↓
Ethylcellulose	Polyisobutylene	
Polyamide 11	Polyvinyl butyral	
Polyamide 6-6	Formo-phenolique, hardened	
Melanime formol	Polychlorobutadiene	
Wool, knitted	Natural rubber	
Silk,woven	Polyacrylonitrile	
Cellulose	Polybisphenol carbonate	
Cellulose acetate	Polystyrene	
Polyethylene glycol adipate	Polyethylene	
Polydiallyl phthalate	Polypropylene	
Cotton, woven	Polyimide	
Wood	Polyethylene terephthalate	
Hard rubber	Polyvinyl chloride	
Acetate, rayon	Polytrifluorochloroethylene	
Polyvinyl alcohol	Polytetrafluoroethylene	

Figure 4.20 A triboelectric series on electrostatic charges.

1 https://owlsmag.wordpress.com/2010/01/20/a-natural-history-devin-corbin
2 https://enwikipedia.org/wiki/Van-de-Graaff-generator

Materials tending to the "positive" direction are more likely to lose electrons, and materials tending to the "negative" direction are more likely to get electrons. This means that the farther apart the two materials are in the sequence, the larger the charge transfer is. In addition to the material in the sequence, the surface morphology of the material can be modified by microfabrication, and nanomaterials can be prepared to improve the specific surface area of the material, thus effectively improving the contact area and triboelectric electrification effect. Surface functionalization can greatly change the surface potential. The introduction of nanostructures on the surface can change the local contact characteristics and can also prepare composite materials. It not only changes the electrical properties of materials but also changes the dielectric properties of materials, thus affecting the electrostatic induction process.

4.3.3 Triboelectric Fiber Devices

In recent years, fiber-shaped, cloth-based, and other triboelectric fiber devices have been developed. Wang and coworkers [80] reported the first batch of fiber-shaped triboelectric NGs. The generator is made of two modified cotton threads woven together: a cotton thread coated with CNTs and a cotton thread coated with tetrafluoroethylene and CNTs. Fiber-shaped TENG mainly has two structural designs. One is a single-fiber shape composed of an inner fiber electrode and an outer wrapper, where the outer wrapper serves as both a dielectric layer and a protective layer. As for the working principle, only SE-type TENG and CS-type TENG have been realized in the triboelectric fiber devices.

SE-mode fiber-shaped TENG is usually designed with a single-fiber structure. The other is usually designed as a core-sheath structure with a gap between the inner core column and the outer tube. Generally, the fiber TENG based on the vertical CS mode is usually designed as a core-sheath structure. In this section, we will respectively introduce the related research of TENG based on single fiber in SE mode and CS mode.

With the merits of simple single-fiber structure, good performance, and versatile applications, the SE-mode fiber-shaped TENGs have been widely designed and reported. The simplest structure of single-fiber-shaped TENGs is that dielectric polymers are directly wrapped around core fiber electrodes, which single energy harvesting fiber was fabricated by coating PDMS or silicone rubber on the surface of conductive fiber [81]. For a typical example, in 2017, Wang and coworkers [82] presented a new type of single-thread-based triboelectric energy harvester, which was constructed from a multitwisted stainless steel thread as the conducting electrode and supersoft silicone rubber as the triboelectric material (Figure 4.21a). As a result, the short-circuit current (I_{sc}) of 5 cm in length reached to ~7 μA. Then, they took stainless steel wire coated with silicone rubber as the base and integrated it into the fabric in a folding way to prepare stretchable and foldable triboelectric NG fabric. The generating electric output reached up to 200 V and 200 μA. However, the surface-coated polymers will not only increase the overall diameter of fibers but also cause itching of human skin. Therefore, artificial commercial fibers are gradually

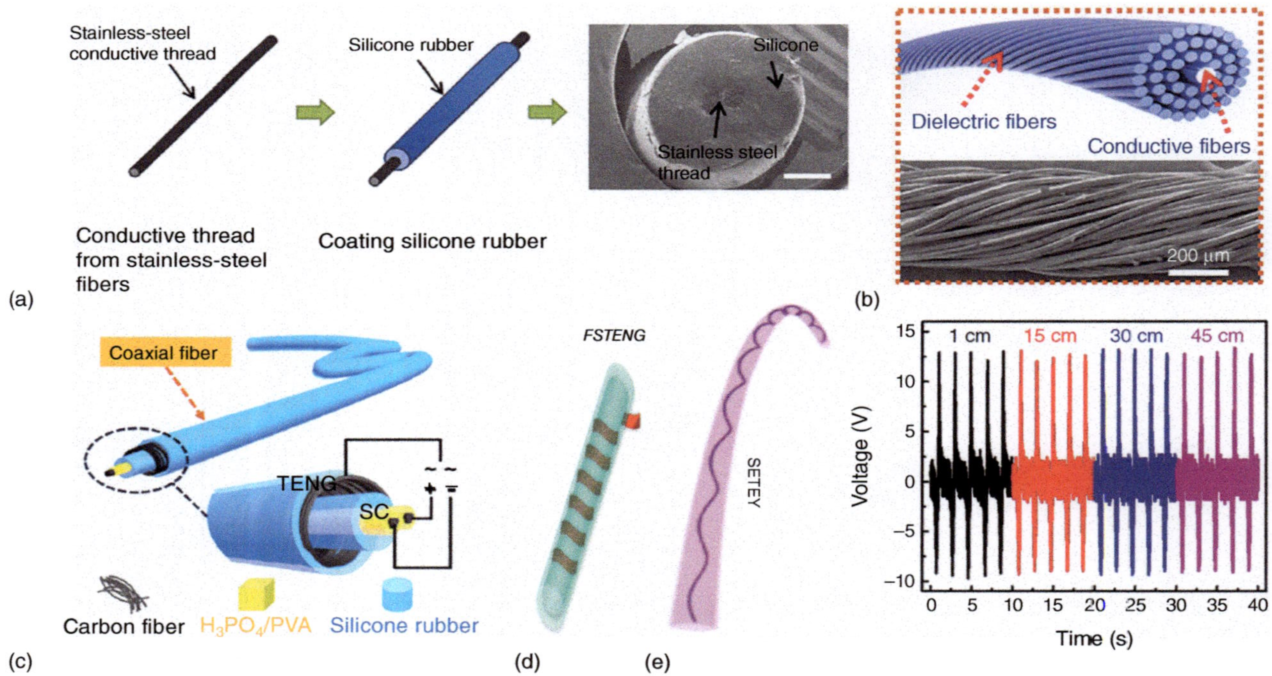

Figure 4.21 Structure and performance of typical SE mode single fiber-shaped TENGs. (a) A highly stretchable and scalable TENG textile realized by sewing the energy harvesting fiber that is composed of silicone rubber-coated stainless steel fiber into a serpentine shape on an elastic textile. Source: Adapted with permission from Lai et al. [82]. Copyright (2016) by WILEY-VCH Verlag GmbH & Co. KGaA, Weinheim. (b) A core-shell-yarn-based TENG applied by twining artificial polymer fibers tightly around core conductive fibers. Source: Adapted with permission from Yu et al. [83]. Copyright (2017) by American Chemical Society. (c) Fabrication process of the coaxial TENG and SC fiber. Source: Yang et al. [84]. © 2018, American Chemical Society. (d) The FSTENG is composed of silicone rubber and conductive thread. Source: Park et al. [85]. © 2017, Royal Society of Chemistry. (e) Schematic diagram of single-electrode triboelectric yarn and the output performances of single-electrode triboelectric yarn measured under different depth in water. Source: Gong et al. [86]. Licensed under CC BY 4.0.

used to replace polymers. Another of the structure of a coaxial energy harvesting fiber was developed by tightly twining artificial fibers around a conductive core fiber covering yarn, which could be further woven or knitted into fabrics (Figure 4.21b) [83]. The core-shell-yarn-based TENG textile (4.5 cm × 8 cm) exhibits motion relative to a common polyester (PET) textile in SE mode; an open-circuit voltage (V_{oc}) of ~75 V and short-circuit current (I_{sc}) of ~1.2 μA can be generated.

The SE-mode single-fiber-shaped TENG can be fabricated by using CNTs as a conductive core and silk fibroin as a dielectric sheath [87]. Thanks to its low cost, high efficiency, and mature technology, it is very promising to realize the large-scale production of this energy harvesting fiber. However, stretchability is absent in the above SE-mode single-fiber-shaped TENGs due to the existence of straight core fiber. It can be obtained by either selecting stretchable materials or applying stretchable structure. The fiber-shaped stretchability TENG can be achieved by twining fiber electrodes on the surface of stretchable fibers. Silicone rubber also acts as the triboelectric material for the TENG and as the encapsulation material for the coaxial fiber. Subsequently, the TENG in the coaxial fiber was fabricated by winding carbon fiber bundles as the electrode material over the supercapacitor (SC) fiber and by coating silicone rubber outside as the triboelectric material (Figure 4.21c) [84]. Park et al. [85] proposed a new structure of fiber-based single-electrode triboelectric nanogenerator (FSTENG). The proposed FSTENG uses silicone rubber as the negative electrode and conductive wires as the electrode of TENG. The prototype of the proposed FSTENG shows an electrical output of 28 V and 0.56 μA when in contact with human skin and shows a high strain up to 100% (Figure 4.21d). Another structure of achieving stretchability is that conductive fiber with built-in helical structure is inserted into elastic silicone rubber tubes, as illustrated in Figure 4.22e [86, 94]. Gong et al. [86] reported a continuous and scalable method for spinning an underwater-operational SE triboelectric yarn, which consists of intrinsically elastic silicone rubber tubes and built-in helical-structure stainless steel yarns, showing super stretchability and flexibility, as well as the output performances of SE triboelectric yarn measured under different depth in water (Figure 4.21e). It shows high performance in both air and liquid. In addition, energy harvesting fibers made up of stretchable components are also certainly stretchable. For example, a superstretchable SE-mode single-fiber-shaped TENG was obtained by injecting nontoxic liquid metals, e.g. Galinstan and EGaIn into stretchable silicone rubber tube [95].

Vertical CS mode is usually designed as a core-sheath structure, which can be classified into four main types:

(1) Dielectric layer wrapped around inner electrode as the core and only outer electrode as the shell, which can be denoted as the type-I fiber-shaped TENGs.
(2) Dielectric layer wrapped around inner electrode as the core and the same configuration as the shell, which can be denoted as the type-II fiber-shaped TENGs.
(3) Inner electrode wound on dielectric fiber as the core and the same configuration as the shell, which can be denoted as the type-III fiber-shaped TENGs.
(4) The outer surface of the type III further covered by dielectric or encapsulation layer, which can be denoted as the type-IV fiber-shaped TENGs.

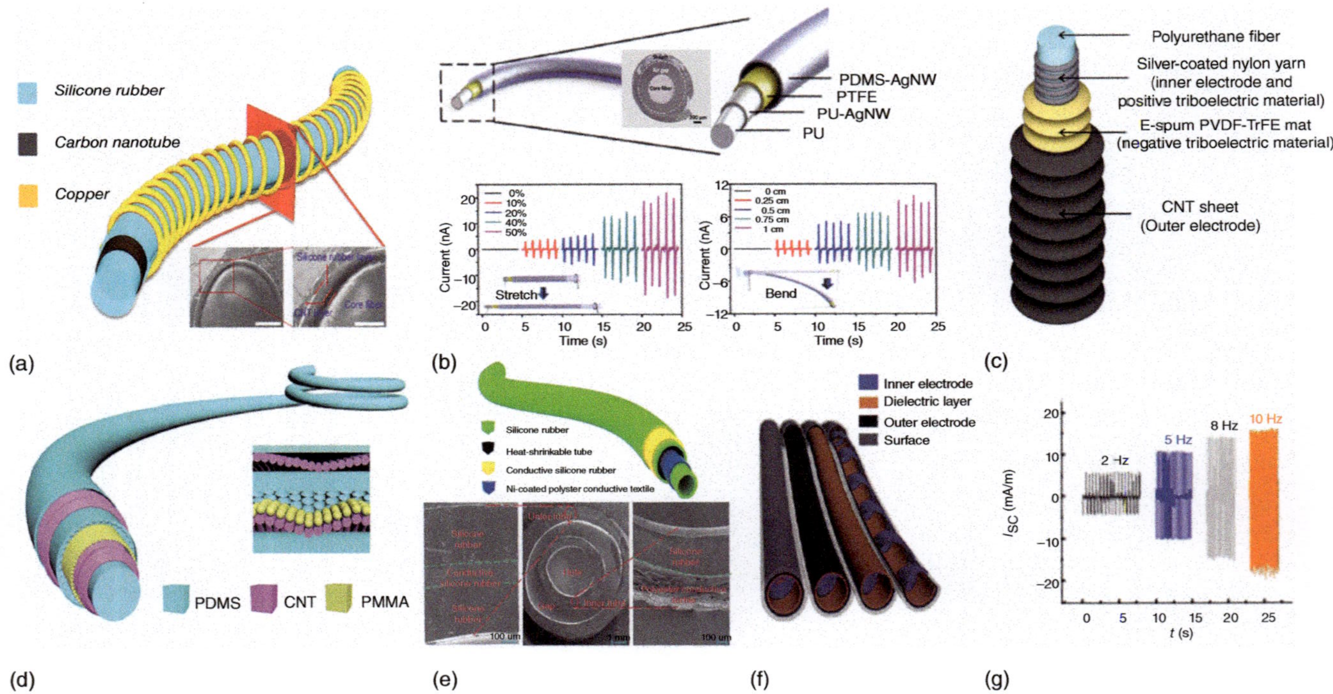

Figure 4.22 Structure and performance of vertical contact separation mode triboelectric fibers. (a) Structure schematic of the fiber-shaped TENG. Insets show the SEM image and magnified image of the fiber section. Silicone rubber layer, 200 μm; CNT layer, 80 μm; core fiber, 3.5 mm; scale bar (left and right), 1 mm and 500 μm, respectively. Source: Adapted with permission from He et al. [88]. Copyright (2016) by WILEY-VCH Verlag GmbH & Co. KGaA, Weinheim. (b) The structural schematic diagram of the FNG device. The output short-circuit current of the FNG under mechanical input of stretching (strain of 0-50%) and bending (deflection of 0-1 cm). Source: Adapted with permission from Cheng et al. [89]. Copyright (2017) by Elsevier. (c) Schematic diagram of stretchable triboelectric structure. Source: Sim et al. [90]. Licensed under CC BY 4.0. (d) A coaxial triboelectric fiber developed by using aligned CNT sheets as inner and outer electrodes and designing porous structures in triboelectric polymers of PDMS and PMMA. Source: Yu et al. [91]. © 2017, Royal Society of Chemistry. (e) A core-shell coaxially structured fiber-shaped TENG designed with conductive textile-wrapped silicone rubber tube as the core and CNT sandwiched between two silicone rubber layers as the shell. Source: Tian et al. [92]. Licensed under CC BY 4.0. (f) A high-output tube-like TENG proposed by attaching a belt-like inner electrode to the interior surface of a silicone rubber tube, which is further covered by outer electrode and outmost encapsulation layer and short-circuit current of tube-like TENG at various pressing frequencies. Source: Adapted with permission from Wang et al. [93]. Copyright (2019) by Springer Nature Limited.

No matter what type is selected, one of the basic design principles is that different triboelectric polarities are chosen for the outer-side materials of the inner core column and the inner-side materials of the outer shell tube.

For the type-I fiber-shaped TENGs, in 2017, Wang and coworkers [88] prepared the functional layer of TENG by coating on silicone rubber wire. Finally, a highly stretchable fiber triboelectric nanogenerator was formed by winding with copper wire. The fiber triboelectric nanogenerator can still drive when the stretch reaches 70%, as presented in Figure 4.22a. The stretching mode can be easily realized by stretch motions and can generate a maximum V_{oc} of 142.8 V and a maximum Q_{sc} of 61 nC per stretching cycle in a single-fiber-shaped TENG. Its high stretchability was achieved by convolving Cu wire onto the inner core column.

For the type-II fiber-based TENGs, a core-shell triboelectric fiber-based TENG was prepared with PTFE-coated PU/Ag NWs as the core and AgNWs/PDMS as the shell, as shown in Figure 4.22b [89]. The resistance of the sheath electrode without prestrain keeps increasing monotonically along with the tensile strain, by 3.5 times at strain of 50%, and retains a residual resistance increase of 150% even though the stretch was completely released. The electric conductivity of the sheath electrode remained exceedingly stable during the bending, twisting, and pressing, with a negligible resistance increase of within 5%.

For the type-III fiber-shaped TENGs, a nanostructured fiber-shaped TENG was developed by inserting Au-coated ZnO NWs grown on Al wires into Al foil-wrapped nanostructured PDMS tube [96]. The stretchable triboelectric fibers can be reversibly stretched up to 50% in tensile direction while generating voltage output proportional to the applied tensile strain. The reversible distance change induced by the Poisson's ratio difference between the core fiber (silver-coated nylon/polyurethane) and the shell (wrinkled polyvinylidene fluoride-*co*-trifluoroethylene/CNT layer) during tensile deformation is the key working principle for electrical generation (Figure 4.22c) [90]. The stretchable triboelectric fiber showed good sensitivity and stability, and strain could be detected up to 50% at various frequencies up to 10 Hz.

For the type-IV fiber-shaped TENGs, it is the most commonly used structure for the single-fiber-shaped TENGs, which has been widely reported. As shown in Figure 4.22d, a coaxial triboelectric fiber was developed with CNT-coated PDMS as the core and CNT sandwiched between two PDMS layers as the shell [91]. Polymethyl methacrylate (PMMA) microspheres were deposited on the inner CNT electrode in order to increase effective contact between the core and shell. With the similar structure, a triboelectric fiber was fabricated with silicone rubber tube covered by Ni-coated PET fabric as the core and CNTs layer sandwiched between two silicone rubber layers as the shell (Figure 4.22e) [92]. A superflexible tube-like TENG with a high surface charge density of 250 μA/cm^2 was designed by inserting a helix conductive belt into the shell tube consisting of carbon black/CNT layer sandwiched between two silicone rubber layers (Figure 4.22f) [93]. A high short-circuit current of 11 mA and open-circuit voltage of 380 V can be obtained from a the TENG tubes (CSTN) with a length of 6 cm, corresponding to a high peak power of 1.638 mW at a load resistance of about 10 MU (Figure 4.22g). As for type IV, conductive fibers are usually wound around dielectric fibers in order to provide stretchability. A varnished wire consisting of conductive Cu and insulation varnish at both ends acting as the shell was spirally wound around a core fiber, which was fabricated with CNT/polyaniline (PANI)-coated silicone rubber [97].

4.4 Thermoelectric Fibers

Thermal energy is also a common energy near the human body, and the use of heat energy around human body is very promising for driving wearable electronics. Thermoelectric generators (thermoelectric devices) can continuously generate electricity using the temperature difference between the human body and the surrounding environment without any transmission components and are expected to provide permanent power for mobile electronic products. Existing TE devices mostly possess the flat structure, which are too hard and bulky to be easily integrated into clothing. Flexible prototype devices that use out-of-plane thermal gradients are too fragile for practical wearing applications. Fiber-shaped smart configuration is undoubtedly the most effective way to achieve abrasion resistance of thermoelectric equipment. Fiber-shaped thermoelectric devices are usually fabricated by coating thermoelectric materials on a conductive fiber substrate, which can better adapt to the human skin and be wearable.

4.4.1 Introduction of Thermoelectric Effect

The thermoelectric effect is a phenomenon in which the electrons (holes) in the heated object move from the high-temperature region to the low-temperature region with the temperature gradient. In 1821, Seebeck, a German scientist, discovered that the magnetic needle placed next to the wire would rotate when heated at one end of the circuit of two wires. The ratio of voltage difference to temperature difference was later named Seebeck coefficient. Thermoelectric material is a kind of material that can directly convert electric energy to thermal energy. The efficiency of thermoelectric material is mainly determined by the thermoelectric merit value ZT:

$$ZT = S^2 T\sigma / k$$

S is Seebeck coefficient, T is absolute temperature, σ is conductivity, and k is thermal conductivity. Compared with other energy materials, thermoelectric materials have many advantages: (i) small and light, no mechanical rotation, so no working noise; (ii) accurate temperature controllability and accuracy of within (+0.1 C); (iii) no use of Freon-containing substances, so no pollution to the environment; and (iv) fast response speed, long use time, and easy to control (Figure 4.23).

However, the device efficiency of thermoelectric materials is still far from that of traditional generators, so how to improve the efficiency of thermoelectric materials is the ultimate goal of almost all thermoelectric materials research. Generally speaking, there are two ways to increase the ZT value: one is to increase the power factor ($S^2\sigma$), the most important of which is to increase the Seebeck coefficient, and the other is to reduce the heat conduction coefficient (K). Among the parameters

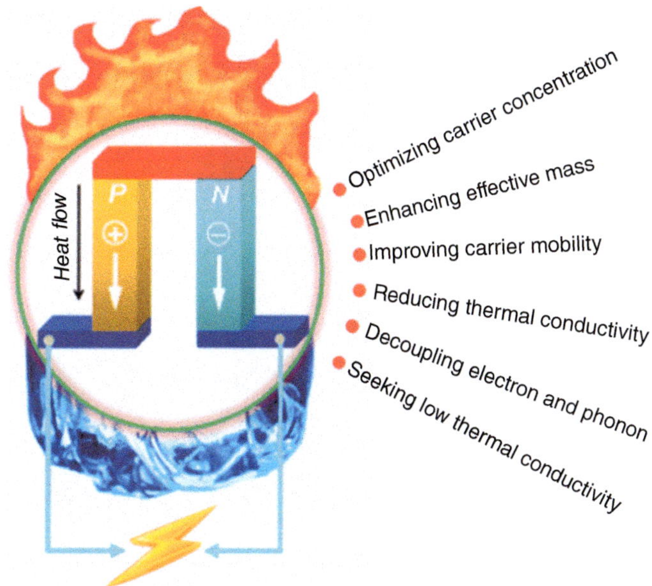

Figure 4.23 The working principle of the thermoelectric device. Source: Tan et al. [97]. © 2016, American Chemical Society.

affecting the power factor (scattering parameters, energy density, carrier mobility, and Fermi level), only Fermi level can be adjusted by changing the doping concentration to obtain high *ZT* value.

4.4.2 TE Materials for Wearable Thermoelectric Devices

The development of thermoelectric devices is closely related to the performance of TE materials, which are capable of converting heat energy and electric energy. This requires TE materials with appropriate carrier concentration and carrier mobility with high TE energy conversion efficiency (high *ZT* value) or by minimizing the Joule heating effect and maintaining a large temperature gradient when the power factor increases to fulfill. These materials should also have the expected mechanical properties (high flexibility and durability) and eco-friendly and stable performance near room temperature. The TE materials required in the preparation of thermoelectric fiber devices mainly include inorganic type, carbon-based type, conductive polymer type, composite type, etc.

As one of the most promising TE materials, the inorganic materials bearing the brunt, including Bi_2Te_3, PbTe, and SnSe, etc., have achieved rapid development in the past few decades, as reviewed by Zhao and coworkers (Figure 4.24) [97]. For the application of wearable energy supply, the TE materials need to efficiently convert human body heat into electrical energy around room temperature. Te-based materials are considered to be one of the ideal inorganic TE materials for wearable TE energy conversion since the TE performance was reported in the Bi-Te system, due to their stability near room temperature, ease of preparation, low thermal

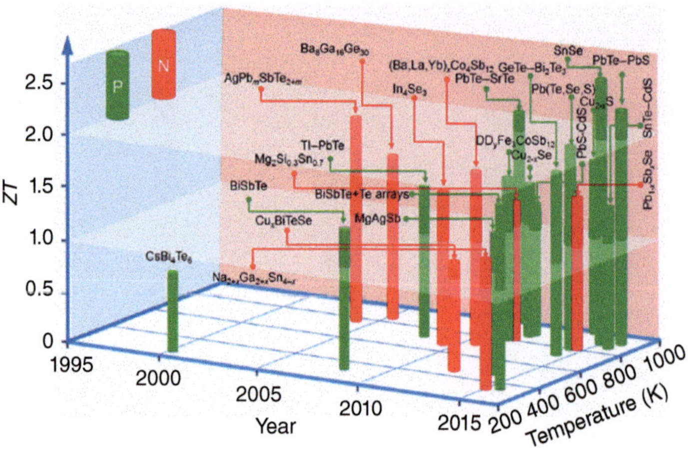

Figure 4.24 Current state-of-the-art inorganic TE materials. The TE ZT as a function of temperature and year illustrating important milestones. Source: Tan et al. [97]. © 2016, American Chemical Society.

conductivity, and high electrical conductivity [98]. The molecular dynamics of electron and phonon transport was systemically studied in Te-based system by Huang and coworker in 2008, indicating the potential of low-dimensional Te-based nanomaterials for high-performance TE materials [99]. Kanatzidis group aims to synthesize new bulk materials with high *ZT*s greater than the historic threshold of $ZT = 1$ with two general approaches. The endotaxial SrTe nanoprecipitates inhibit the heat flow in the system, but the crystallographic alignment of the SrTe and PbTe does not affect the hole mobility, allowing a large power factor to be achieved concurrently with a very low thermal conductivity. This decoupling of phonon and electron transport allows the system to reach $ZT = 2.2$ at 900 K (Figure 4.25a) [100]. Zhao Lidong and He Jiaqing of Nanke University in Science reported that the maximum *ZT* of n-type tin selenide (SnSe) crystals was $\sim 2.8 \pm 0.5$ at 773 K (Figure 4.25b) [101]. The thermal and electrical properties of tin selenide in the wide temperature range of 300-773 K can be greatly improved by applying the special electronic band structure and multi-valley effect of tin selenide. The *ZT* value can be increased from 0.1-0.9 to 0.7-2.0.

Carbon-based TE materials are also widely used in the preparation of thermoelectric fiber-shaped devices, such as graphene, graphene nanoribbons (GNR), CNTs, etc., and also have attracted much attention in TE energy harvesting due to their extraordinary physical, chemical, and mechanical properties. The electrical conductivity of graphene can be described by the relativistic Dirac equation: $E \pm (q) \approx \pm v_F |q|$, where q is momentum relating to the Dirac point and v_F is Fermi velocity. High carrier mobility of 200 000 $cm^2/V{\cdot}s$ and electrical conductivity of 106 S/cm have been reported by Du et al. with a low carrier concentration at room temperature [102]. Thus, it makes graphene a potential TE material that is more promising than the state-of-the-art TE materials near room temperature, such as bismuth telluride. However, the thermal conductivity of graphene is dangerous

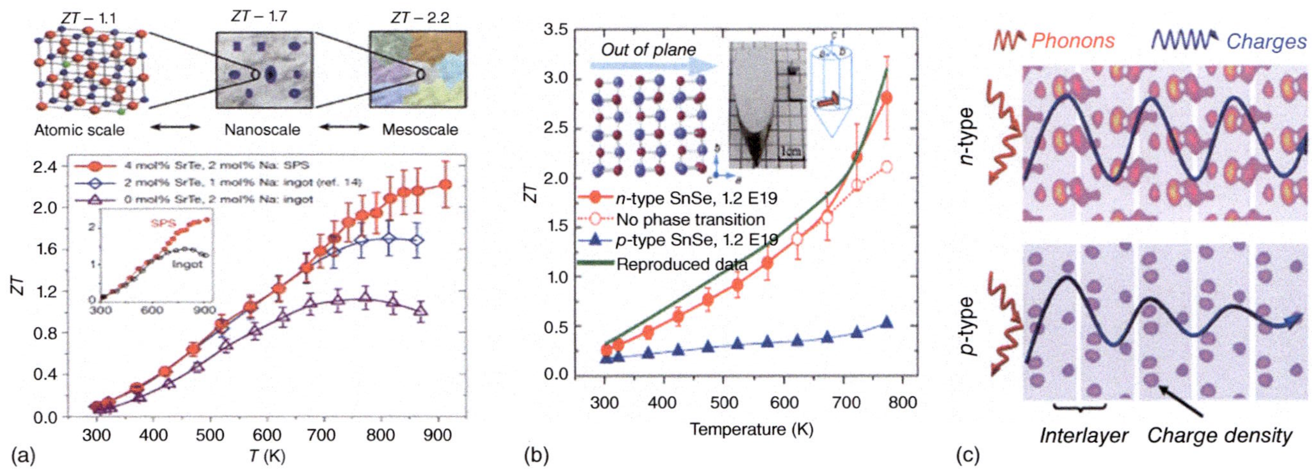

Figure 4.25 Seebeck coefficient (TEP). (a) TEP of endotaxial SrTe nanoprecipitates [100]. Source: Biswas et al. [100]. © 2012, Springer Nature. (b) TEP of n-type tin selenide (SnSe) crystals [101]. Source: Adapted with permission from Zhao et al. [101]. Copyright (2014) by Springer Nature Limited.

for good TE materials. The usually thermal transport in graphene system is caused by the quantum effect on heat transport, such as localization caused by quantum interference. According to this principle, Seol et al. have reported a reduced thermal conductivity in a range of 2000-3000 Wm/K at room temperature by experimental studies [103], but this value is also too high to achieve an ideal *ZT* value for graphene. The improvements in energy conversion efficiency for graphene materials have been explored by structural modifications and band structure modulation. GNR as the quasi-1D structure material achieved by patterning graphene in a particular direction exhibits similar mechanical and electronic properties for 2D graphene. The formation of 1D GNRs can effectively reduce its lattice thermal conductivity and improve its TE performance while maintaining the electrical conductivity [104]. Compared with GNRs, CNTs can be supposed to be curled from the monolayer graphene, which also have been proposed as potential functional materials for wearable application, due to its excellent mechanical properties (lightweight, high strength), chemical properties, physical properties, and surface structure [105]. The TE performance of CNTs and their electron transport properties have been effectively adjusted by the changing in defect types, degree of defects, strain, and diameter of CNTs. In addition, the effects of humidity for TE performance of CNT films also have systematically been researched by Daisuke et al. [106], and their Seebeck coefficient and electrical conductivity are effectively changed in humid environment, thus achieving a threefold increases in power factor of CNT films combined with the nonequilibrium Green's function theory. Yu et al. [107] uncovered the interesting phenomenon that the transfer types of the carriers in CNTs can be changed from p type to n type after doping of electron-donating dopants of polyethyleneimine (PEI) in 2012; this promoted the development of flexible p–n-type TE energy conversion devices. The advances in n-type doping CNTs including methods, analysis, and application have been reviewed by Brownlie and Shapter [108], who reviewed the recent research into n-type doping of CNTs in TE materials under the banner of flexible electronics and pointed out the great potential of n-type CNTs toward wearable energy conversion devices.

Conducting polymers (CPs) as the promising organic materials [109], displaying high TE properties close to inorganic materials and considering to be the next-generation TE materials for wearable devices arising from their unique advantages including their low thermal conductivity (1-3 orders of magnitude lower than inorganic materials), lightweight, high flexibility and stretchability, good film formation, low cost-effectiveness, etc. In the past few decades, a large number of CPs have been explored for potential TE materials, such as polyacetylene (PA), PANI, polypyrrole (PPy), poly(3,4-ethylenedioxythiophene) (PEDOT), and other CPs. PEDOT is considered as the candidate of organic TE materials in these reported CPs, owing to its high electrical conductivity, outstanding environmental stability, and low density. However, the solubility and film formation of PEDOT limited its application. This discovery greatly increased the range of applications for PEDOT, and electrical conductivity of the formed PEDOT:PSS can be increased by several orders of magnitude after doping with dielectric solvents, such as DMSO, ethylene glycol (EG), N-methyl-2-pyrrolidone (NMP), etc. [110], which is ascribed to the fact that the aligned PEDOT chains increase the mobility and improve carrier hopping by reducing hopping distance. Generally, the CPs show the low Seebeck coefficient from ~10 to ~20 μV/K near room temperature. Xiong and co-workers first reported the significantly enhanced electrical conductivity for PEDOT:PSS (1500 S/cm) sub-microfilm using a dilution filtration with these dielectric solvents, and the ZT value reached to ~0.1 [111]. Kim et al. [112] effectively improved Seebeck coefficient of PEDOT:PSS films from 12.6 to 54.8 μV/K after over-dropping with a mixture of DMSO/NaOH ethanol mixture, owing to the precise control of de-doping level of PEDOT chains. This shows good potential for low-power wearable electronic equipment owing to the increased demand for flexible energy generation, although this *ZT* value is still lower than the inorganic materials.

Composite TE materials have also been developed to improve the performance of thermoelectric devices. For example, the incorporation between nanoscale in organic materials and CP matrices has been considered as an effective method to improve their TE performance. A large number of organic-inorganic composites have been prepared through controlling the interfacial energy barriers (energy filtering effect) to optimize their TE performance in recent years, which present opportunities simultaneously harnessing tailored interface between the two classes of materials to create composite system. For the first time, See et al. [113] reported one promising strategy of in situ composite method to optimize the organic-inorganic composite materials ($S^2\sigma \sim 70\,\mu W/m{\cdot}K$ at room temperature) through the effectively interface control, and their TE performance was further improved by a simple chemical treatment ($S^2\sigma \sim 284\,\mu W/m{\cdot}K$ at room temperature) [114], which promotes the studies of interface control for improving TE performance of organic-inorganic composites. Thus, composite materials reveal the good potential for wearable TE energy conversion devices by the effective combination of organic and inorganic materials.

4.4.3 Fiber-Shaped Thermoelectric Devices

Most thermoelectric devices are flat-panel structures, which are not well adapted to human wear. Flexible fiber-shaped thermoelectric devices have shown excellent tensile, bending, and shear properties and thus are more suitable for wearable energy conversion. A large number of fiber-shaped devices have been reported in recent years. For high-performance TE fiber devices, the fabrication process involves the deposition of high-performance TE materials on the commercial fibers, yarns, or fabrics, for which surface modification or coating methods for fiber-shaped electrodes are very important.

Many TE materials have been applied in the fiber-shaped thermoelectric device. As early as 2008, Yadav et al. [115] deposited nickel and silver as adjacent strips on flexible silicon fibers, which attracted widespread interests in wearable TE energy conversion, promoting the development of flexible TE fibers. These fibers can be woven into energy harvesting fabrics or can be interwoven with structural fibers such as carbon fibers to create an energy harvesting composite. An open-circuit voltage of 19.6 μV/K per thermocouple junction is measured for Ni-Ag thin films, and a maximum power of 2 nW for seven couples at $\Delta T = 6.6$ K is measured. Heat transfer analysis is used to project performance for several other material systems, with a predicted power output of 1 μW per couple for Bi_2Te_3/Sb_2Te_3-based fiber coatings with a hot junction temperature of 100 °C.

Considering the performance of woven thermoelectric cloths or fiber composites, relevant properties and dimensions of individual thermoelectric fibers are optimized. Then, Liang et al. [116] reported that 300 nm-thick PdTe nanocrystals were deposited on the surface of glass fibers, which exhibited excellent TE performance and flexibility (their ZT value at 400 K was up to 0.72, and bending curvature was 84.5 °C). They reported an example using a scalable solution-phase deposition method to coat thermoelectric nanocrystals onto the surface of flexible glass fibers (Figure 4.26a,b). Their investigation of the thermoelectric properties yields high performance comparable with the state of the art from the bulk crystals, and proof-of-concept demonstration also suggests the potential of wrapping the thermoelectric fibers on the industrial pipes to improve the energy efficiency. The InSe alloy core thermoelectric fibers with glass cladding were fabricated by a reactive molten core drawing method (Figure 4.26c) [117]. The In_4Se_3 crystals were obtained in the fiber core by annealing the as-drawn fibers. The Seebeck coefficient of the In_4Se_3 core fibers (=347 μV/K) was enhanced by ~85% compared with that of the as-drawn InSe core fibers (=189 μV/K). Wei and coworkers [118] demonstrate flexible and ultralong TE fibers using thermal drawing technology to seamlessly integrate high-performance and intrinsically crystalline inorganic TE micro/nanowires in a fiber-like carrier (Figure 4.26d). The resulting thermoelectric fibers are intrinsically crystalline, highly flexible, ultralong, and mechanically stable while maintaining high thermoelectric properties as their bulk counterparts. Two types of thermoelectric generators covered on different curved surfaces are constructed to provide mW/cm^2-level output power density. Additionally, a wearable 2D cooling textile is assembled to achieve a maximum cooling of 5 °C. Recently, Ag_2Te nanocrystals were coated on nylon mesh fibers followed with a low annealing process (just about 150 °C), which display a high Seebeck coefficient and electrical conductivity, and promoting the fabrication of fiber-shaped TE conversion devices toward wearable energy conversion. However, the surface modification or coating also has its own disadvantages such as variance in smoothness and uniformity of these coatings, which are waiting to be solved for the further researches [119].

Various methods have been employed for preparing thermoelectric fiber-shaped devices, such as electrostatic spinning, dry spinning, wet spinning, etc. Among them, the direct preparation of fiber-shaped materials via the simple wet spinning attracts more and more attentions toward wearable energy conversion, which mainly focus on CNTs, graphene, or partially CPs. Jalili and coworkers successfully prepared the flexible PEDOT:PSS fibers with high TE performance by using this method in 2011 [120], achieving the 30-fold enhancements in electrical conductivity from 9 to 264 S/cm after a post-spinning treatment with EG, owing to an improved molecular ordering of the PEDOT chains in the direction of the fiber axis and the consequential enrichment of linear (or expanded coil-like) conformation to preference bipolaronic electronic structures. However, the flexible PEDOT:PSS fibers are sensitive to humidity conditions. In order to maintain the stability in

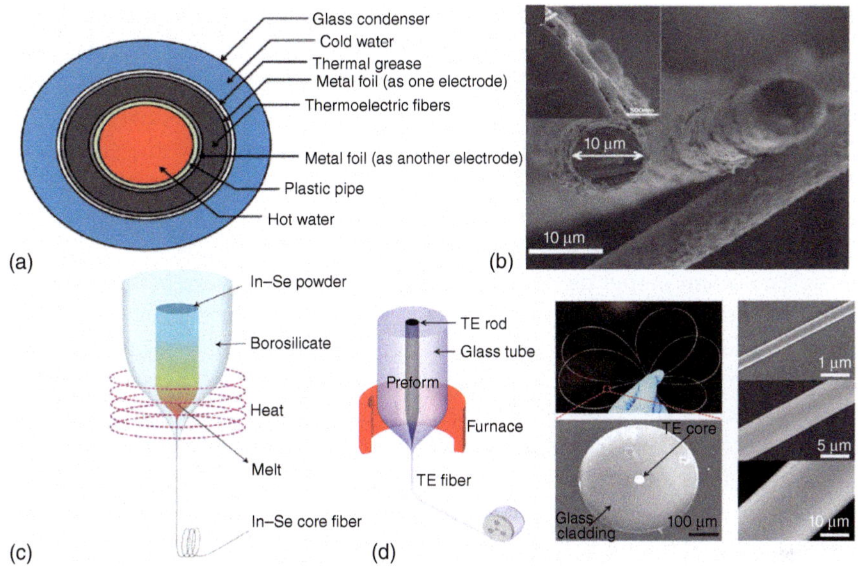

Figure 4.26 Demonstration of using flexible thermoelectric fibers for harvesting energy from industrial pipes. (a) Scheme of the proof-of-concept design. Source: Liang et al. [116]. © 2012, American Chemical Society. (b) SEM image of PbTe nanocrystal-coated glass fibers. Source: Adapted with permission from Liang et al. [116]. Copyright (2012) by American Chemical Society. (c) Sketch of the In-Se core fiber drawing process. Source: Sun et al. [117]. © 2018, Elsevier. (d) Flexible and ultralong TE fiber fabricated by thermal drawing. Source: Adapted with permission from Zhang et al. [118]. Copyright (2017) by Elsevier.

surrounding environment, Liu et al. used PDMS to seal the PEDOT:PSS fibers for practical application in wearable energy conversion [121]. Recently, the uniform CNT fibers with the diameter of ~50 mm were prepared by Ito et al. based on their past experience, which exhibited high electrical conductivity more than ~500 S/cm and promising flexibility toward wearable energy harvesting [122]. Interestingly, the conductive type of CNT fibers can be changed from p to n type accompanying the Seebeck coefficient changed from 47.8 to −49.1 μV/K after doping with 1-butyl-3-methylimidazolium hexafluorophosphate ($[BMIM]PF_6$), which demonstrated great potential toward wearable TE energy harvesting devices with all-carbon materials. Although the fabrication of fiber-based materials offers more opportunities for the application as wearable energy storage devices, the preparation of fiber-shaped materials with high performance and accurate measurement of cross-sectional area (that is electrical conductivity) and thermal conductivity are still challenges. Thus, the development of effective preparation methods and more accurate measurement are being imminent for the promotion and evaluation of these fiber-shaped TE materials.

4.5 Conclusions and Outlook

This chapter summarizes common energy harvesting fiber devices, which include photovoltaic fibers, piezoelectric fibers, triboelectric fibers, thermoelectric fibers, etc. The fiber-shaped energy harvesting device can be more widely used and can effectively obtain environmental energy or human movement energy in a continuous, convenient, and environmentally friendly way, providing a feasible power supply for wearable electronic products. For energy harvesting fiber devices, FPSCs give a PCE of 10%, and fiber-shaped TENG device can give a power output of 100 μW. In this way, a variety of functional applications such as motion tracking, health monitoring, pressure pixel array, and human-computer interaction can be realized without the need for additional power.

Significant progress has been made so far through continuous research on them. However, current output power, energy density, stability, etc. are particularly low, requiring further work. To improve the power outputs of fabric-based energy conversion devices, especially under on-body conditions, we should focus on several aspects, such as increasing the effective contact area, improving flexibility of nanostructure function materials, introducing tandem structure for multiple composited devices, and so on.

Although the current fiber energy harvesting device still needs a lot of work to improve its performance, the development of these wearable energy devices is expected to broadly apply in medical and health, military, education, entertainment, and other fields.

References

1 Baps, B., Eber-Koyuncu, M., and Koyuncu, M. (2002). Ceramic based solar cells in fiber form. *Key Engineering Materials* 206 (213): 937–940.

2 Liu, J., Namboothiry, M., and Carroll, D.L. (2007). Optical geometries for fiber-based organic photovoltaics. *Applied Physics Letters* 90 (13): 138.

3 Qiu, L., Deng, J., Lu, X. et al. (2014). Integrating perovskite solar cells into a flexible fiber. *Angewandte Chemie International Edition* 53: 10425–10428.

4 Fan, X., Chu, Z., Wang, F. et al. (2008). Wire-shaped flexible dye-sensitized solar cells. *Advanced Materials* 20 (3): 592–595.

5 Lee, M.R., Eckert, R.D., Forberich, K. et al. (2009). Solar power wires based on organic photovoltaic materials. *Science* 324: 232–235.

6 Deng, J., Qiu, L., Lu, X. et al. (2015). Elastic perovskite solar cells. *Journal of Materials Chemistry A* 3 (42): 21070–21076.

7 Zou, D., Wang, D., Chu, Z. et al. (2010). Fiber-shaped flexible solar cells. *Coordination Chemistry Reviews* 254 (9–10): 1169–1178.

8 Xu, W., Choi, S., and Allen, M.G. (2010). Hairlike carbon-fiber-based solar cell. *2010 IEEE 23rd International Conference on Micro Electro Mechanical Systems (MEMS).*

9 Zhang, L., Song, L., Tian, Q. et al. (2012). Flexible fiber-shaped $CuInSe_2$ solar cells with single-wire-structure: design, construction and performance. *Nano Energy* 1 (6): 769–776.

10 He, R., Day, T.D., Krishnamurthi, M. et al. (2013). Silicon p-i-n junction fibers. *Advanced Materials* 25 (10): 1461–1467.

11 Martinsen, F.A., Smeltzer, B.K., Nord, M. et al. (2014). Silicon-core glass fibres as microwire radial-junction solar cells. *Scientific Reports* 4: 6283.

12 Li, G., Zhu, R., and Yang, Y. (2012). Polymer solar cells. *Nature Photonics* 6 (3): 153–161.

13 Liu, D., Zhao, M., Li, Y. et al. (2012). Solid-state, polymer-based fiber solar cells with carbon nanotube electrodes. *ACS Nano* 6 (12): 11027–11034.

14 Chen, T., Qiu, L., Li, H., and Peng, H. (2012). Polymer photovoltaic wires based on aligned carbon nanotube fibers. *Journal of Materials Chemistry* 22 (44): 23655–23658.

15 Zhang, Z., Yang, Z., Wu, Z. et al. (2014). Weaving efficient polymer solar cell wires into flexible power textiles. *Advanced Energy Material* 4 (11): 1301750.

16 Zhang, Z., Yang, Z., Deng, J. et al. (2015). Stretchable polymer solar cell fibers. *Small* 11 (6): 675–680.

17 O'Connor, B., Pipe, K.P., and Shtein, M. (2008). Fiber based organic photovoltaic devices. *Applied Physics Letters* 92 (19): 172.

18 Liu, D., Li, Y., Zhao, S. et al. (2013). Single-layer graphene sheets as counter electrodes for fiber-shaped organic polymer solar cells. *RSC Advances* 3: 13720–13727.

19 Bedeloglu, A., Demir, A., Bozkurt, Y., and Sariciftci, N.S. (2010). A photovoltaic fiber design for smart textiles. *Textile Research Journal* 80: 1065–1074.

20 Cao, Y., Liu, Y., Zakeeruddin, S.M. et al. (2018). Direct contact of selective charge extraction layers enables high-efficiency molecular photovoltaics. *Joule* 2 (6): 1108–1117.

21 Meng, Q.B., Takahashi, K., Zhang, X. et al. (2003). Fabrication of an efficient solid-state dye-sensitized solar cell. *Langmuir* 19 (9): 3572–3574.

22 O'Regan, B., Schwartz, D.T., Zakeeruddin, S.M., and Grätzel, M. (2000). Electrodeposited nanocomposite n-p heterojunctions for solid-state dye-sensitized photovoltaics. *Advanced Materials* 12 (17): 1263–1267.

23 Fu, Y., Lv, Z., Hou, S. et al. (2011). Conjunction of fiber solar cells with groovy micro-reflectors as highly efficient energy harvesters. *Energy & Environmental Science* 4: 3379–3383.

24 Feng, H., Tao, S., Zhang, X. et al. (2014). From wires to veins: wet-process fabrication of light-weight reticulation photoanodes for dye-sensitized solar cells. *Chemical Communications* 50 (26): 3509.

25 Fan, X., Zhang, X., Zhang, N. et al. (2015). Wet-process fabrication of low-cost all-solid wire-shaped solar cells on manganese-plated electrodes. *Electrochimica Acta* 161: 358–363.

26 Yang, Z., Sun, H., Chen, T. et al. (2013). Photovoltaic wire derived from a graphene composite fiber achieving an 8.45% energy conversion efficiency. *Angewandte Chemie International Edition* 52 (29): 7545–7548.

27 Chen, L., Dai, H., Zhou, Y. et al. (2014). Porous, single crystalline titanium nitride nanoplates grown on carbon fibers: excellent counter electrodes for low-cost, high performance, fiber-shaped dye-sensitized solar cells. *Chemical Communication* 50 (92): 14321–14324.

28 Chen, L., Zhou, Y., Dai, H. et al. (2015). One-step growth of $CoNi_2S_4$ nanoribbons on carbon fibers as platinum-free counter electrodes for fiber-shaped dye-sensitized solar cells with high performance: polymorph-dependent conversion efficiency. *Nano Energy* 11: 697–703.

29 Yang, Z., Deng, J., Sun, X. et al. (2014). Stretchable, wearable dye-sensitized solar cells. *Advanced Materials* 26 (17): 2643–2647.

30 Yan, J., Uddin, M.J., Dickens, T.J. et al. (2014). 3D wire-shaped dye-sensitized solar cells in solid state using carbon nanotube yarns with hybrid photovoltaic structure. *Advanced Materials Interfaces* 6 (1): 1400075.

31 Weintraub, B., Wei, Y., and Wang, Z. (2009). Optical fiber/nanowire hybrid structures for efficient three-dimensional dye-sensitized solar cells. *Angewandte Chemie International Edition* 47 (121): 8981–8985.

32 Zhang, S., Ji, C., Bian, Z. et al. (2011). Single-wire dye-sensitized solar cells wrapped by carbon nanotube film electrodes. *Nano Letters* 8 (11): 3383–3387.

33 Sun, H., Li, H., You, X. et al. (2014). Quasi-solid-state, coaxial, fiber-shaped dye-sensitized solar cells. *Journal of Materials Chemistry A* 2 (2): 345–349.

34 Liang, J., Zhang, G., Sun, W., and Dong, P. (2015). High efficiency flexible fiber-type dye-sensitized solar cells with multi-working electrodes. *Nano Energy* 12: 501–509.

35 Guo, W., Xu, C., Wang, X. et al. (2012). Rectangular bunched rutile TiO_2 nanorod arrays grown on carbon fiber for dye-sensitized solar cells. *Journal of the American Chemical Society* 134 (9): 4437–4441.

36 Zhu, B., Huang, Y., Fan, L. et al. (2016). Novel fuel cell with nanocomposite functional layer designed by perovskite solar cell principle. *Nano Energy* 19: 156–164.

37 Sum, T.C. and Mathews, N. (2014). Advancements in perovskite solar cells: photophysics behind the photovoltaics. *Energy & Environmental Science* 7: 2518–2534.

38 Lee, M.K., Yohan, J., and Yongseok (2015). Efficient fiber-shaped perovskite photovoltaics using silver nanowires as top electrode. *Journal of Materials Chemistry A* 3 (38): 19310–19313.

39 Hu, H., Jun, Y., Peng, M. et al. (2016). Fiber-shaped perovskite solar cells with 5.3% efficiency. *Journal of Materials Chemistry A* 4 (10): 3901–3906.

40 He, S., Qiu, L., Fang, X. et al. (2015). Radically grown obelisk-like ZnO arrays for perovskite solar cell fibers and fabrics through a mild solution process. *Journal of Materials Chemistry A* 3 (18): 9406–9410.

41 Qiu, L., He, S., Yang, J. et al. (2016). Fiber-shaped perovskite solar cells with high power conversion efficiency. *Small* 12 (18): 2419–2424.

42 Qiu, L., He, S., Yang, J. et al. (2016). An all-solid-state fiber-type solar cell achieving 9.49% efficiency. *Journal of Materials Chemistry A* 4 (26): 10105–10109.

43 Hu, H., Dong, B., Chen, B. et al. (2017). High-performance fiber-shaped perovskite solar cells based on lead acetate precursor. *Sustainable Energy Fuels* 2: 79–84.

44 Woon, S.Y., Byung-Wook, P., Eui, H.J. et al. (2017). Iodide management in formamidinium-lead-halide-based perovskite layers for efficient solar cells. *Science* 356 (6345): 1376–1379.

45 Liu, H., Zhong, J., Lee, C. et al. (2018). A comprehensive review on piezoelectric energy harvesting technology: materials, principles, and applications. *Applied Physics Reviews* 5 (4): 041306.

46 Wang, A., Wu, C., Pisignano, D. et al. (2018). Polymer nanogenerators: opportunities and challenges for large-scale applications. *Journal of Applied Polymer Science* 135 (24): 45674–45690.

47 Wang, X., Guo, R., and Liu, J. (2019). Liquid metal based soft robotics: materials, designs, and applications. *Advanced Materials Technologies* 4 (2): 1800549–1180063.

48 Dong, K., Peng, X., and Wang, Z. (2019). Fiber/fabric-based piezoelectric and triboelectric nanogenerators for flexible/stretchable and wearable electronics and artificial intelligence. *Advanced Materials* 32 (5): e1902549.

49 Kingon, A.I. and Srinivasan, S. (2005). Lead zirconate titanate thin films directly on copper electrodes for ferroelectric, dielectric and piezoelectric applications. *Nature Materials* 4 (3): 233–237.

50 Dimitroula, M., Savvas, V., and Derman, V.B. (2018). Piezoelectric textile fibers for wearable energy harvesting systems. *Materials Research Express* 5: 065508.

51 Fang, J., Niu, H., Wang, H. et al. (2013). Enhanced mechanical energy harvesting using needleless electrospun poly(vinylidene fluoride) nanofiber webs. *Energy & Environmental Science* 6 (7): 2196.

52 Wang, Z. and Song, J. (2006). Piezoelectric nanogenerators based on zinc oxide nanowire arrays. *Science* 312 (5771): 242–246.

53 Wang, Z. (2007). Nanopiezotronics. *Advanced Materials* 19 (6): 889–892.

54 Wang, Z. (2010). Piezopotential gated nanowire devices: piezotronics and piezo-phototronics. *Nano Today* 5 (6): 540–552.

55 Lin, P., Pan, C., and Wang, Z. (2018). Two-dimensional nanomaterials for novel piezotronics and piezophototronics. *Materials Today Nano* 4 (2018): 17–31.

56 Qin, Y., Wang, X., and Wang, Z. (2008). Microfibre-nanowire hybrid structure for energy scavenging. *Nature* 451 (7180): 809–813.

57 Lee, M., Chen, C., Wang, S. et al. (2012). A hybrid piezoelectric structure for wearable nanogenerators. *Advanced Materials* 24 (13): 1759–1764.

58 Cauda, V., Canavese, G., and Stassi, S. (2014). Nanostructured piezoelectric polymers. *Journal of Applied Polymer Science* 132 (13): 41667.

59 Li, Z. and Wang, Z. (2010). Air/liquid-pressure and heartbeat-driven flexible fiber nanogenerators as a macro/nano-power source or diagnostic sensor. *Advanced Materials* 23 (1): 84–89.

60 Zhang, L., Bai, S., Su, C. et al. (2015). A high-reliability kevlar fiber-ZnO nanowires hybrid nanogenerator and its application on self-powered UV detection. *Advanced Functional Material* 25 (36): 5794–5798.

61 Du, Y., Fu, C., Gao, Y. et al. (2017). Carbon fibers/ZnO multimaterial piezoelectric fibers nanowires hybrid nanogenerator based on an insulating interface barrier. *RSC Advances* 7 (35): 21452–21458.

62 Shin, K.Y., Lee, J.S., and Jang, J. (2016). Highly sensitive, wearable and wireless pressure sensor using free-standing ZnO nanoneedle/PVDF hybrid thin film for heart rate monitoring. *Nano Energy* 22: 95–104.

63 Choi, M., Murillo, G., Hwang, S. et al. (2017). Mechanical and electrical characterization of PVDF-ZnO hybrid structure for application to nanogenerator. *Nano Energy* 33: 462–468.

64 Gao, H., Minh, P.T., Wang, H. et al. (2018). High-performance flexible yarn for wearable piezoelectric nanogenerators. *Smart Material and Structure* 27: 095018.

65 Lund, A., Rundqvist, K., Nilsson, E. et al. (2018). Energy harvesting textiles for a rainy day: woven piezoelectrics based on melt-spun PVDF microfibers with a conducting core. *NPJ Flexible Electronics* 2 (1): 9.

66 Egusa, S., Wang, Z., Chocat, N. et al. (2010). Multimaterial piezoelectric fibers. *Nature Materials* 9 (8): 643–648.

67 Sim, H.J., Choi, C., Lee, C.J. et al. (2015). Flexible, stretchable and weavable piezoelectric fiber. *Advanced Engineering Materials* 17 (9): 1270–1275.

68 Kim, M. and Yun, K.S. (2017). Helical piezoelectric energy harvester and its application to energy harvesting garments. *Micromachines* 8 (4): 115.

69 Kechiche, M.B., Bauer, F., Harzallah, O., and Drean, J.Y. (2013). Development of piezoelectric coaxial filament sensors P(VDF-TrFE)/copper for textile structure instrumentation. *Sensors and Actuators A: Physical* 204: 122–130.

70 Baniasadi, M., Huang, J., Xu, Z. et al. (2015). High-performance coils and yarns of polymeric piezoelectric nanofibers. *ACS Applied Material & Interfaces* 7 (9): 5358–5366.

71 Lu, X., Qu, H., and Skorobogatiy, M. (2017). Piezoelectric micro and nanostructured fibers fabricated from thermoplastic nanocomposites using a fiber drawing technique: comparative study and potential applications. *ACS Nano* 11 (2): 2103–2114.

72 Lee, Y.B., Han, J.K., Noothongkaew, S. et al. (2016). Toward arbitrary-direction energy harvesting through flexible piezoelectric nanogenerators using perovskite $PbTiO_3$ nanotube arrays. *Advanced Materials* 29 (6): 1604500.

73 Mokhtari, F., Foroughi, J., Zheng, T. et al. (2019). Triaxial braided piezo fibers energy harvesters for self-powered wearable technologies. *Journal of Materials Chemistry A* 7: 8245.

74 Wen, Z., Yang, Y., Sun, N. et al. (2018). A wrinkled PEDOT:PSS film based stretchable and transparent triboelectric nanogenerator for wearable energy harvesters and active motion sensors. *Advanced Functional Materials* 28: 1803684.

75 Wang, P., Liu, R., Ding, W. et al. (2018). Complementary electromagnetic- triboelectric active sensor for detecting multiple mechanical triggering. *Advanced Functional Materials* 28 (11): 1705808.

76 Chen, J. and Wang, Z. (2017). Reviving vibration energy harvesting and self-powered sensing by a triboelectric nanogenerator. *Joule* 1 (3): 480–521.

77 Zhang, L., Zhang, B., Chen, J. et al. (2015). Lawn structured triboelectric nanogenerators for scavenging sweeping wind energy on rooftops. *Advanced Materials* 28 (8): 1650–1656.

78 Xu, M., Zhao, T., Wang, C. et al. (2019). High power density tower-like triboelectric nanogenerator for harvesting arbitrary directional water wave energy. *ACS Nano* 13: 1932.

79 Luo, J. and Wang, Z. (2019). Recent advances in triboelectric nanogenerator based self-charging power systems. *Energy Storage Materials* 23: 617–628.

80 Zhong, J., Zhang, Y., Zhong, Q. et al. (2014). Fiber-based generator for wearable electronics and mobile medication. *ACS Nano* 8 (6): 6273–6280.

81 Liu, M., Cong, Z., Pu, X. et al. (2019). High-energy asymmetric supercapacitor yarns for self-charging power textiles. *Advanced Function Material* 41 (29): 1806298.

82 Lai, Y., Deng, J., Zhang, S. et al. (2016). Single-thread-based wearable and highly stretchable triboelectric nanogenerators and their applications in cloth-based self-powered human-interactive and biomedical sensing. *Advanced Functional Materials* 27 (1): 1604462.

83 Yu, A., Pu, X., Wen, R. et al. (2017). Core-shell-yarn-based triboelectric nanogenerator textiles as power cloths. *ACS Nano* 11 (12): 12764–12771.

84 Yang, Y., Xie, L., Wen, Z. et al. (2018). Coaxial triboelectric nanogenerator and supercapacitor fiber based self-charging power fabric. *ACS Applied Materials Interfaces* 10: 42356.

85 Park, J., Choi, A.Y., Lee, C.J. et al. (2017). Highly stretchable fiber-based single-electrode triboelectric nanogenerator for wearable devices. *RSC Advances* 7 (86): 54829–54834.

86 Gong, W., Hou, C., Zhou, J. et al. (2019). Continuous and scalable manufacture of amphibious energy yarns and textiles. *Nature Communications* 10 (1): 868.

87 Zhang, M., Zhao, M., Jian, M. et al. (2019). Printable smart pattern for multifunctional energy-management e-textile. *Matter* 1 (1): 20–21.

88 He, X., Zi, Y., Guo, H. et al. (2016). A highly stretchable fiber-based triboelectric nanogenerator for self-powered wearable electronics. *Advanced Functional Materials* 27 (4): 1604378.

89 Cheng, Y., Lu, X., Hoe Chan, K. et al. (2017). A stretchable fiber nanogenerator for versatile mechanical energy harvesting and self-powered full-range personal healthcare monitoring. *Nano Energy* 41: 511–518.

90 Sim, H.J., Choi, C., Kim, S.H. et al. (2016). Stretchable triboelectric fiber for self-powered kinematic sensing textile. *Scientific Reports* 6: 35153.

91 Yu, X., Pan, J., Zhang, J. et al. (2017). A coaxial triboelectric nanogenerator fiber for energy harvesting and sensing under deformation. *Journal of Materials Chemistry A* 5: 6032.

92 Tian, Z., He, J., Chen, X. et al. (2018). Core–shell coaxially structured triboelectric nanogenerator for energy harvesting and motion sensing. *RSC Advances* 8: 2950.

93 Wang, J., Li, S., Yi, F. et al. (2016). Sustainably powering wearable electronics solely by biomechanical energy. *Nature Communication* 7: 12744.

94 Xie, L., Chen, X., Wen, Z. et al. (2019). Spiral steel wire-based fiber-shaped stretchable and tailorable triboelectric nanogenerator for wearable power source and active gesture sensor. *Nano-Micro Letters* 11 (1): 39.

95 Yang, Y., Sun, N., Wen, Z. et al. (2018). Liquid-metal-based super-stretchable and structure-designable triboelectric nanogenerator for wearable electronics. *ACS Nano* 12 (2): 2027–2034.

96 Kim, K.N., Chun, J., Kim, J.W. et al. (2015). Highly stretchable 2D fabrics for wearable triboelectric nanogenerator under harsh environments. *ACS Nano* 9 (6): 6394–6400.

97 Tan, G., Zhao, L., and Kanatzidis, M.G. (2016). Rationally designing high-performan cebulk thermoelectric materials. *Chemical Review* 116: 12123–12149.

98 Bed, P., Hao, Q., Ma, Y. et al. (2008). High-thermoelectric performance of nanostructured bismuth antimony telluride bulk alloys. *Science* 320: 634–638.

99 Huang, B. and Massoud, K. (2008). Abinitio and molecular dynamics predictions for electron and phonon transport in bismuth telluride. *Physical Review B* 77: 125209.

100 Biswas, K., He, J., Blum, I.D. et al. (2012). High-performance bulk thermoelectrics with all-scale hierarchical architectures. *Nature* 489 (7416): 414–418.

101 Zhao, L., Lo, S., Zhang, Y. et al. (2014). Ultralow thermal conductivity and high thermoelectric figure of merit in SnSe crystals. *Nature* 508 (7496): 373.

102 Du, X., Ivan, S., Anthony, B., and Eva, Y.A. (2008). Approaching ballistic transport in suspended graphene. *Nature Nano Technology* 3: 491–495.

103 Seol, J.H., Jo, I., Moore, A.L. et al. (2010). Two-dimensional phonon transport in supported graphene. *Science* 328: 213–216.

104 Ma, W., Liu, Y., Yan, S. et al. (2018). Chemically doped macroscopic graphene fibers with significantly enhanced thermoelectric properties. *Nano Research* 11: 741–750.

105 Takahiro, F., Tsuyohiko, F., and Naotoshi, N. (2015). Development of n-type cobaltocene-encapsulated carbon nanotubes with remarkable thermoelectric property. *Scientific Report* 5: 7951.

106 Daisuke, H., Tomohiro, U., Yusuke, N. et al. (2016). Thermoelectric properties of single-wall carbon nanotube films: effects of diameter and wet environment. *Applied Physical Express* 9: 025102.

107 Yu, C., Murali, A., Choi, K., and Ryu, Y. (2012). Air-stable fabric thermoelectric modules made of N- and P-type carbon nanotubes. *Energy & Environmental Science* 5: 9481.

108 Brownlie, L. and Shapter, J. (2018). Advances in carbon nanotube n-type doping: methods, analysis and applications. *Carbon* 126: 257–270.

109 Chen, G., Xu, W., and Zhu, D. (2017). Recent advances in organic polymer thermo electric composites. *Journal of Materials Chemistry C* 5: 4350.

110 Xiong, J., Jiang, F., Zhou, W. et al. (2015). Highly electrical and thermoelectric properties of a PEDOT:PSS thin-film via direct dilution-filtration. *RSC Advances* 5: 60708–60712.

111 Zhu, Z., Liu, C., Jiang, Q. et al. (2015). Optimizing the thermo electric properties of PEDOT:PSS films by combining organic co-solvents with inorganic base. *Journal of Electronic Material* 26: 8515–8521.

112 Kim, G.-H., Shao, L., Zhang, K., and Pipe, K. (2013). Engineered doping of organic semiconductors for enhanced thermoelectric efficiency. *Nature Materials* 12: 719–723.

113 See, K.C., Feser, J.P., Chen, C.E. et al. (2010). Water-processable polymer-nanocrystal hybrids for thermoelectrics. *Nano Letters* 10: 4664–4667.

114 Eun, J.B., Young, H.K., Kwang-Suk, J., and Song, Y.C. (2016). Enhancement of thermoelectricproperties of PEDOT:PSS and tellurium-PEDOT:PSS hybrid composites bysimple chemical treatment. *Scientific Reports* 6: 18805–18814.

115 Yadav, A., Pipe, K.P., and Shtein, M. (2008). Fiber-based flexible thermoelectric power generator. *Journal of Power Sources* 175: 909–913.

116 Liang, D., Yang, H., Finefrock, S.W., and Wu, Y. (2012). Flexible nanocrystal-coated glass fibers for high-performance thermoelectric energy harvesting. *Nano Letters* 12: 2140–2145.

117 Sun, M., Tang, G., Qian, G. et al. (2018). In_4Se_3 alloy core thermoelectric fibers. *Materials Letters* 217: 13–15.

118 Zhang, T., Li, K., Zhang, J. et al. (2017). High-performance, flexible, and ultralong crystalline thermoelectric fibers. *Nano Energy* 41: 35–42.

119 Linden, A., Steven, H., and Trisha, L.A. (2017). Towards seamlessly-integrated textile electronics: methods to coat fabrics and fibers with conducting polymers for electronic applications. *Chemical Communications* 53: 7182–7193.

120 Innis, P.C., Razal, J.M., Wallace, G.G., and Jalili, R. (2011). One-step wet-spinning process of poly (3,4-ethylenedioxythiophene):poly(styrenesulfonate) fibers and the origin of higher electrical conductivity. *Advanced Functional Materials* 21: 3363–3370.

121 Liu, J., Jia, Y., Jiang, Q. et al. (2018). Highly conductive hydrogel polymer fibers toward promising wearable thermoelectric energy harvesting. *ACS Applied Materials Interfaces* 10: 44033–44040.

122 Mitsuhiro, I., Takuya, K., Hirotaka, K. et al. (2017). From materials to device design of a thermoelectric fabric for wearable energy harvesters. *Journal of Materials Chemistry A* 5: 12068–12072.

5

Energy Storage Fibers

Energy storage is still a serious global challenge. The modern society and the increasing emerging ecological problem badly require low-cost and environmentally friendly energy storage devices with high performance [1]. In recent years, the prosperity of portable electronics and electric vehicle market stimulates the development of flexible energy storage devices toward higher power density and energy density. In order to meet the requirements of flexible wearing, energy storage devices with different types of structures, such as two-dimensional (2D) plane, one-dimensional (1D) fiber, and interwoven textile, have been fabricated, which show promise for such applications compared with their three-dimensional (3D) counterparts.

For energy storage devices with different structures, 1D energy storage fibers have got much attention due to their higher structure freedom for wearing. Compared with 2D energy storage devices, fibrous electronic devices can satisfy the above application needs since they can be easily woven into any desired shape or be knitted into fabrics. Therefore, the development of fiber-type energy storage devices (energy storage fibers) is highly demanded and may open up a new avenue in the area of energy storage devices.

Supercapacitors (SCs), batteries, and phase-transit devices are three main types of energy storage devices. SC fibers and battery fibers are two typical energy storage fibers, which is composed of two electrodes, separator, and electrolyte. SCs have a higher power density than traditional capacitors. Compared with SCs, batteries have a lower power density but much higher energy density. Besides SC fibers and battery fibers, phase-transit fibers (PTFs) are energy storage devices that are able to store and release a large amount of heat within a relatively narrow range of temperature variation.

This chapter mainly discussed SC fibers, battery fibers, and PTFs.

5.1 Supercapacitor Fibers

SCs have two electrodes, which are connected via electrolyte and segregated by a separator (ion permeable, e.g. ceramic, cardboard, paper or plastic, and glass)

Textile-Based Energy Harvesting and Storage Devices for Wearable Electronics, First Edition.
Xing Fan, Nannan Zhang, and Yi Wang.

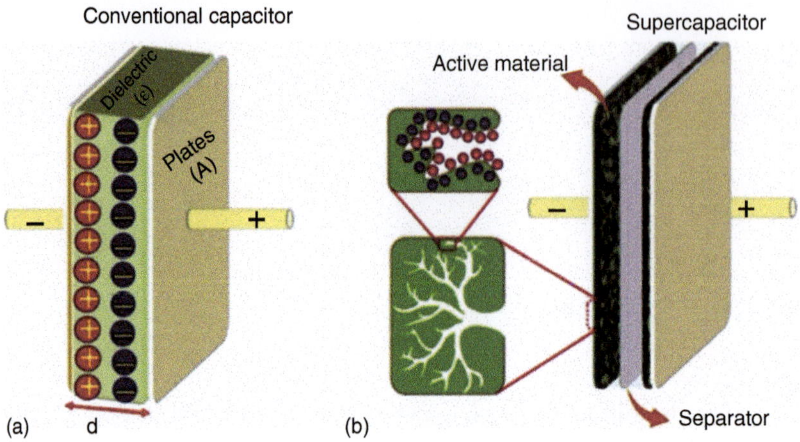

Figure 5.1 Scheme structure of different energy storage devices. (a) Traditional capacitors. (b) SCs. Source: Adapted with permission from Khan et al. [2]. © 2019, Elsevier.

(Figure 5.1). In SCs, the charge accumulated via segregation of charges is in Helmholtz double layers at the electrode-electrolyte interface (i.e. the electrical or electrochemical double layer capacitance) or by Faradaic charge transfer via redox reactions (i.e. pseudocapacitance). In recent SCs, electrolyte is intrinsically capable to go through Faradaic electron transfer reactions or redox-additive electrolyte, where redox active species are incorporated to allow Faradaic electron transfer reactions at electrolyte-electrode interfaces. Another part is separator, which is porous membrane, sandwiched in the electrodes. The separator stops immediate self-discharge and electrical short circuits while allows ionic transport, which is necessary for closing the circuit.

Commonly, SCs were fabricated in a basic planar or 2D configuration. The flexibility of planar SCs highly depends on the electrode substrate. Different kinds of planar-type electrode substrates have been used to develop flexible SCs, such as metal coated plastics, conductive papers, carbon material coated cloth, and conductive carbon fiber (CF) cloths or fabrics [3]. Unfortunately, these electrode substrates do not provide sufficient flexibility to satisfy the demands of future flexible wearable electronic devices. To tackle these problems, remarkable efforts in the design of SCs have shifted from planar (2D) to fiber (1D) format with lightweight and more flexibility.

SC fibers are usually similar to 1D wires, with diameters ranging from a few microns to millimeters. They have reasonable flexibility, that is, they can be bend repeatedly without significantly changing their energy storage performance. Their fibrous structure makes them suitable for compact or curved spaces, and they can be woven or knitted into wearable textiles. SC fibers also comprise three parts: two electrodes sandwiching the electrolyte and separator. It is undoubtedly that capacitive materials play an essential role to the device performance since the reign charge storage and transportation. Capacitive materials include carbon materials, transition metal oxides, and conducting polymers. However, the assembly of capacitive materials for SC fibers is more difficult than that for traditional SCs, which will be broken easily under deformation.

According to different type of capacitive materials, supercapacitor fibers can be divided into two types, supercapacitor fibers with carbon-based capacitive materials and supercapacitor fibers with composited capacitive materials, which will be discussed as follows.

5.1.1 Supercapacitor Fibers with Carbon-Based Capacitive Materials

SCs with carbon-based capacitive materials (or carbon-based SCs) means the two electrodes were coated by carbon materials (i.e. carbon nanotube [CNT], graphene). Generally speaking, carbon-based SC stores charges or ions statically by forming an electrical double layer (EDL) at the electrode-electrolyte interface, so this carbon-based SC is called an electrical double-layer capacitor (EDLC). As capacitive materials, carbon materials have the advantages of large specific surface area, good electrical conductivity, high rate capacity, and excellent cycle stability. In fiber SCs, carbon-based fibers such as single-walled or multi-walled carbon nanotube fibers, graphene fibers, CFs, and carbon-carbon composite fibers are most attractive, owing to their 1D, flexible, and conductive features.

CNTs are the most commonly used materials in SC fibers. They are used as promising electrode materials for SCs due to their unique 1D structure with good electrical conductivity that facilitates rapid charge transport and high mechanical strength and lightweight. SC fibers with carbon-based capacitive materials can be divided into common SC fibers, and functional SC fibers depends on whether they have additional function.

For common SC fibers, Yang et al. [4] developed a novel family of highly stretchable, fiber-shaped, high-performance SC for the first time in 2013 (Figure 5.2). CNT sheets that are sequentially wrapped on an elastic fiber serve as two electrodes. The use of aligned CNT sheets offers combined remarkable properties including high flexibility, tensile strength, electrical conductivity, and mechanical and thermal stability. As a result, the fiber-shaped SC maintains a high specific capacitance of approximately 18 F/g after stretch by 75% for 100 cycles.

For functional SC fibers, the SC fibers were endowed with extra function like resistance to torsional deformations or shape memory function. They have attracted abundant attention in recent years. For example, Meng et al. [5] reported a flexible torsionable fiber-shaped SC, which is very stable even under torsional deformation (Figure 5.3). The torsional SCs were prepared with CNT films. When the membrane is integrated with the elastic fiber as an electrode, it can withstand torsional levels of up to 20 000 rad/m and has relatively stable electrical conductivity. As a result, a fibrous SC made from such a twistable electrode can retain almost the same capacitance during the twisting process.

In addition, Deng et al. [6] developed a shape memory SC fiber by winding CNT sheets on a shape memory polyurethane substrate (Figure 5.4). Despite its flexibility and stretchability, the deformed shape under bending and stretching can "freeze" as

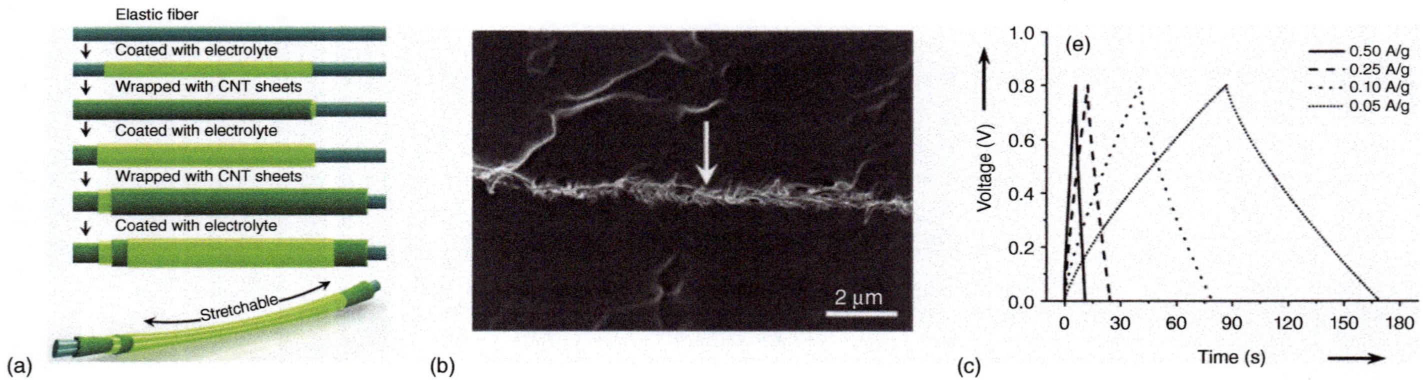

Figure 5.2 Schematic diagram, SEM image, and performance of CNT-based SC fibers. (a) Illustration to the fabrication of highly stretchable SC fibers. (b) SEM image of a fiber-shaped supercapacitor. (c) Galvanostatic charge-discharge curves of a fiber-shaped supercapacitor at different current densities. Source: (a, c) Yang et al. [4]. © 2013, John Wiley & Sons, (b) Adapted with permission from Yang et al. [4]. Copyright (2013) by WILEY-VCH Verlag GmbH & Co. KGaA, Weinheim.

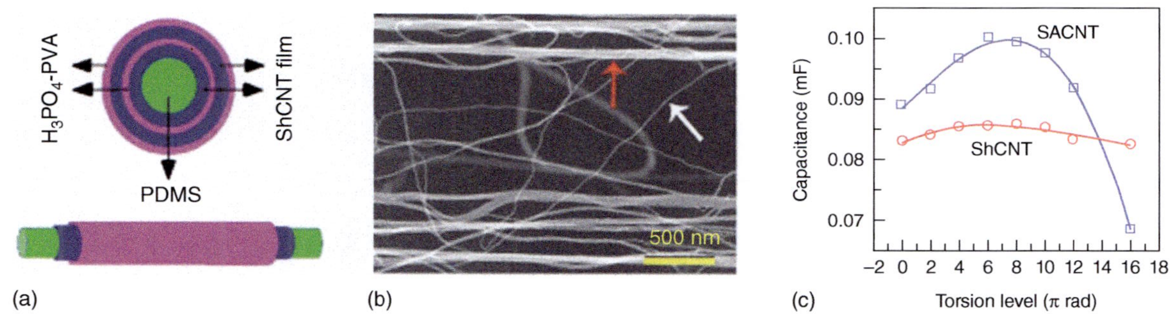

Figure 5.3 Schematic diagram, SEM image, and performance of functional CNT-based SC fibers. (a) The structure of a torsionable supercapacitor. (b) SEM image of a single-layer ShCNT film. (c) The capacitances of supercapacitors based on the ShCNT film and SACNT film. Source: (a, c) Meng et al. [5]. © 2017, Royal Society of Chemistry, (b) Adapted with permission from Meng et al. [5]. Copyright (2017) by Royal Society of Chemistry.

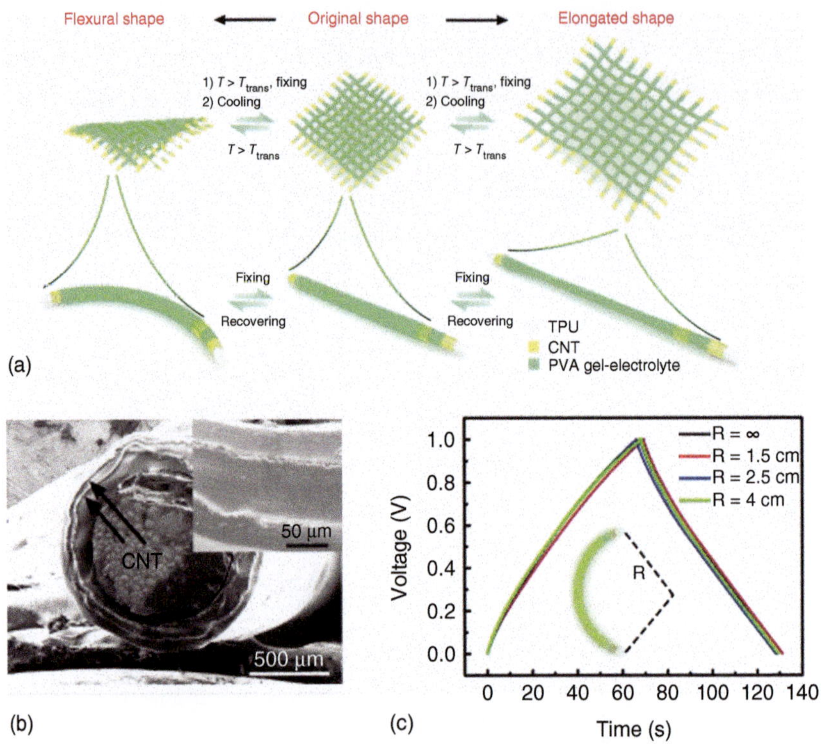

Figure 5.4 Schematic diagram, SEM image, and performance of the SFSC. (a) Schematic diagram of the SFSC and the resulting textile that are reversibly transformed into flexural or elongated states and recovered to the original shape. (b) Cross-sectional image of the SFSC (inset, high magnification). (c) Galvanostatic charge-discharge curves of the SFSC at different flexural shapes. Source: (a, c) Deng et al. [6]. © 2015, John Wiley & Sons, (b) Adapted with permission from Deng et al. [6]. Copyright (2015) by WILEY-VCH Verlag GmbH & Co. KGaA, Weinheim.

expected and return to its original state as needed. The electrochemical properties of the SC fiber are well maintained in the deformation process.

Graphene is another choice of lightweight capacitive material for SC applications, which has favorable properties such as good electrical conductivity even at room temperature, good mechanical flexibility, high surface area (2675 m^2/g), and low preparation cost. It can offer a maximum theoretical specific capacitance of 550 F/g. It can be prepared via different methods such as laser irradiation, mechanical exfoliation, liquid-phase exfoliation, hydrothermal reactions, chemical or electrochemical reduction, chemical vapor deposition (CVD), and so on, where graphene oxide (GO) is the major source to obtain graphene. Hu et al. [7] designed and fabricated a new RGO-GO-RGO SC fiber (Figure 5.5). A large-scale spinning method was employed to make GO fibers from a GO suspension. This all-in-one SC fiber inherits the mechanical flexibility of graphene fiber while maintaining the high performance in capacitance. The capacitance of the RGO-GO-RGO fiber can reach a value as high as 1.2 mF/cm^2 when a current density of 80 $\mu A/cm^2$ is applied and maintained a considerable value of 0.45 mF/cm^2 at a current density of 200 $\mu A/cm^2$.

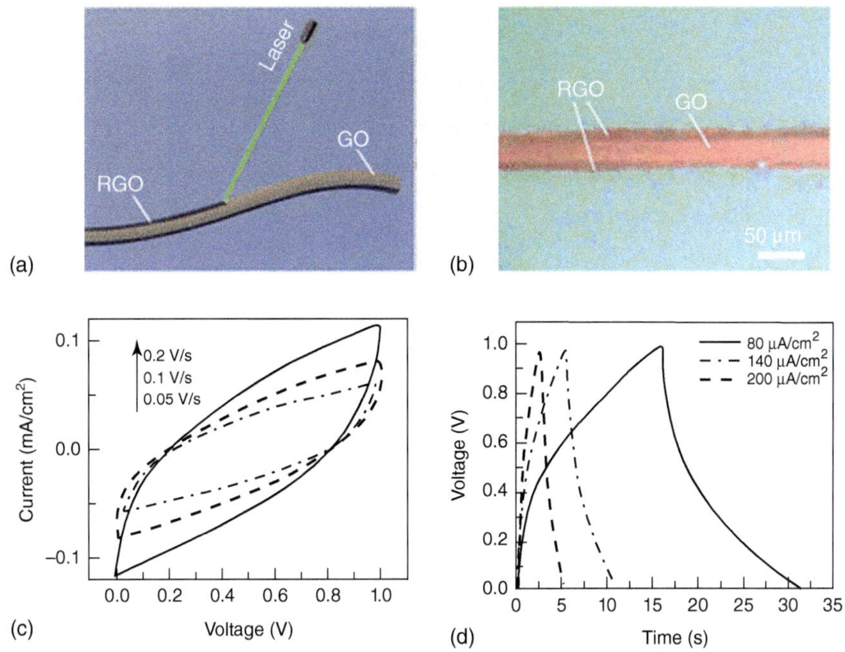

Figure 5.5 Schematic diagram, SEM image, and performance of RGO-GO-RGO fiber: (a) Schematic diagram of the laser reduction of GO fiber for formation of RGO-GO-RGO fiber. (b) The digital image of the RGO-GO-RGO fiber under optical microscopy. (c) CV curves of RGO-GO-RGO fiber. (d) Charge–discharge curves of RGO-GO-RGO fiber. Source: (a, c, d) Hu et al. [7]. © 2014, Royal Society of Chemistry, (b) Adapted with permission from Hu et al. [7]. Copyright (2014) by Royal Society of Chemistry.

Hybrid carbon-based SC means two or more carbon materials are used in the SC. Hybrid carbon materials are gaining increasing interests because they combine the advantages of the two individual materials, beneficial for high performance of SCs. CNTs were the most commonly used materials for hybrid carbon-based SC. The most widely studied hybrid carbon-based SCs are CNT/graphene SC fibers and CNT/carbon microfiber SC fibers. For CNT/graphene SC fibers, Sun et al. [8] reported the synthesis of a new family of graphene/carbon composite fibers. The CNTs serve as an effective bridge to enhance charge transfer, thanks to powerful π-π interactions between graphene sheets and composite fibers that simultaneously achieve high electrical conductivity and electro-catalytic activities (Figure 5.6). Graphene/carbon composite fibers are used as electrodes to produce supercapacitors with specific capacitance of up to 31.50 F/g (4.97 mF/cm^2 or 27.1 F/cm).

For CNT/carbon microfiber SC fibers, Le et al. [9] reported a coaxial SC fiber, which consists of carbon microfiber bundles coated with multi-walled CNTs as core electrodes and carbon nanofiber paper as external electrodes (Figure 5.7). The electrode volume ratio was determined by the half-cell test of each electrode. The high capacitance, energy, and power density obtained were attributed to the high surface area of the coaxial structure and the high conductivity of the carbon

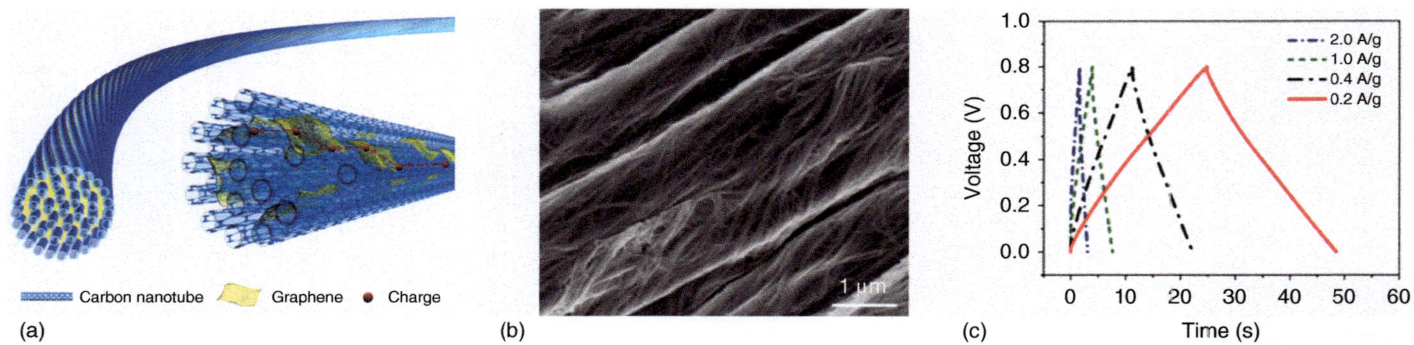

Figure 5.6 Schematic diagram, SEM image, and performance of graphene/CNT composite fiber. (a) Schematic illustration of the structure of the graphene/CNT composite fiber. (b) SEM image of graphene/CNT composite fiber. (c) Galvanostatic charge–discharge curves of graphene/CNT composite fiber. Source: (a, c) Sun et al. [8]. © 2014, John Wiley & Sons, (b) Adapted with permission from Sun et al. [8]. Copyright (2014) by WILEY-VCH Verlag GmbH & Co. KGaA, Weinheim.

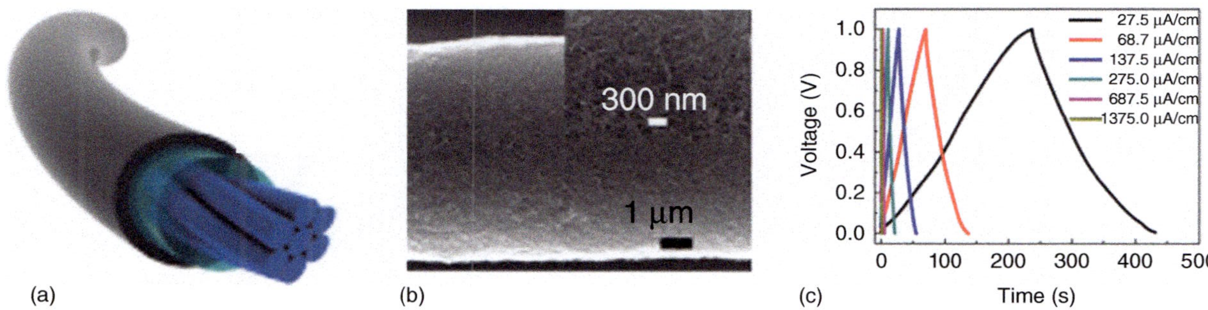

Figure 5.7 Schematic diagram, SEM image, and performance of CMF coated with MWCNTs. (a) Schematic photo of a coaxial fiber supercapacitor. (b) SEM image of CMF coated with MWCNTs. (c) Charge–discharge curves of CMF at different current densities. Source: (a, c) Le et al. [9]. © 2013, American Chemical Society, (b) Adapted with permission from Le et al. [9]. Copyright (2013) by American Chemical Society.

material, especially multi-walled carbon nanotubes (MWCNTs)/carbon microfiber (CMF). When the SC fiber was bent 180°, the cyclic voltammetry characteristic changes can be ignored.

5.1.2 Supercapacitor Fibers with Composited Capacitive Materials

Besides carbon materials, pseudocapacitive materials (conducting polymers and metal oxides) can also be used as capacitive materials in SC. In pseudocapacitive SCs, charges are stored through fast faradic redox reactions (charge transfer reactions) on the surface of the electrode materials similar to the case in batteries. The main advantage of pseudocapacitive materials is that they can provide higher specific capacitances than carbon materials due to the involvement of multi-electron redox reactions within the operating potential window. Carbon materials alone tend to have limited EDLC, while pseudocapacitor materials (conducting polymers and metal oxides) alone have higher specific capacitance but poor electrical conductivity. It is an effective way to obtain high-performance electrode materials by combining carbon materials with pseudocapacitor materials as composited capacitive materials.

For SC with composited capacitive materials, the two electrodes were usually coated by two types of composited capacitive materials, including carbon materials/conducting polymers and carbon materials/metal oxides. Among all the carbon materials, CNT were the most commonly used materials. In the following parts, SC fibers with CNT/conducting polymers and SC fibers with CNT/metal oxides will be discussed.

CNT/conducting polymers such as CNT/polyaniline (PANI), CNT/polypyrrole (Ppy), and CNT/poly(styrenesulfonate) (PEDOT:PSS) composites are a good choice for achieving high-performance SCs. For traditional SCs, adding pseudocapacitive materials such as metal oxides and conductive polymers directly into the carbon electrode is considered to be an effective method to improve the capacitance and energy density. The same approach has also been explored for SC fibers.

For SC fibers with CNT/PANI composite, Wang et al. [10] prepared a two-ply yarn SC constructed from CNT yarns and PANI nanowires (Figure 5.8). In the SC, a pure CNT yarn was used both as active material, as well as current collector for the threadlike SC. Conducting PANI was used as electrode material because of its high energy storage capability and flexibility. The threadlike SCs have excellent electrochemical capacity and flexibility. When woven or knitted into wearable electronic device, the two-ply yarn SC can be stretched repeatedly without significant loss of capacitance.

For SC fibers with CNT/PPy composite, Shang et al. [11] demonstrated a self-stretchable fiber-shaped SC based on helical CNT yarns or CNT/PPy composite yarns without using external substrates (Figure 5.9). The helical CNT yarn SCs can work stably under many extreme deformation conditions such as super-elongation (tensile strain up to 150%), arbitrary shape change (entangling), and high-frequency stretching (up to 10 Hz over 10 000 cycles). By designing and manufacturing particular yarn structures, the fiber-shaped SCs with unprecedented properties

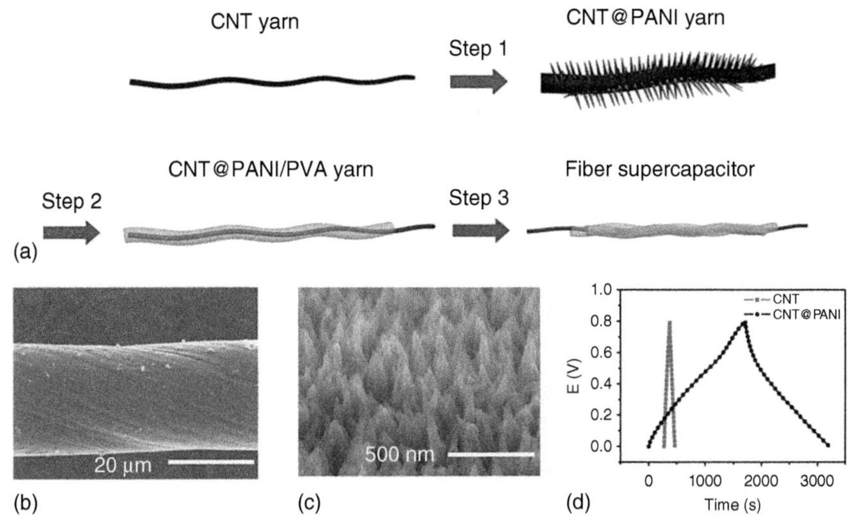

Figure 5.8 Schematic diagram, SEM images, and performance of CNT@PANI composite yarn. (a) Schematic diagram of the preparation procedures for the two-ply yarn supercapacitor. (b, c) SEM images of CNT@PANI composite yarn. (d) Galvanostatic charge-discharge curves of the two-ply yarn supercapacitors. Source: (a, d) Wang et al. [10]. © 2013, John Wiley & Sons, (b, c) Adapted with permission from Wang et al. [10]. Copyright (2013) by WILEY-VCH Verlag GmbH & Co. KGaA, Weinheim.

and high performance could be developed, which have potential applications in portable, wearable, and body-integrated energy storage devices.

For SC fibers with CNT/PEDOT:PSS composite, a high-performance CNT/PEDOT:PSS yarn SC has been constructed from CNT yarn or its composites using a simple process [12] (Figure 5.10). The CNT yarn was first treated with gamma irradiation in the presence of oxygen and then coated with conductive polymer PEDOT:PSS to maximize the energy storage capacity of the as-prepared SC. The coating of PEDOT:PSS brings further improvements to the energy storage capacity of the SC. The CNT/PEDOT:PSS yarn-based SC is strong and flexible and has good cyclic charge–discharge stability.

CNT/metal oxides or CNT/metal hydroxide composites are also good choice for achieving high-performance SCs. For CNT/metal oxides, in order to overcome the structural drawbacks of those reported electrodes without impairing their high electrochemical performance, Choi et al. [13] developed MWCNTs yarn electrode with high internal porosity (Figure 5.11). Due to the 3D porosity inside the yarn, manganese dioxide (MnO_2) can be effectively trapped in the pores during the deposition process, forming a good mixing zone composed of nanoscale MnO_2 and well-arranged CNT bundles. The specific capacitance (25.4 F/cm^3) and the energy and average power density (3.52 mWh/cm^3 and 127 mW/cm^3) were obtained in the micro-diameter all-solid-state SC at the scan rate of 10 mV/s.

For CNT/metal hydroxides, Lu et al. [14] proposed a method of preparing asymmetric fibrous solid supercapacitor (AFSSC) by electrodeposition of CNT-modified CF beam, in which CF bundle@CNT-NiCo(OH)$_x$ (CF-@CNC) was the positive

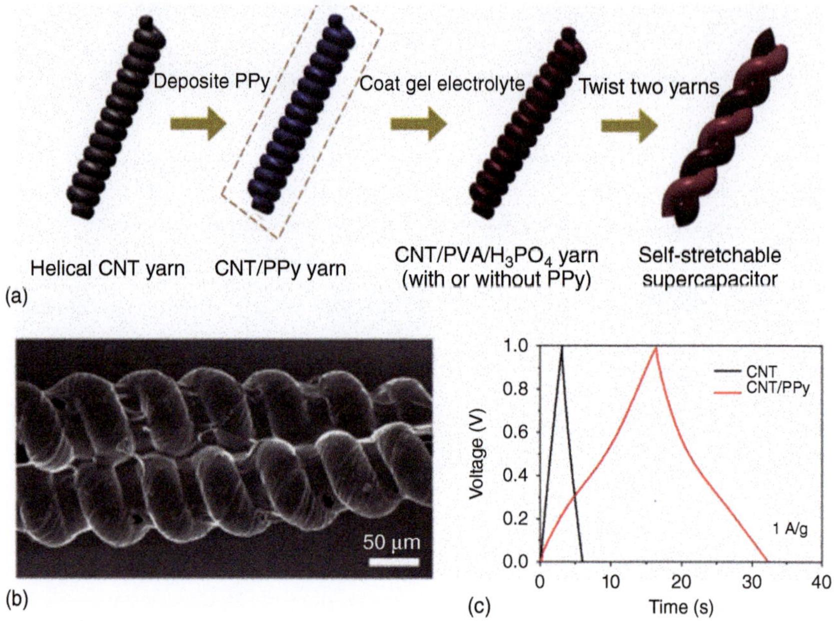

Figure 5.9 Schematic diagram, SEM image, and performance of CNT/PPy supercapacitor. (a) Illustration of the process to prepare the double-helix supercapacitor. (b) SEM image of the double-helix supercapacitor. (c) Galvanostatic charge-discharge curves for a CNT and CNT/PPy supercapacitor at a current density of 1 A/g. Source: (a, c) Shang et al. [11]. © 2015, Elsevier, (b) Adapted with permission from Shang et al. [11]. Copyright (2014) by Elsevier.

electrode and activated carbon (AC)-coated CFs (CF-@AC) was the negative electrode (Figure 5.12). In order to overcome the inherent disadvantages of CFs, such as hydrophobicity, poor electrical conductivity, and low surface area, CNT modification of CF@CNC and air plasma treatment were adopted. These two strategies work together to establish a close interaction between CF and electrodeposited porous $NiCo(OH)_x$. The method not only improve ion/electron transport at the interface but also prevent $NiCo(OH)_x$ deposition from shedding during the reaction.

In summary, compared with traditional SCs, the research of fiber SCs is still in its infancy. A common research goal is to develop flexible fiber SCs that maintain or even exceed their electrochemical properties compared with conventional SCs. In the field of microdevices, the continuous miniaturization of electronic products requires small energy storage units. Limited life cycles are particularly problematic for some applications, such as implantable sensors, and low power density impedes applications that require high power. In the miniaturized electronic products, SC fibers can replace the micro-battery as a small energy storage unit. In the field of macroscale equipment, flexible energy storage units need a wide range of new applications, such as artificial skin, scalable sensors, and intelligent textiles. Therefore, SC fiber has unique 1D structure and mechanical flexibility, which can be used in the form of yarn or fabric and can be integrated into intelligent textiles, which has great application potential in this field.

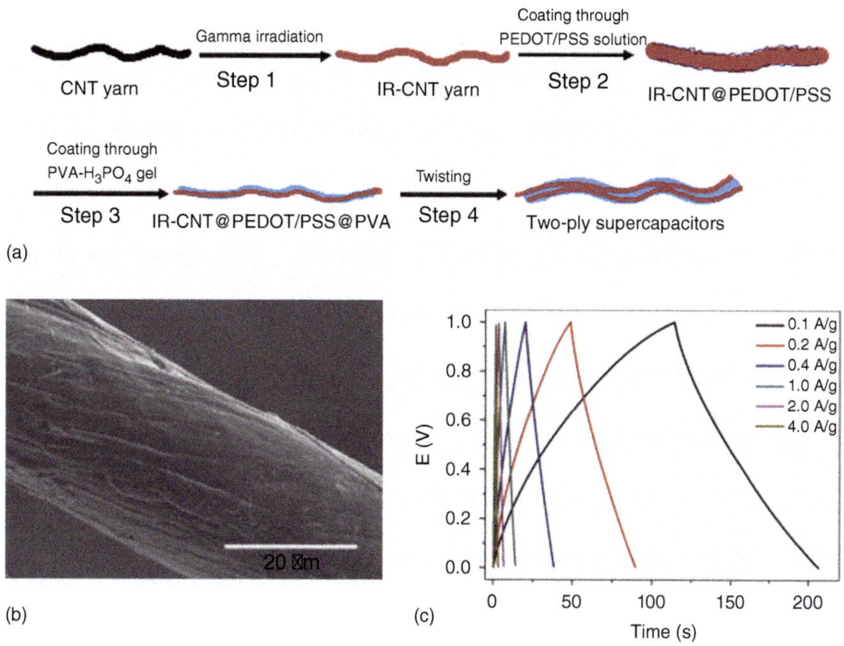

Figure 5.10 Schematic diagram, SEM image, and performance of CNT@PEDOT:PSS: (a) Schematic diagram of the preparation procedures for the two-ply yarn supercapacitors. (b) SEM image of CNT@PEDOT:PSS. (c) Galvanostatic charge-discharge curves measured at different current densities. Source: (a, c) Su et al. [12]. © 2014, Elsevier, (b) Adapted with permission from Su et al. [12]. Copyright (2014) by Elsevier.

5.2 Battery Fibers

Battery refers to a device containing electrolyte solution and metal electrodes that convert chemical energy into electricity. Battery can be divided into primary battery and secondary battery. Primary battery is a battery that cannot be recharged and restored after being discharged, such as alkaline zinc-manganese battery, while secondary battery can be recharged. Typical secondary batteries include lithium-ion batteries (LIBs), lithium-sulfur batteries, metal-air batteries, etc.

As the commercially available batteries are typically bulk and rigid, which cannot satisfy the standards of flexible/wearable electronics that require portability and flexibility, developing batteries with flexibility and stretchability are urgently needed. Recently, numerous flexible batteries have been demonstrated successfully. In recent years, flexible battery fibers have been successfully prepared for achieving a high performance. Compared with SC fibers, battery fibers have a higher energy density. Compared with the traditional bulky or planar structure, 1D fibrous structure can make the battery deformation in all directions. Primary battery fibers such as alkaline zinc-manganese battery fibers, LIB fibers, lithium-sulfur battery fibers, and metal-air battery fibers were most studied, and they will be discussed in the following parts.

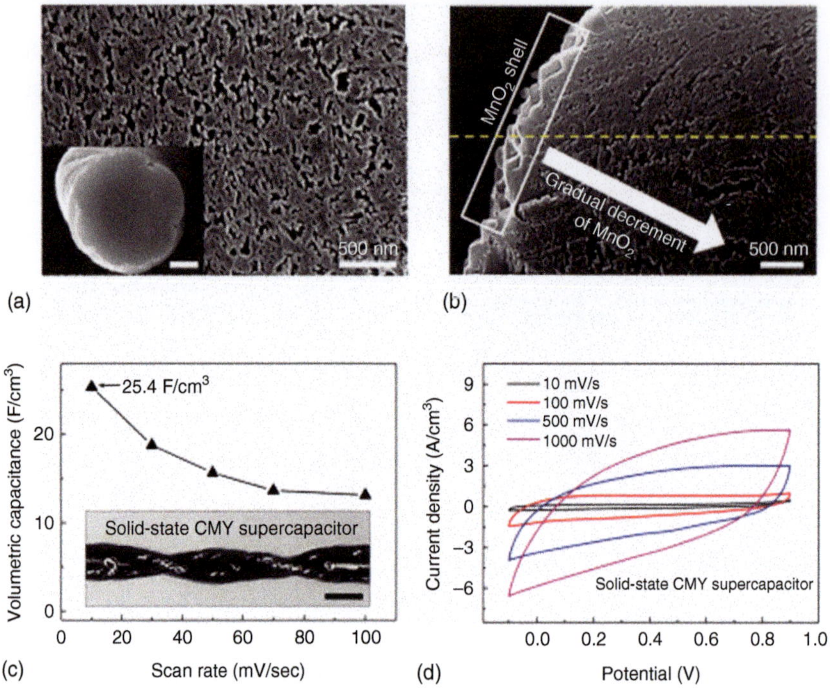

Figure 5.11 SEM images and performance of CNT/MnO_2 fiber electrode. (a) Cross-sectional SEM image of the CNT/MnO_2 fiber electrode (inset) and its magnified image, showing high internal porosity. (b) Magnified edge part of the CNT/MnO_2 fiber electrode structure and direction of gradual concentration decrement of MnO_2. (c) Volumetric capacitance as a function of scan rate and optical image of the CMY supercapacitor. (d) CV curves with various scan rates. Source: (a, b) Adapted with permission from Choi et al. [13]. Copyright (2013) by WILEY-VCH Verlag GmbH & Co. KGaA, Weinheim. (c, d) Choi et al. [13]. © 2014, John Wiley & Sons.

5.2.1 Primary Battery Fibers

As discussed earlier, primary battery is a battery that cannot be recharged and restored after being discharged. Alkaline zinc-manganese battery is a kind of typical primary battery (Figure 5.13). The negative electrode of alkaline zinc-manganese battery is mainly composed of zinc (Zn), and the positive electrode is composed of MnO_2 and graphite powder, while potassium hydroxide solution is used as electrolyte.

In 2013, Yu et al. [15] first reported a high-performance fiber zinc–manganese battery (Figure 5.14). In this battery, lightweight, low-cost commercial CFs are used as collectors, coated with anode or cathode materials. The anode and cathode are sealed in a transparent flexible polymer plastic tube filled with electrolyte. The resulting open-circuit voltage of the battery is 1.5 V, which is close to the theoretical voltage. When discharge density is 70 mA/g or 1.3 mA/cm^2, discharge capacity is 158 mAh/g or 1.50 mA/cm^2 (calculated by mass of positive MnO_2). The fiber battery also showed good flexibility; when the bending radius was between 3.0 and 0.7 cm, there was

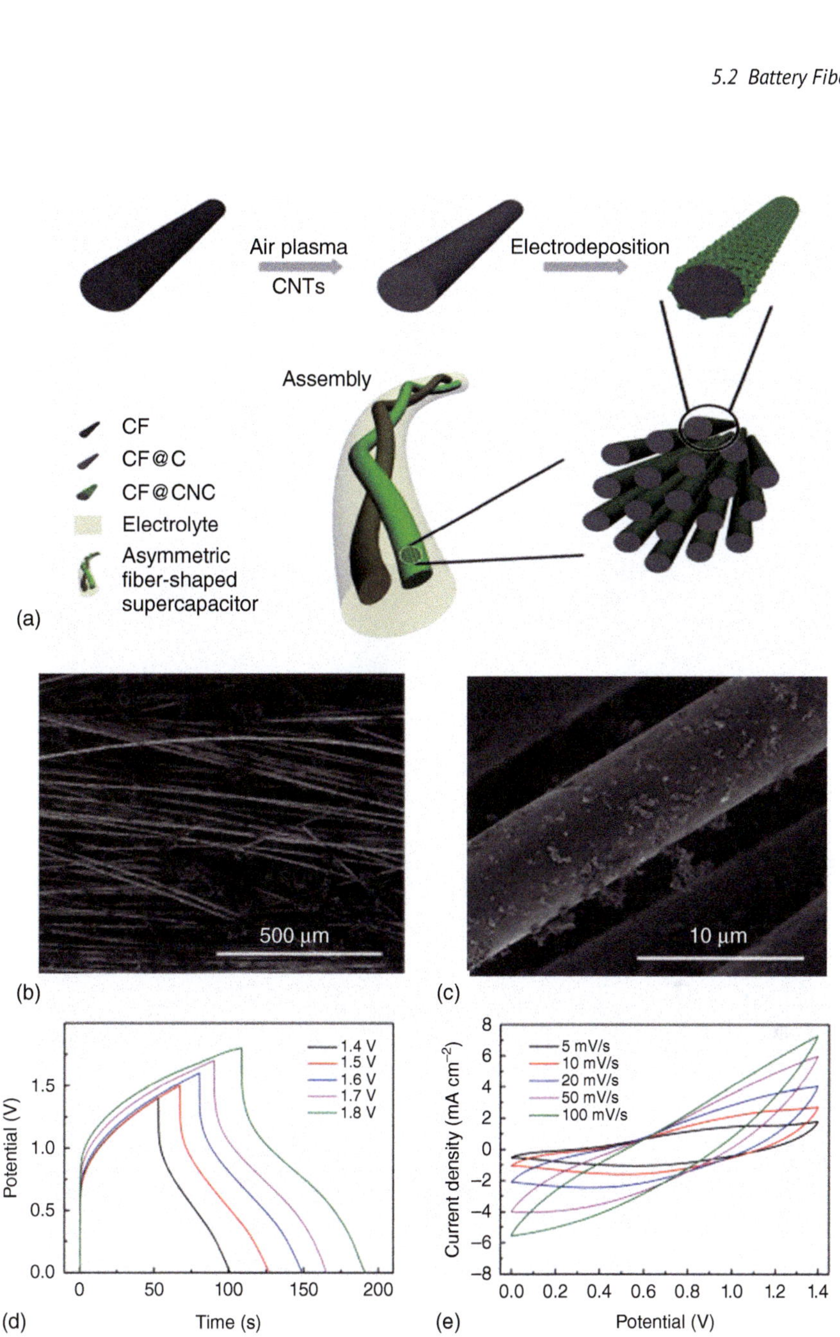

Figure 5.12 Schematic diagram, SEM images, and performance of CF@AC fiber electrode: (a) Schematic illustration of the synthesis procedure for the CF@CNC and fabrication of the asymmetric fiber-shaped solid-state supercapacitor. (b, c) SEM images of the CF@AC at different magnifications. (d) GCD curves of the AFSSC with different operating potential windows. (e) CV curves with scan rates between 5 and 100 mV/s. Source: (a, d, e) Lu et al. [14]. © 2016, Royal Society of Chemistry, (b, c) Adapted with permission from Lu et al. [14]. Copyright (2016) by Royal Society of Chemistry.

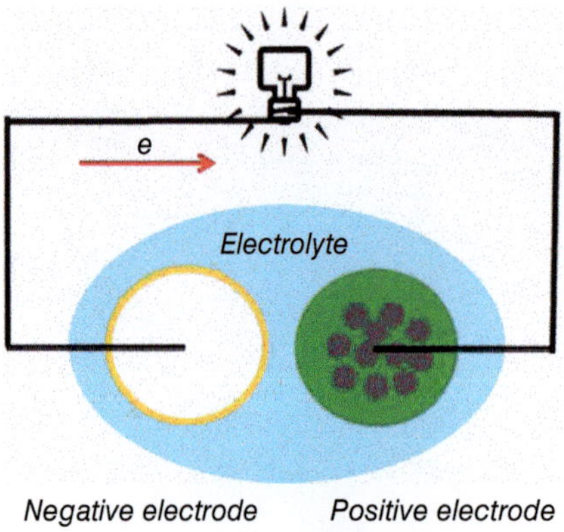

Figure 5.13 Schematic illustration of the alkaline Zn–MnO_2 battery system and its working principle. Source: Adapted with permission from Yu et al. [15]. © 2013, Elsevier.

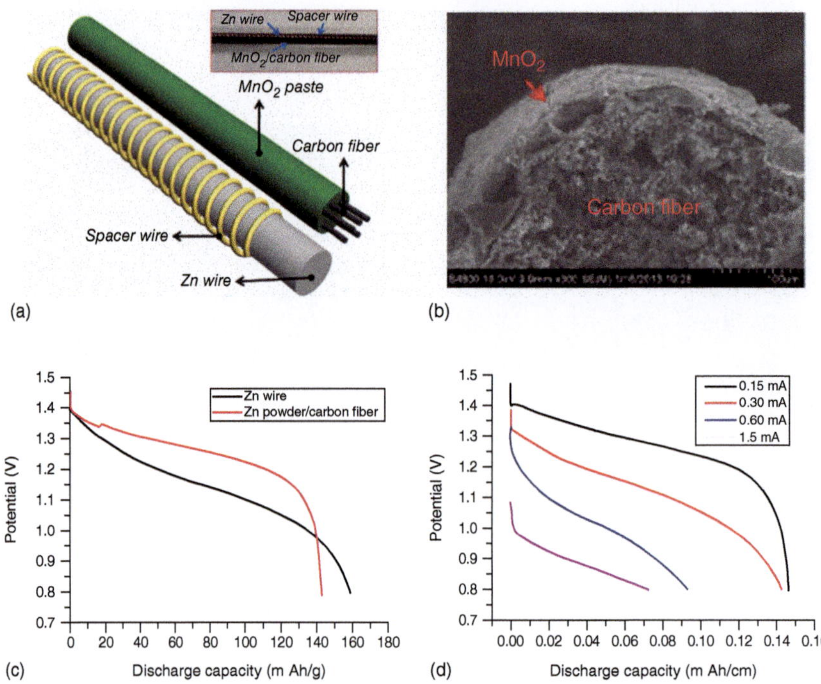

Figure 5.14 Schematic diagram, SEM image, and performance of zinc–manganese battery. (a) The schematic illustration of the fiber battery based on Zn wire and MnO_2/carbon fiber. (b) Scanning electron microscopy image of the MnO_2/carbon fiber electrode. (c) The discharge curves of fiber battery with Zn wire or Zn powder/carbon fiber as negative electrode. (d) The discharge curves of all carbon fiber-based battery with different discharge current densities. Source: (a, c, d) Yu et al. [15]. © 2013, Elsevier, (b) Adapted with permission from Yu et al. [15]. Copyright (2013) by Elsevier.

no loss of capacity. When the battery fiber length increased from 2.0 to 8.0 cm, the discharge capacity remained basically unchanged. The results show that fiber battery can be easily produced on a large scale by increasing the battery length without affecting the battery performance.

Because the primary battery fibers cannot be charged, so much attention have been paid to secondary battery fibers, such as LIB fibers, lithium-sulfur battery fibers, and metal-air battery fibers, especially.

5.2.2 Lithium-Ion Battery Fibers

LIBs are the most widely used secondary batteries all over the world. Conventional LIB consists mainly of anode and cathode, which is electronically isolated by electrolyte (Figure 5.15). The anode is the negative electrode, which is oxidized during the electrochemical reaction to donating electrons to the external circuit. The cathode is the positive electrode, which is reduced during the electrochemical reaction and accepts electrons from the external circuit. The electrolyte is a solid or liquid medium that allows the Li-ions to move between the anode and cathode and force the electrons to move through the external circuit.

LIBs were introduced by Sony in 1991. They exhibit a theoretical specific energy of 400 Wh/kg and have been scrutinized by researchers all over the world for the last two decades to improve the energy density due to advances in the portable electronics industry along with the requirements of vehicle electrification and grid energy storage. As one of the most promising electrochemical systems, LIBs have been widely used in portable electronics due to their high energy density, high safety, and long cycling life. However, conventional LIBs cannot meet the needs of high flexibility and breathability for wearable electronics. Wire-shaped electrochemical storage devices that are flexible can be woven into textiles have emerged and thrived in response.

To get a higher performance, many attempts have been made to get fibrous LIBs. In 2012, Kwon et al. [17] introduced an unprecedented concept of battery

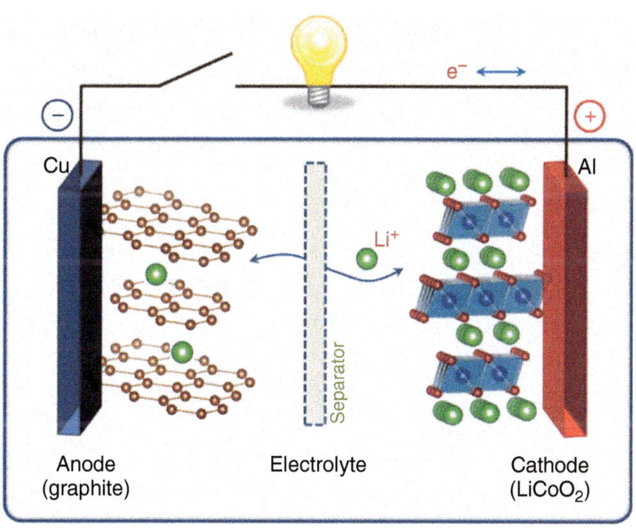

Figure 5.15 Schematic illustration of the Li-ion battery system and its working principle. Source: Adapted with permission from Goodenough et al. [16]. © 2013, American Chemical Society.

structure, a cable-type LIB that can achieve unprecedented mechanical flexibility and may indeed bring necessary breakthroughs in the field of flexible electronics (Figure 5.16a). The linear shape and omnidirectional flexibility of the linear battery allowed designers to break away from the traditional constraints, because the battery can be placed anywhere and in any shape. In addition, it can be worn on the wrist, neck, or any other part of the human body, instead of installing the battery inside the device, thus gaining the maximum freedom in the device design and greatly promoting the implementation of practical wearable electronic devices. In the same year, Peng Huisheng's group [18] developed a micro-battery with high performances by using aligned MWCNT fibers as electrodes (Figure 5.16b). The wire-shaped battery was produced by twisting an aligned MWCNT fiber and a lithium wire that was functioned as positive and negative electrodes, respectively. The specific capacity was calculated as 94.37 mAh/cm^3 or 174.40 mAh/g at 2×10^{-3} mA. In 2014, a stretchable wire-shaped LIB is produced from two aligned MWCNT/lithium oxide composite yarns as the anode and cathode without extra current collectors and binders [19] (Figure 5.16c). The two composite yarns can be well paired to keep a safe battery with superior electrochemical properties, such as energy densities of 27 Wh/kg or 17.7 mWh/cm^3 and power densities of 880 W/kg or 0.56 W/cm^3. Later in 2015, an aligned CNT/MoS_2 nanosheet hybrid fiber was synthesized to display combined remarkable mechanical, electronic, and electrochemical properties (Figure 5.16d) [20]. It was used to fabricate LIBs with high specific capacitance of 135 F/cm^3 and high specific capacity of 1298 mAh/g, respectively. In 2016, Peng Huisheng's group [21] developed a new type of fibrous aqueous lithium-ion battery (FAL) with polyimide/CNT composite fiber as the positive electrode and $LiMn_2O_4$/CNT composite fiber as the negative electrode (Figure 5.16e). The output power density of the battery is 10 217.74 W/kg, more than most SCs, and the energy density is 48.93 Wh/kg, equivalent to thin film LIB. Compared with the traditional planar structure, the shape of the fiber also has unique advantages such as 3D deformability.

5.2.3 Lithium-Sulfur Battery Fibers

Lithium-sulfur battery is another widely used battery all over the world. A conventional Li/S system where sulfur-based composite, lithium metal, and a nonaqueous (mostly organic solution) Li-ion containing salt solution act as a cathode, anode, and electrolyte, respectively, was in charged state after assembled (Figure 5.17). The elemental sulfur that displays orthorhombic crystal structure and cyclooctasulfur (S_8) molecular structure at room temperature needs 16 mol of electrons to reduce 1 mol of S_8. During discharging, Li metal anode release electrons to form Li-ions that are later adsorbed and react with the S_8 of cathode and reduce S_8 to lithium sulfide (Li_2S). Due to the high theoretical capacity (1675 mAh/g) and energy density (2600 Wh/g), Li-S batteries have attracted growing research interests and are regarded as a promising successor to conventional LIBs. In recent years, some groups have successfully prepared fiber-shaped Li-S battery.

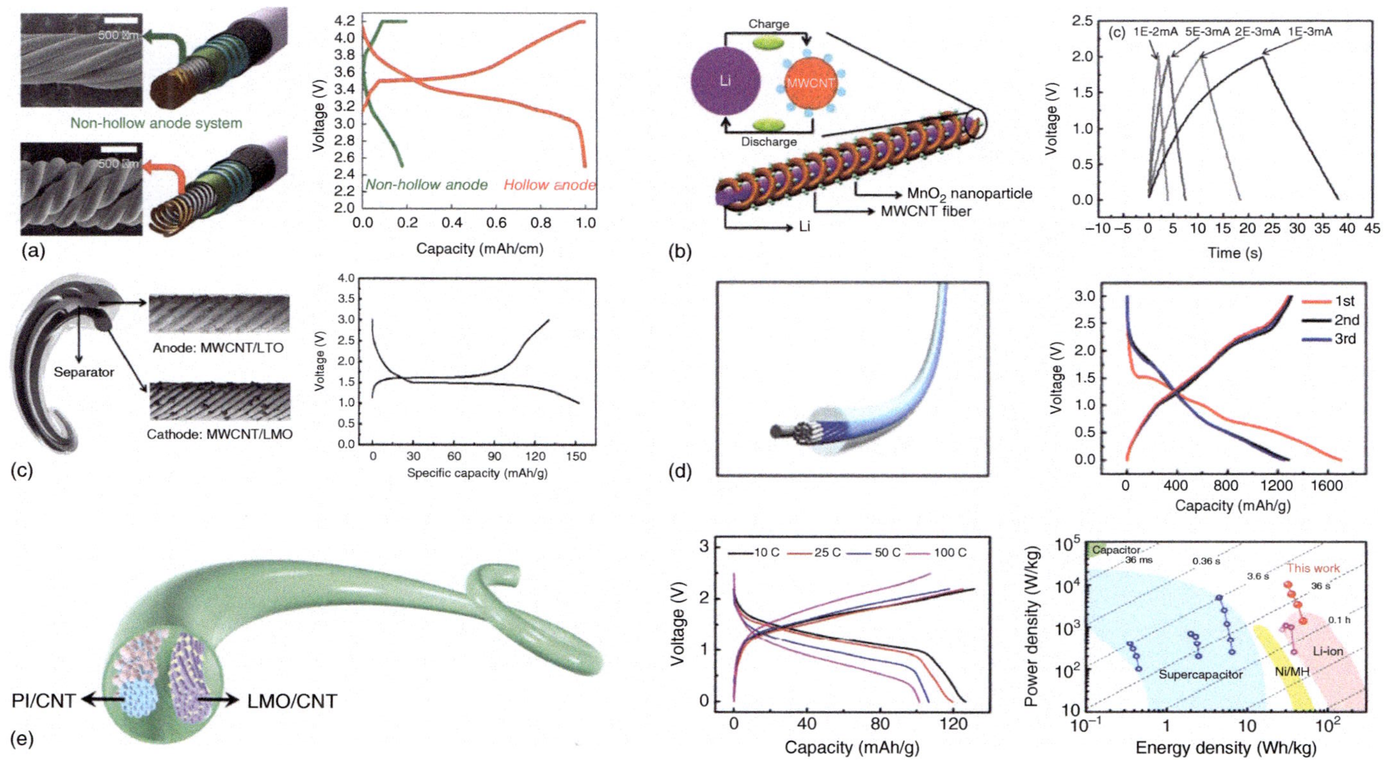

Figure 5.16 Schematic illustration and performance of lithium-ion battery fibers. (a) Images of the cable batteries with non-hollow and hollow anode system and first charge–discharge profiles. Source: Adapted with permission from Kwon et al. [17]. Copyright (2012) by WILEY-VCH Verlag GmbH & Co. KGaA, Weinheim. (b) Schematic illustration of the wire-shaped lithium ion battery and galvanostatic charge–discharge curves. Source: Ren et al. [18]. © 2013, John Wiley & Sons. (c) Structure of the flexible wire-shaped lithium-ion battery and charge–discharge curves at 0.02 mA. Source: Ren et al. [19]. © 2014, John Wiley & Sons. (d) Schematic illustration of the synthesis of the aligned CNT/MoS_2 hybrid fiber and first three charge-discharge curves at 0.2 A/g. Source: Luo et al. [20]. © 2015, Royal Society of Chemistry. (e) Schematic illustration, charge-discharge curves, and energy and power densities of the FAL. Source: Zhang et al. [21]. © 2016, Royal Society of Chemistry.

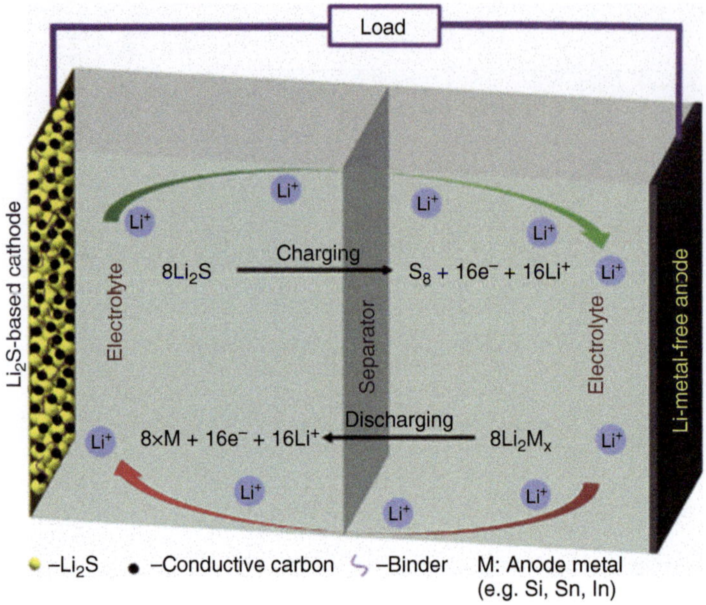

Figure 5.17 Schematic illustration of the lithium–sulfur battery system and its working principle. Source: Kaiser et al. [22]. © 2019, Elsevier.

The first fiber-shaped Li-S battery was reported by Fang et al. in 2015, in which carbon-nanostructured fibers with strengths ranging from 31 to 34 MPa were used as both the sulfur support and current collector (Figure 5.18a) [23]. After combining the hybrid cathode and a lithium wire anode, the as-prepared cable-shaped Li-S battery delivered a high specific capacity of 400-700 mAh/g, three to five times higher than those of the previously reported fiber-shaped LIBs. As a demonstration, the battery with only 1.8 mg of hybrid fiber (10 cm) could power a red light-emitting diode (LED) for 30 min. Moreover, the cable-shaped Li-S battery presented excellent flexibility, with the open-circuit voltage remaining unchanged in different bending states. Then in 2016, Chong et al. explored a facile wet-spinning technique to prepare the first-ever graphene-based rGO/CNT/S hybrid fibrous cathodes, and a cable-like lithium-sulfur batteries prototype was fabricated [24] (Figure 5.18b). The 1D fibrous cathode delivered a high initial capacity of 1255 mAh/g, owing to its high electrical conductivity of 11.7 S/cm and a superior areal energy density compared with other types of LSBs. The high stability of the electrochemical performance under cyclic bending and excellent mechanical flexibility of the current LSBs highlight great potential of graphene-based fiber assemblies in the quest for shape-compliant electrode materials. One year later in 2017, a highly flexible and mechanically robust fibrous cathode has been fabricated by depositing rGO/sulfur composites onto strain sensing fabric (SSF) yarn under the capillary-driven self-assembly as shown in Figure 5.18c [25]. The sulfur component is well reserved by the stacking of rGO nanosheets and tightly connected to the highly conducive SSF current collector. Such a dip-coating manufacturing route using industrially available SSF

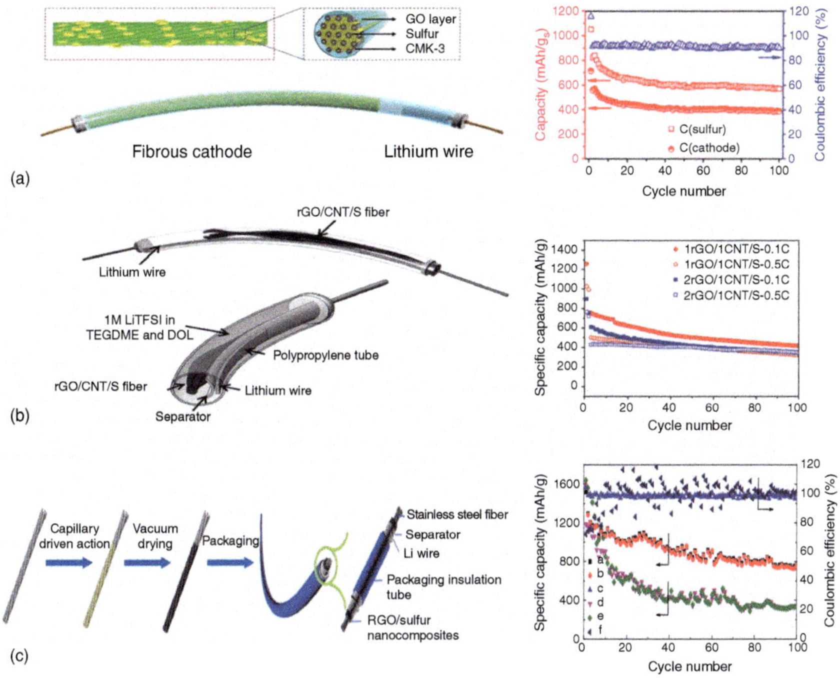

Figure 5.18 Schematic illustration and performance of lithium-sulfur battery fibers. (a) Schematic illustration, discharge capacities, and coulombic efficiencies over 100 cycles at 0.1 C of a cable-shaped Li-S battery. Source: Fang et al. [23]. © 2016, John Wiley & Sons. (b) Schematic illustration and galvanostatic cyclic stability of 1rGO/1CNT/S and 2rGO/1CNT/S fiber electrodes for 100 cycles. Source: Chong et al. [24]. © 2017, John Wiley & Sons. (c) Scheme of synthesis process and cycle performance of rGO/S/SSF hybrid fiber electrode at scanning rate of 0.2 mV/s. Source: Liu et al. [25]. © 2017, Elsevier.

yarns as skeleton allows the continuous and large-scale production of fibrous electrodes for wire-shaped batteries. The fabricated wire-shaped lithium-sulfur battery shows high electrochemical performances, excellent flexibility, waterproofness, and wearability.

5.2.4 Metal-Air Battery Fibers

Metal-air battery (typically, Li-air battery; Figure 5.19) is viewed as a promising candidate for LIB and can satisfy the ever-growing demand for high energy density, low cost, and environmental friendly power sources for mobile electronics and electric vehicles. In general, the metal-air battery is composed of porous cathode, electrolyte, and metal anode. Metals such as Li, Na, Zn, etc. are appropriately used as anode materials in metal-air batteries. Compared with a flexible fibrous LIB with a closed system, the flexible metal-air battery fibers have a unique semi-open system, which mainly uses oxygen from the ambient air to achieve energy storage. The air electrode has the minimized mass and volume and thus can increase the energy density of the battery. The energy storage mechanism of metal-air battery is mainly

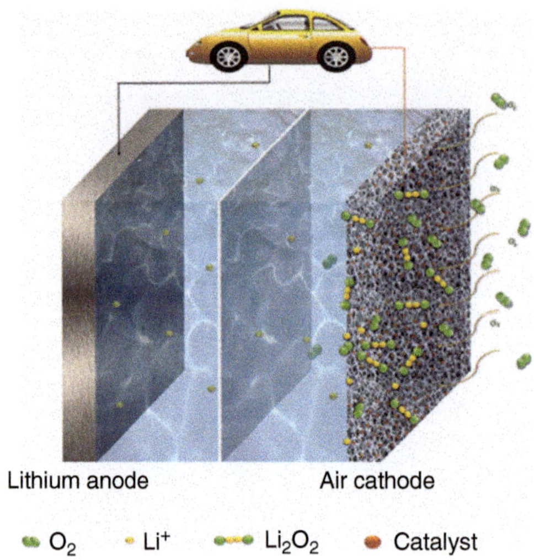

Figure 5.19 Schematic presentation of a nonaqueous aprotic Li-air battery composed of a Li metal anode, aprotic electrolyte, a separator, and a porous air cathode. Source: Li et al. [26]. © 2017, John Wiley & Sons.

achieved through the dissolution or deposition of metals on the negative electrodes and oxygen reduction or oxygen evolution reactions on positive electrodes. The common metal-air battery fibers include lithium-air battery fibers, zinc-air battery fibers, aluminum-air battery fibers, silicon-air battery fibers, etc.

For lithium-air battery fibers, Yin et al. [27] designed and manufactured a highly safe flexible Li-O_2 (SFLO) battery, in which the innovative central ventilation structure and the polyimide and polyimide (vinyl-hexafluoropropylene) (PIPV) composite diaphragm played a crucial role (Figure 5.20a). The results show that the PIPV separator has good water resistance, excellent thermal stability, and good ionic conductivity. In addition, the SFLO batteries have excellent water resistance and fire resistance, effectively avoiding leakage of liquid electrolytes. Surprisingly, the resulting SFLO batteries exhibit excellent flexibility and structural stability during deformation. Importantly, the assembled SFLO battery can work well in water. In addition, the SFLO battery has the characteristics of long cycle life, large capacity, and fast speed.

For zinc-air battery fibers, Peng Huisheng's group [28] developed a flexible, stretchable, and rechargeable fibrous zinc-air battery with an air cathode consisting of CNT sheets arranged and cross-stacked for oxygen reduction reaction (ORR), RuO_2-based catalyst for oxygen evolution reaction (OER), Zn anode, and independent hydrogel polymer electrolyte (Figure 5.20b). In the discharge process, O_2 is reduced to OH^- on the CNT sheet, and on the RuO_2-based catalyst layer, OH^- is oxidized to O_2 in the charging process. Apparently, the CNT layer also acts as a diffusion layer, absorbing oxygen from the atmosphere and a current collector to transfer electrons. The CNT-based air cathode makes the zinc-air battery have stable electrochemical performance under bending and stretching conditions, stable discharge at 10 A/g, and good discharge-charge performance at 2 A/g.

For aluminum-air battery fibers, Peng Huisheng's group [29] developed a new type of fibrous aluminum–air battery, which was coated on an aluminum substrate

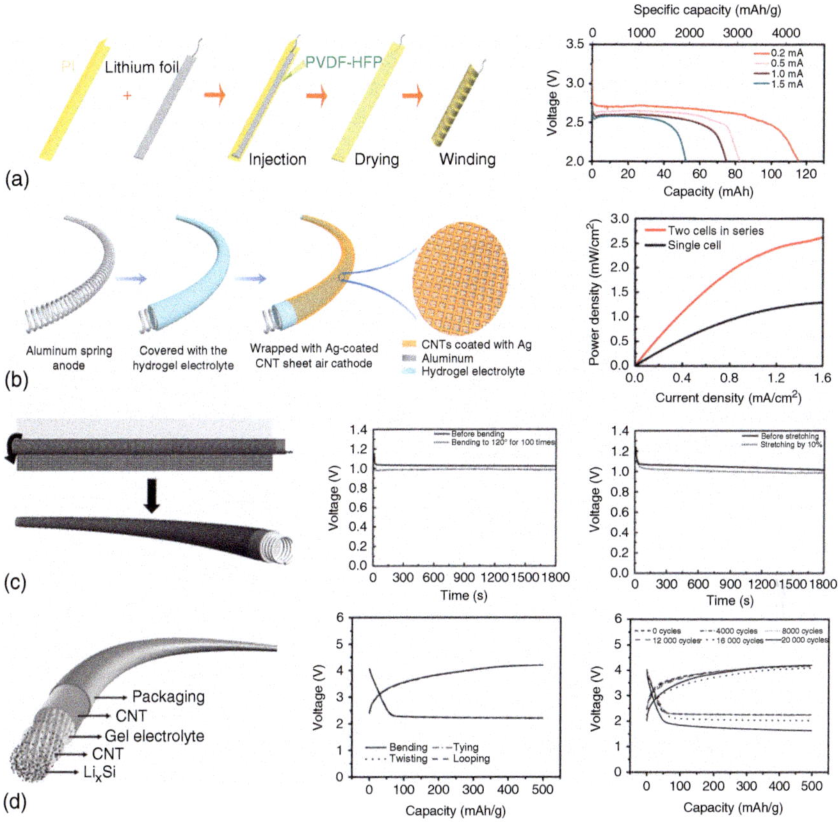

Figure 5.20 Schematic illustration and performance of metal-air battery fibers. (a) Preparation of PIPV separator and the rate capacity of SFLO battery. Source: Yin et al. [27]. © 2018, John Wiley & Sons. (b) Fabrication and power density curves of the fiber-shaped Al-air battery. Source: Xu et al. [28]. © 2015, John Wiley & Sons. (c) Schematic illustration and discharge curves of the fiber-shaped Zn-air battery with a length of 5 and 10 cm. Source: Xu et al. [29]. © 2016, John Wiley & Sons. (d) Schematic illustration and charge-discharge curves of the SOB fiber. Source: Zhang et al. [30]. © 2017, John Wiley & Sons.

by a continuous layer of CNT/silver nanoparticles as an air cathode on gel electrolyte (Figure 5.20c). Due to the high theoretical energy density of metal-air battery, aluminum-air battery is considered as a promising long-term electronic power source. However, the available energy density of aluminum-air batteries is much lower than expected and is limited by the current electrode material. They developed a new type of all-solid fiber aluminum-air battery with a specific capacity of 935 mAh/g and energy density of 1168 Wh/kg. The electrode composed of CNTs/silver nanoparticles has excellent electrochemical properties. The shape of the fibers also gives aluminum-air batteries a unique advantage. For example, they are flexible and stretchable and can be woven into a variety of textiles for large-scale applications.

For silicon-air battery fibers, Peng Huisheng's group [30] have discovered a silicon-oxygen battery (SOB) fiber with high energy density and ultrahigh flexibility by designing a solid-state coaxial architecture with the lithiated silicon/CNT hybrid fiber as inner anode, a polymer gel as middle electrolyte, and a bare CNT sheet as outer cathode (Figure 5.20d). The full SOB fiber exhibited an energy density of 512 Wh/kg based on the total weight of the two electrodes. The as-fabricated lithiated silicon/CNT hybrid fiber not only avoided dendrite formation and the safety issue of lithium metal but also showed ultrahigh flexibility. The SOB fiber could effectively work after 20 000 bending cycles.

5.2.5 Other Battery Fibers

In addition to the abovementioned batteries, many other battery systems, including zinc–ion battery (ZIB), sodium–ion battery (SIB), and all-MOF (metal-organic framework) battery, have been made into fiber shape.

Taking ZIB as an example, recently, growing attention has been paid to develop flexible fiber-shaped aqueous ZIB (Figure 5.21). To date, the reaction mechanisms of ZIB involved remain a topic of discussion and are underdeveloped. According to the literature, the redox reactions in aqueous ZIB systems mainly involve three mechanisms: Zn^{2+} insertion/extraction, chemical conversion reaction, and H^+/Zn^{2+} insertion/extraction. Li et al. [32] presented a quasi-solid washable and customizable elastic yarn ZIB composed of a double–helix wire electrode and a polyacrylamide (PAM)-based polymer electrolyte (Figure 5.22a). Using double-helix CNT as the substrate of MnO_2 cathode and Zn anode, the strength and robustness of the electrode under different deformation conditions are effectively improved, and the

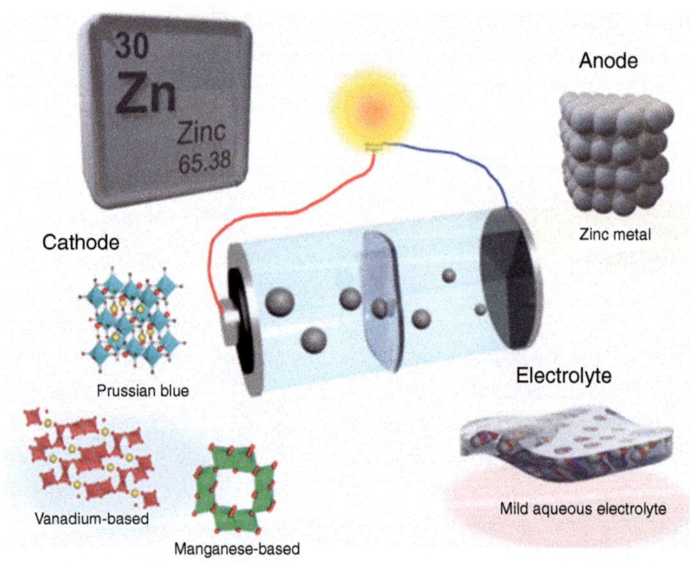

Figure 5.21 Schematic diagram of ZIBs. Source: Adapted with permission from Konarov et al. [31]. Copyright (2018) by American Chemical Society.

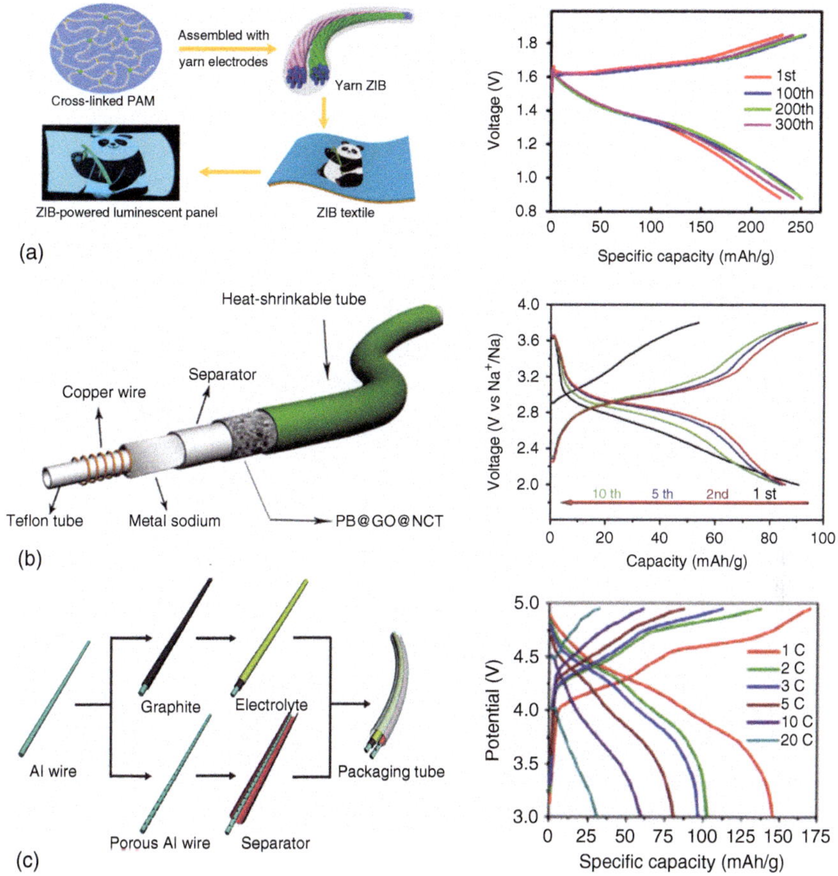

Figure 5.22 Schematic illustration and performance of other battery fibers. (a) Schematic fabrication process and discharge curves of the yarn ZIB after different bending cycles. Source: Li et al. [32]. © 2017, John Wiley & Sons. (b) Schematic illustration and charge-discharge profiles of tube-type flexible SIBs. Source: Zhu et al. [33]. © 2017, John Wiley & Sons. (c) The schematic illustration of the fabrication and charge-discharge curves of the FSDIB. Source: Song et al. [34]. © 2019, Elsevier.

electrolyte wettability of the electrode surface was significantly improved. The prepared quasi-solid yarn ZIB has a high specific capacity and volumetric energy density (302.1 mAh/g and 53.8 mWh/cm^3, respectively) and good cyclic stability (a capacity retention rate of 98.5% after 500 cycles under the high current of 2 A/g).

Similar to ZIB, SIB is also a promising alternative for the state-of-art LIBs. It is composed of cathode, anode, electrolyte, and separator. Upon discharging the SIB, Na-ions deintercalate from the lattice sites of anode (oxidation), while Na-ions intercalate into the lattice sites of cathode (reduction), and electrons moves from the anode to cathode in the external circuit. Zhu et al. [33] prepared a tube-type flexible and wearable sodium-ion batteries by developing a simple and economical strategy to effectively revive electroless nickel plating wastewater and the Ni-coated cotton textile (NCT) waste toward a novel electrode substrate (Figure 5.22b). Unexpectedly, the obtained NCT has the advantages of high mechanical strength/flexibility, good electrical conductivity, and good electrochemical stability. In addition, as a proof-of-concept application, the unbonded PB-@GO-@NTC electrode was successfully obtained with superior electrochemical performance, with a high capacity of more than 110 mAh/g at a high rate of 30 C and an extremely long cycle life of 1800 cycles.

Dual-ion battery (DIB) is a novel type of battery developed in recent years, with the merits of high safety, wide voltage window, and high specific capacity. Song et al. [34] proposed a novel design of fiber-shaped DIBs with omnidirectional porous Al wire as the anode and graphite on the Al wire as the cathode (Figure 5.22c). During charge-discharge processes, the Li^+ and $PF6^-$ ions moved back and forth, accompanied by reversible $PF6^-$ anion intercalation/deintercalation into/from the graphite interlayers. The fiber-shaped DIB delivered a high specific capacity of 145.7 mAh/g at 1 C and could maintain a capacity of over 130.3 mAh/g after 200 cycles. Employing liquid metal anodes instead of the current solid anodes may be a positive approach to overcoming the dendrite problem of metallic electrodes. Besides the above battery fibers, other battery fibers including all-MOF battery fiber [35], Ni/Fe battery fiber [36], and liquid metal-air battery fiber [37] have also been prepared successfully.

In summary, the development of flexible, wearable, and intelligent electronic devices has promoted the research of battery fibers. Although significant progress has been made in the research of batteries with excellent electrochemical properties, the research of flexible battery fibers is still in the laboratory stage. In order to meet the needs of developing wearable electronic products, we should focus on the practical application of flexible battery fibers.

To satisfy the requirements on the energy and power densities, operation safety and good cyclic stability of flexible energy storage devices, great efforts must be made to meet the development needs of the next generation of flexible battery fibers, especially from the following four aspects: developing new high-performance flexible fibrous electrode materials or emerging battery systems including high-energy battery fibers; optimizing battery components, including reducing the weight of inactive components (collector, binder, conductive additive) and improving the energy density of the whole battery unit; using advanced technologies to prepare flexible fibrous electrodes with high-quality loaded active materials; and the volume energy density is also an important factor that cannot be ignored, when battery fibers are applied to flexible electronic devices.

Besides, in order to realize large-scale and low-cost production of battery fibers, it is urgent to develop simple, fast, and controllable flexible fiber electrode manufacturing process and assembly technology. For example, emerging technologies such as advanced 3D printing offer a promising solution for the manufacture of fiber electrodes and flexible fiber cells at low cost and with high flexibility.

5.3 Phase-Transit Fibers

Different from SCs and batteries, phase-transit materials (PTMs) store and release thermal energy while SCs and batteries store and release electric energy.

Traditionally, PTM refers to a certain type of substance featuring a great latent heat of fusion with regard to melting or solidifying at a nearly constant temperature. In the process of phase transition, PTM is able to store and release a large amount of heat within a relatively narrow range of temperature variation (Figure 5.23). For example, relative to the conventional thermal storage materials, such as water or rocks, an equivalent volume of PTM can store or release 5-14 times more heat. To this end, PTMs have been widely explored for use as thermal materials for heat management. It should be pointed out that the storage or release of energy is only one reflection of the changes in intermolecular interactions (e.g. van der Waals forces and hydrogen bonding) in PTM during phase transition. The change in intermolecular interactions also causes drastic variations in other physicochemical parameters, such as mobility and density. For example, when butter is melted upon heating in a pan, its mobility will be increased substantially to allow flow. The increase in mobility during a solid-to-liquid transition can be leveraged to control the release of payloads trapped in a solid matrix, leading to a temperature-responsive release system.

Traditionally, a typical shape-stable PTM consists of a supporting component (inorganic porous/laminar material or polymer) and a latent heat storage (LHS) component (solid-liquid PTMs). PTF, as a stable phase change material in fiber form, has attracted wide attention in academia and industry due to its application prospect in intelligent fabrics and fabrics in recent 30 years. According to LHS component, PTFs can be divided into three types, PTFs based on hydrocarbons and fatty acids, PTFs based on fatty alcohols, and PTFs based on other kinds of phase-transit materials, which will be discussed respectively as follows.

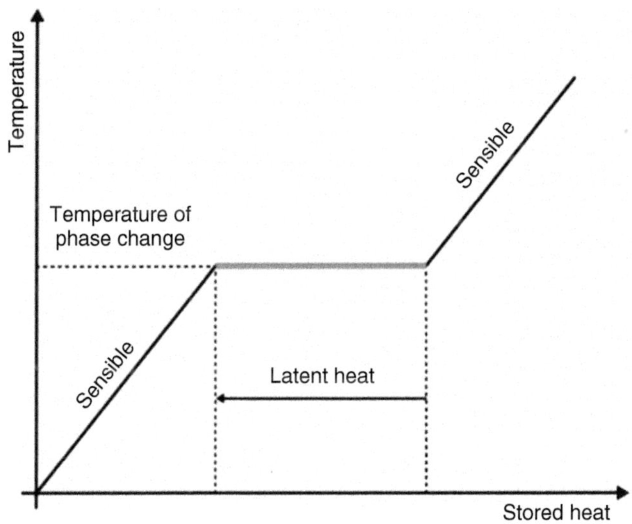

Figure 5.23 A representative phase diagram showing the temperature of a PTM as a function of the stored heat. Source: Qiu et al. [38]. © 2020, John Wiley & Sons.

5.3.1 Phase-Transit Fibers Based on Hydrocarbons and Fatty Acids

Hydrocarbons (e.g. dodecane, tetradecane, hexadecane, octadecane, and eicosane) refer to hydrocarbons with the basic properties of aliphatic compounds. A fatty acid is a carboxylic acid composed of a long aliphatic chain, either saturated or unsaturated. Long-chain fatty acids and their eutectics have also been researched extensively and applied as typical PTMs in various areas due to the unique advantages of suitable melting temperature, high latent heat of fusion, ready availability, nontoxicity, and noncorrosiveness.

For PTFs based on hydrocarbons, Xia's group [39] first proposed a method of preparing long-chain hydrocarbon core and composite sheaths of phase change nanofibers in 2006 (Figure 5.24a). Shape-stable phase-transit nanofibers have many potential applications because they absorb, retain, and release large amounts of heat energy to exploit the fusion heat of long-chain hydrocarbons within a given

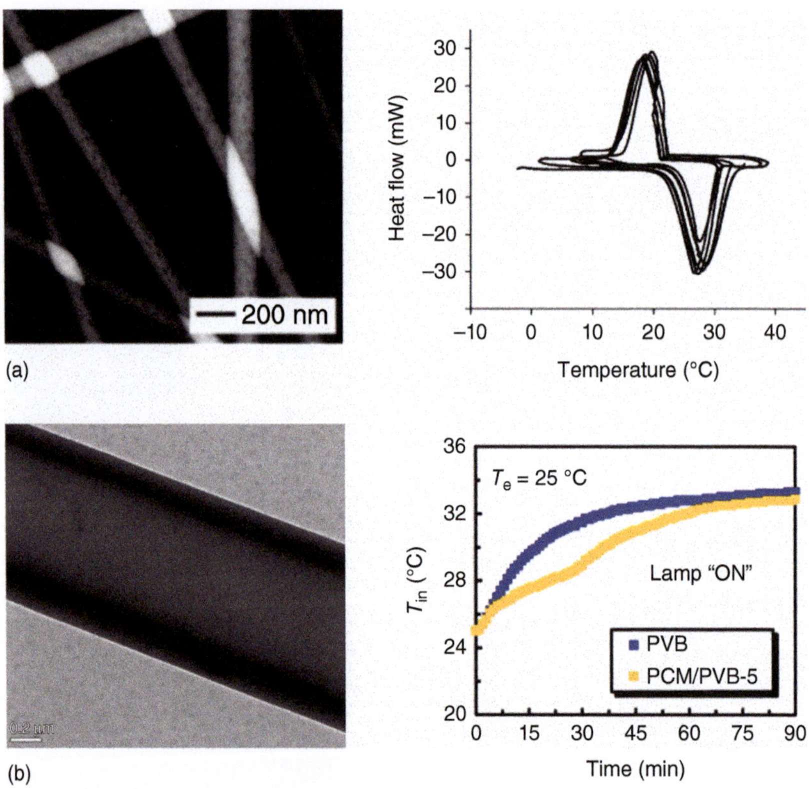

Figure 5.24 TEM image and DSC curves of PTM based on hydrocarbons. (a) TEM image and Lissajous figures derived from DSC measurements of TiO_2-PVP nanofibers. Source: Adapted with permission from McCann et al. [39]. Copyright (2006) by American Chemical Society. (b) TEM image and the time-dependent inner temperature (Tin) of model houses under the periodical simulated solar radiation of PCM/PVB nanofibers. Source: Adapted with permission from Sun et al. [40]. Copyright (2015) by Springer Nature Limited.

temperature range. They have produced thermally stable phase-transit materials up to 45 wt% octadecane. In addition, the obtained fibers exhibit novel core segmented morphology due to the rapid solidification of hydrocarbons driven by evaporative cooling of carrier solution. In 2015, Sun et al. [40] prepared core/shell PTM/polyvinyl butyral (PVB) nanofibers with hydrophobic PVB shell with ethanol as solvent (Figure 5.24b). With the increase of PVB concentration in ethanol solution, the morphology of core/shell PTM/PVB nanofibers became better, and the packing efficiency was higher, but the size uniformity was worse. The higher the shell velocity, the higher the encapsulation efficiency. Under the condition of constant shell velocity such as 3.0 mL/h, the encapsulation rate of PTM/PVB nanofibers increases with the increase of core velocity. PTM/PVB nanofibers show good thermal regulation ability under simulated solar irradiation, and the temperature remains at about 28 °C, close to the comfortable room temperature. The PTM/PVB nanofibers have good thermal regulation and stability during repeated heating/cooling cycles.

For PTFs based on fatty acids, in 2013, Cai et al. [41] prepared polyacrylonitrile (PAN) nanofiber felt and its derived carbon nanofiber felt (Figure 5.25a). PAN nanofiber felt with stable shape was prepared from fatty acid eutectic as raw material for storage and recovery of heat energy. The heat transfer coefficient was improved by converting PAN mat to carbon mat. The maximum adsorption capacity of PAN and CF felt was 99% and 81% wt%, respectively. Differential scanning calorimeter (DSC) results showed that the melting temperature and enthalpy of the prepared composite PTMs were 19.8-31.4 °C and 120.1-165.4 kJ/kg of PAN nanofiber felt, respectively, and the melting temperature and enthalpy of carbon nanofiber felt were 18.8-30.7 °C and 92.1-128.0 kJ/kg, respectively. After that in 2016, a series of ternary fatty acid eutectics were successfully prepared by Ke et al. [42] (Figure 5.25b). Subsequently, silver-plated PAN nanofiber felt with good thermal conductivity was prepared, and the eutectic ternary fatty acid prepared was absorbed by capillary tube, physical surface tension, and nano constraint. The maximum melting enthalpy variation of ternate fatty acid is low eutectic/PAN/Ag to form stable composite adsorption up to 164.8 kJ/kg, and the phase transition temperature can be customized in the range of −0.95 °C to 53.15 °C, which is suitable for slightly lower thermal energy storage (TES) and retrieval applications. In 2017, Ke [43] developed a series of novel quaternary eutectic/PU/Ag stable phase-transit composite fiber membranes (Figure 5.25c). The sputtered silver nanoparticles prepared by magnetron sputtering were successfully deposited on the surface of electrospun polyurethane fibers. The phase transition temperature of quaternary eutectic is lower than that of single fatty acid and binary eutectic. The initial melting point temperature and enthalpy of quarterly eutectic/PU/Ag composite fiber membrane are about 14-29 °C and 95-118 kJ/kg, respectively, which can meet the requirements of building energy conservation, temperature-regulating textiles, and other low-temperature heat storage applications. The maximum adsorption rate of composite fiber membrane to quaternary eutectic was about 91%.

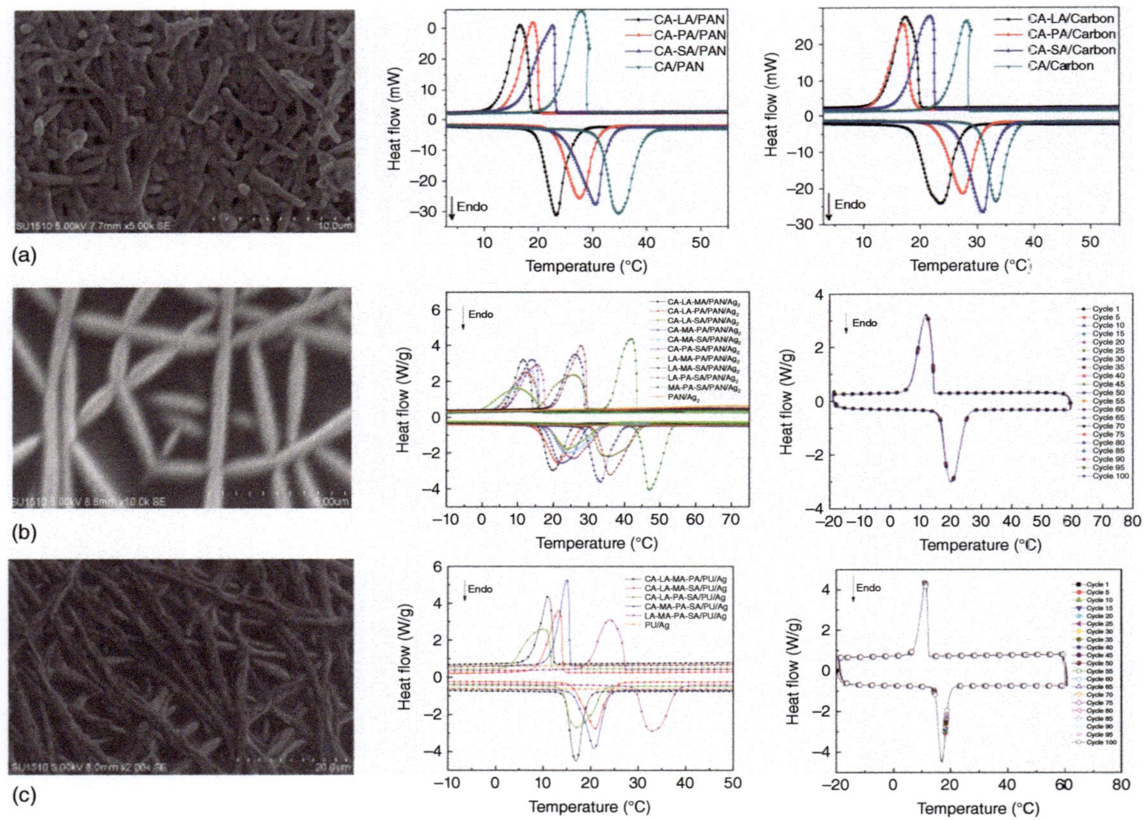

Figure 5.25 SEM image and DSC curves of PTM based on fatty acids. (a) SEM image and DSC curves of composite PCMs consisting of fatty acid eutectics and electrospun PAN carbon and nanofibrous mats. Source: Adapted with permission from Cai et al. [41]. Copyright (2013) by Elsevier. (b) SEM image, DSC curves, and DSC cycle curves of Ag-coated PAN nanofiber membrane. Source: Adapted with permission from Ke et al. [42]. Copyright (2016) by Springer Nature Limited. (c) SEM image, DSC curves, and DSC cycle curves of Ag-coated electrospun PU fibrous membranes. Source: Adapted with permission from Ke et al. [43]. Copyright (2017) by Springer Nature Limited.

5.3.2 Phase-Transit Fibers Based on Fatty Alcohols

Fatty alcohols have a structure similar to that of fatty acids, and they are typically straight-chain primary alcohols. They naturally exist in vegetable oil and animal fat. Fatty alcohols with a chain length of up to C_{18} are known to be biodegradable. To this end, fatty alcohols have also been considered as candidates for controlled release of drugs. Same as fatty acids, the melting points of fatty alcohols generally have a positive correlation with the number of carbon atoms in the chain. Polyethylene glycol (PEG) is a typical PTM with outstanding characteristics of nontoxicity, resistance to corrosion, relatively large enthalpies, and wide selectivity of molecular weight. Thus, PEG and its derivatives are also usually used as LHS component in the form-stable PTMs. PEG/calcium (CA), PEG/poly (dl-lactide) (PLA), PEG/poly(vinylidene fluoride) (PVDF), and PEG/polyamide 6 (PA6) composite fibers were prepared, respectively.

PEG/CA composite fibers were firstly prepared. Chen et al. [44] first prepared cross-linked PEG/CA composite fibers with toluene-2,4-diisocyanate (TDI) as cross-linking agent in 2009 (Figure 5.26a). The cross-linked PEG/CA fibers were cylindrical, smooth, and similar to the original composite fibers in morphology. The results show that cross-linking improves the water resistance and thermal stability of the fiber but decreases the enthalpy of the fiber. Conclusively, as a stable PTM material, cross-linked PGE/CA composite fiber has good energy storage performance in a wide temperature range. And then in 2011, Chen et al. [45] prepared ultrafine PEG/CA composite fibers with different ratios of PEG by electrospinning (Figure 5.26b). By adjusting the content of PEG, the morphology and thermal properties of the composite fibers were significantly changed. The PEG/CA electrospun fibers have good internal temperature regulation ability when the ambient temperature changes. The electrospun PEG/CA composite fibers have reliable thermal properties and are suitable for use as PTFs for heat storage and temperature regulation.

Besides CA, other polymers such as PLA, PVDF, and PA6 were also selected as the fibrous supporting matrix to encapsulate PEG. For PEG/PLA composite, in 2013, Chen et al. [46] prepared a series of PEG/PLA ultrafine phase-transit composite fibers with different mass ratios (Figure 5.27a). The maximum mass ratio of PEG/PLA was 200/100, and the average fiber diameter increased with the increase of PEG content. PEG is randomly distributed on the surface and inside of the composite fiber. DSC and thermogravimetry analysis (TG) results show that the composite fiber has an effective heat storage capacity, and the melting enthalpy ranges from 54.8 to 74.7 J/g, and the degradation temperature of the fiber exceeds 320 °C. After heat treatment, the morphology, thermal properties, and storage and discharge properties of the composite fibers did not change significantly. For PEG/PVDF composite, in 2015, Dang et al. [47] prepared shape-stable PVDF nanofibers (Figure 5.27b). The melting point and maximum enthalpy ratio of PVDF/SiO_2 nanofibers supported by PEG were about 31 °C and 40%, respectively. The temperature of PEG/PVDF core/shell nanofibers was 33 °C and 13.7%, respectively. With the addition of fumed SiO_2 into the PEG/PVDF

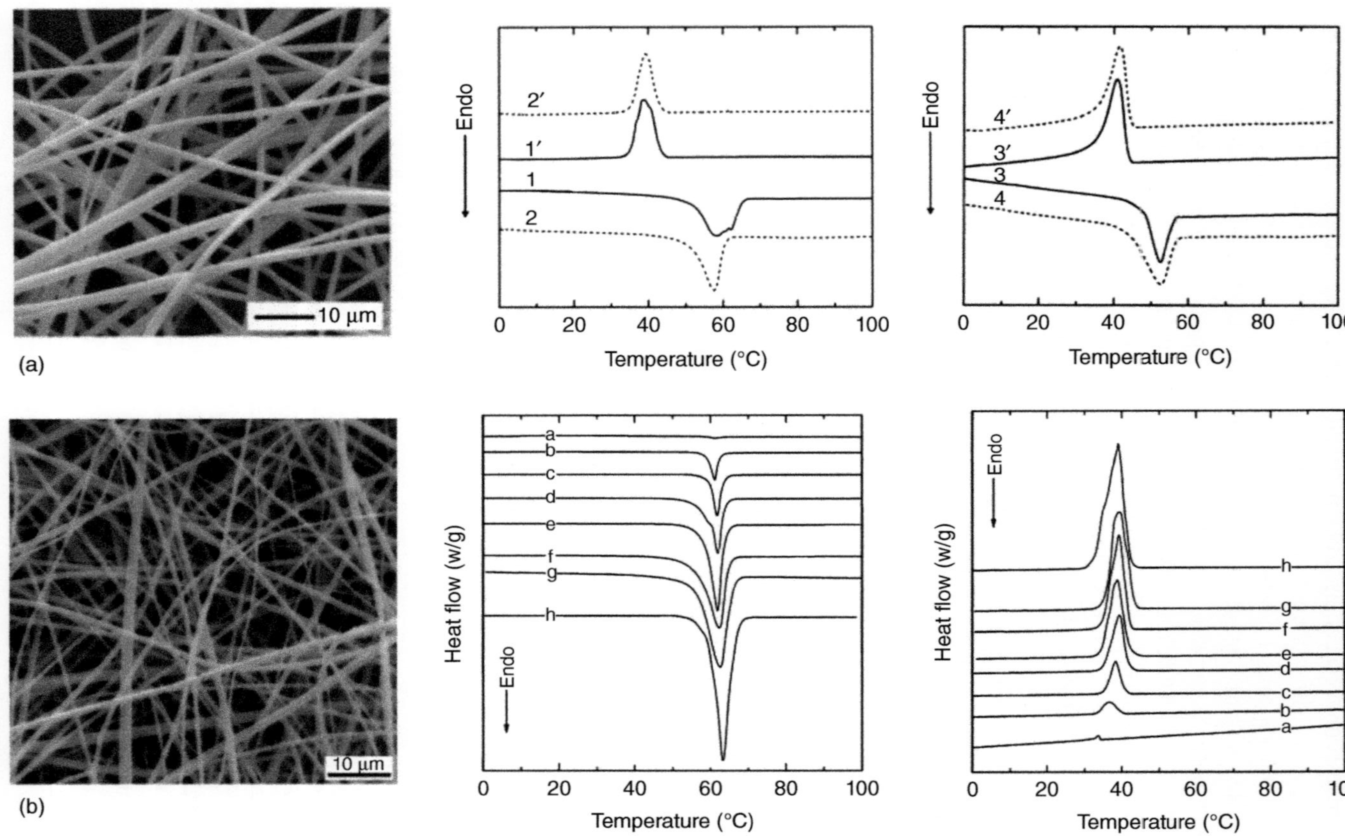

Figure 5.26 SEM image and DSC curves of PTM based on PEG/CA. (a) SEM image and DSC curves of PEG/CA composite fibers before and after 100 heating–cooling cycles. Source: Adapted with permission from Chen et al. [44]. Copyright (2009) by Elsevier. (b) SEM image and heating and cooling curves of PEG/CA fibers with different PEG content. Source: Adapted with permission from Chen et al. [45]. Copyright (2011) by Elsevier.

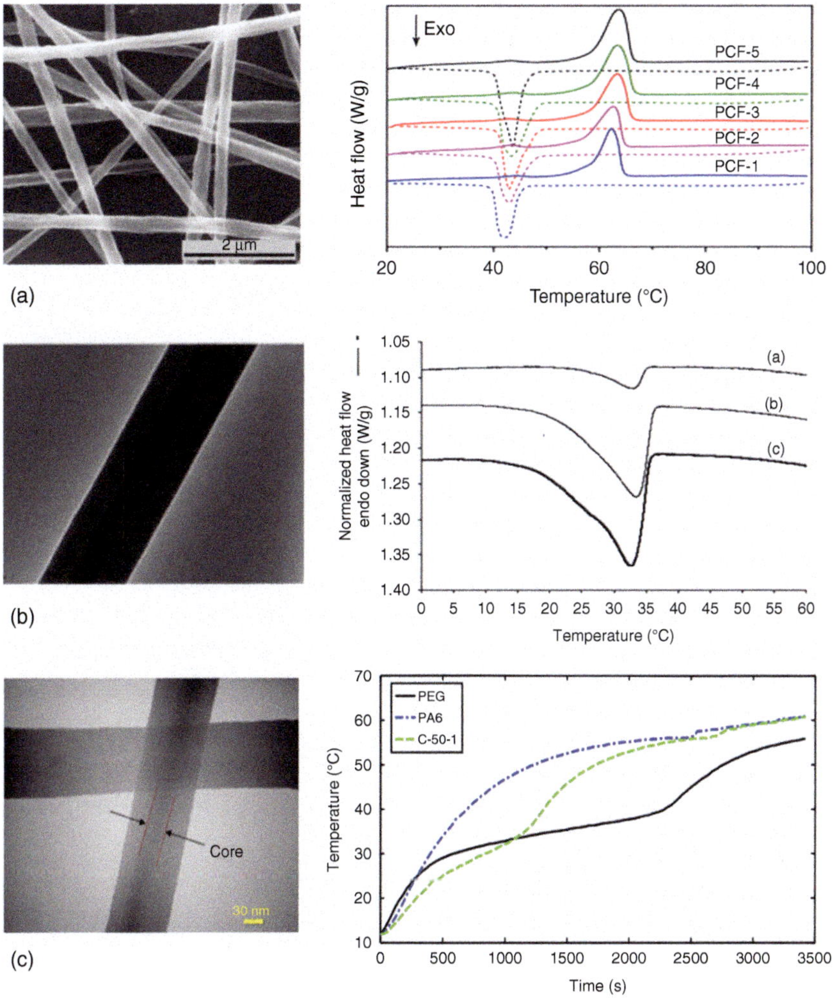

Figure 5.27 Morphology image and DSC curves of PTM based on PEG/PLA, PEG/PA, and PEG/PVDF. (a) SEM image of PLA and DSC thermograms of PEG/PLA composite fibers with different mass ratios. Source: Adapted with permission from Chen et al. [46]. Copyright (2013) by Elsevier. (b) TEM image and DSC curves of PEGs/PVDF. Source: Adapted with permission from Dang et al. [47]. Copyright (2015) by Springer Nature Limited. (c) TEM image of C-60-1 nanofiber and thermal energy storage rates for PEG, PA6, and C-50-1 composite. Source: Adapted with permission from Babapoor et al. [48]. Copyright (2017) by Elsevier.

composite, the PEG-loaded PVDF/SiO_2 composite nanofibers had greater thermal stability and mechanical strength than the PEG-loaded PVDF nanofibers. Both PEG-supported PVDF/SiO_2 composite nanofibers and PEG/PVDF core/shell nanofibers show thermal stability in 100 heating/cooling cycles through DSC. In contrast, the core/shell PEG/PVDF nanofibers, fabricated at core feed rates of 1.0, 2.0, and 3.0 μL/min, were the most form-stable nanofibers in terms of their shape maintenance and their ability to prevent PEG leakage after hot oven tests and water immersion tests. For PEG/PA6 composite, in 2017, Babapoor et al. [48] prepared core/shell PTM nanofibers with PEG1000 (core) and PA6 (shell) as raw materials (Figure 5.27c) by increasing the flow rate of the core solution, slightly thicker fibers can be produced, and the onset temperature of melting is reduced. As the PEG concentration rises, the peak temperature increases, and higher amounts of latent heat enthalpy are achieved. The fabricated core/shell structure has almost resolved the leakage instability normally associated with other types of PTM fibers and, hence, has the potential to improve thermal storage capacity.

5.3.3 Phase-Transit Fibers Based on Other Kinds of Phase-Transit Materials

Besides hydrocarbons, fatty acids, and fatty alcohols, other commercial and synthesized PTMs were also employed as the LHS component of PTFs, such as stearyl stearate (SS), diacid dioctadecyl esters (DADOEs), poly(vinyl alcohol) (PVA), PAN, etc.

SS is a common fatty acid ester that is extensively used as raw material of emulsifiers or oiling agents for foods, spin finishes and textiles, lubricants for plastics, paint and ink additives, and surfactants and base materials for perfumery. It was used as a novel PTM for its high enthalpy, suitable phase change temperature and nontoxicity. Chen et al. [49] synthesized a novel solid-liquid phase-transit material SS, and the SS/polyethylene terephthalate (PET) composite microfiber was used for heat storage (Figure 5.28a). After heat treatment, the morphology of composite fiber does not change significantly, which proves that the material developed has the phase transition characteristics of stable morphology. DSC results showed that the enthalpy of composite fiber increased with the increase of SS content, but the change of SS content had little effect on the fiber phase transition temperature. In addition, after 100 times of heating-cooling thermal cycle, the thermal properties of the fiber remain good, indicating that the composite fiber has good thermal stability and reliability.

DADOE which has good thermal stability was also used in PTMs. In 2012, Chen et al. [50] synthesized a series of DADOEs as new solid-liquid PTMs by esterify of octadecanol, ethane diacid, butane diacid, adipic acid, octane diacid, and sebacic acid, and then ultrafine PTFs of DADOE/PET were fabricated (Figure 5.28b). There was no significant change in the morphology of the fiber after thermal cycling, which confirmed that DADOE/PET fiber had stable phase transition characteristics. DSC curve results showed that DADOE/PET fibers had reversible phase-transit behavior similar to DADOE and the enthalpy of composite fibers increased with the increase of DADOE content (the maximum melting enthalpy of the fibers was greater than 65 J/g). In addition, after the thermal cycle, the thermal properties of the composite fiber can be well maintained, indicating that the composite fiber has good thermal stability and reliability.

PVA, a water-borne polymer, was chosen as the matrix material to prepare PTM/PVA fibers by Zdraveva et al. [51] (Figure 5.28c). PTM was randomly distributed in nanofibers. Higher PTM load in the fiber will lead to increased enthalpy,

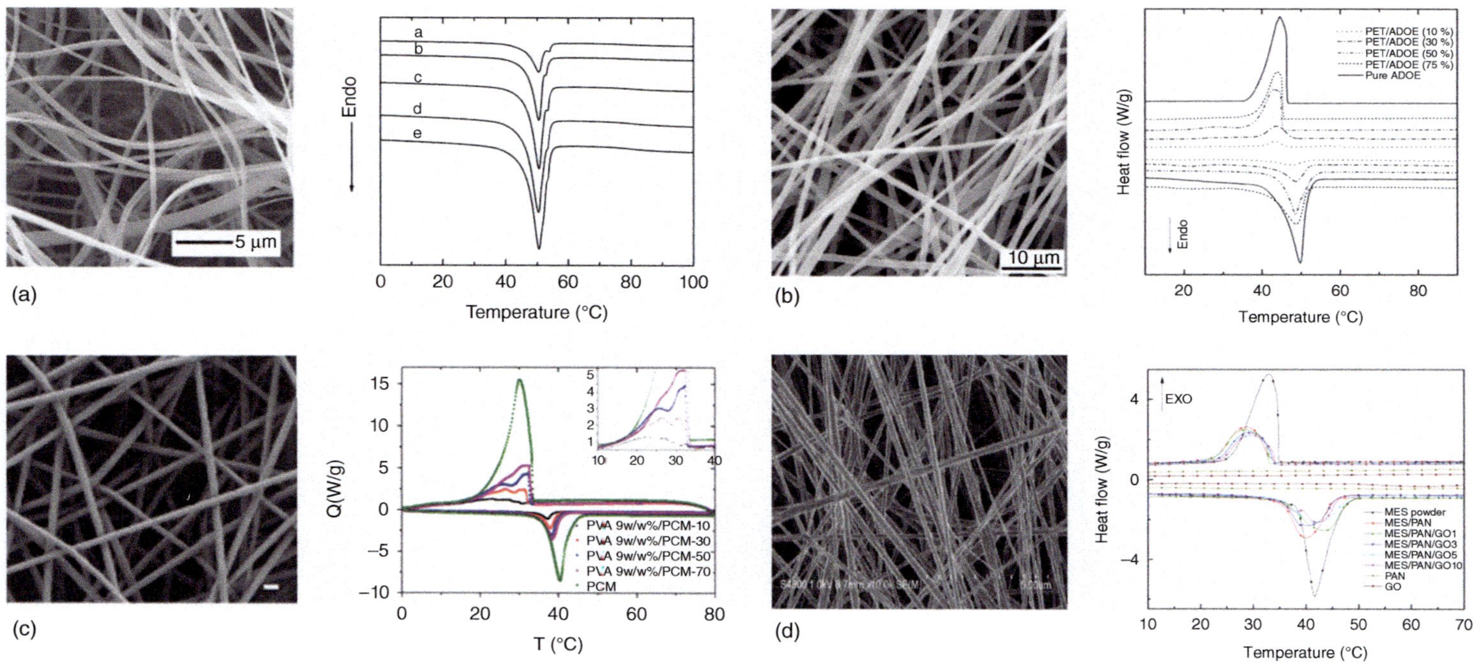

Figure 5.28 SEM image and DSC curves of PTM based on other kinds of phase-transit materials. (a) SEM image and DSC curves of SS/PET. Source: Adapted with permission from Chen et al. [49]. Copyright (2009) by Elsevier. (b) SEM image and DSC curves of DADOE/PET. Source: Adapted with permission from Chen et al. [50]. Copyright (2012) by Elsevier. (c) SEM image and DSC thermograms of PVA/PCM fibers. Source: Adapted with permission from Zdraveva et al. [51]. Copyright (2015) by American Chemical Society. (d) FE-SEM image and DSC curves of MES/PAN. Source: Adapted with permission from Ke et al. [52]. Copyright (2014) by Springer Nature Limited.

grooves on the fiber surface, and irregular cross section. After heat treatment, the integrity of the fiber is maintained even if the PTM melts. PTM/PVA fiber has reliable thermal regulation performance. Due to the polarity mismatch between PTM and PVA, the interaction between PTM and PVA is weak. Therefore, when heat is applied, PTM tends to migrate outward from the hydrophilic matrix. Under an optical microscope, large PTM particles were observed either on the fiber surface or between the fibers.

PAN is an important engineering polymer material that has been widely used to produce electrospun nanofiber membranes for many applications due to its good spinnability, attractive structural and mechanical characteristics, outstanding chemical and thermal stability properties, and low cost. Ke et al. [52] prepared a new type of PAN composite phase-transit material by electrospinning with methyl stearate (MES) wrapped in the support matrix of PAN nanofibers (Figure 5.28d). The average tensile strength of MES/PAN/GO composite nanofibers increased significantly by 573% with the increase of GO's mass fraction, and the maximum elongation at break increased by 107% when GO mass fraction was 3%. DSC results showed that electrospun PAN-based phase-transit composite nanofibers with different GO loads had appropriate phase-transit temperatures, and the latent heat range was between 92 and 109 kJ/kg.

In summary, over the last 30 years, PTFs have been extensively investigated and applied as high-performance nonwoven fabrics and coatings. As a prospective renewable and clean material, PTFs with microscale structure have been successfully prepared by melt/wet spinning for applications in TES and the temperature regulation. With the development of fiber manufacturing techniques, e.g. electrospinning, ultrafine PTFs have been exploited and investigated in the last decade. More efforts are needed to improve the various thermophysical properties of ultrafine PTFs (such as LHS density, thermal conductivity, mechanical strength, etc.) to facilitate their commercial application. Electrospinning ultrafine PTFs also provides a unique opportunity to develop innovative multifunctional fiber materials that combine TES with other useful features.

5.4 Conclusions

Energy storage materials are receiving tremendous attention and research interest due to the increasing concern on the sustainable development of energy, economy, and society, which is closely related to the high efficiency and versatile storage and consumption of energy. To fulfill the newly emerging applications, such as powering portable electronics, advanced energy storage materials with superior integrated performance that enables high energy and power densities and environmentally benign, convenient, and flexible storage of energy are highly demanded. Notable progress has been made in recent years: the capacity, rate capability, and cycling life of electrode materials for SC fibers and battery fibers have been evidently enhanced, and the heat storage and release efficiency of PTFs have been improved. These energy storage fibers with high capacity, long cycling life, good safety, and good reliability will undoubtedly boost the performance of energy storage devices constructed on them and facilitate their wide application. However, challenges remain that hinder the practical applications of 1D devices, in particular, related to their poor performance. More research should focus on exploring new electrode and active materials to enhance the performances of 1D energy harvesting and storage devices. A systematic study of the mechanisms and underlying relationships, such as the dependence of loading density and alignment degree on the curvature derived from the fiber configuration, is desired to improve the energy storage performance by optimizing the interface and structure. Much more manufacturing techniques should be applied to prepare ultrafine PTFs with high performance (LHS density, thermal conductivity, mechanical strength). Energy storage is a severe challenge in the future; meeting the challenge will require the fundamental understanding and scientific breakthroughs in new materials and chemical processes to make possible new energy technologies and performance levels far beyond what is now possible.

References

1 Liu, C., Li, F., Ma, L., and Cheng, H. (2010). Advanced materials for energy storage. *Advanced Materials* 22 (8): E28–E62.

2 Khan, K., Tareen, A.K., Aslam, M. et al. (2019). Going green with batteries and supercapacitor: two dimensional materials and their nanocomposites based energy storage applications. *Progress in Solid State Chemistry* 58: 100254.

3 Qin, S., Seyedin, S., Zhang, J. et al. (2018). Elastic fiber supercapacitors for wearable energy storage. *Macromolecular Rapid Communications* 39 (13): 1800103.

4 Yang, Z., Deng, J., Chen, X. et al. (2013). A highly stretchable, fiber-shaped supercapacitor. *Angewandte Chemie International Edition* 52 (50): 13453–13457.

5 Meng, F., Zheng, L., Luo, S. et al. (2017). A highly torsionable fiber-shaped supercapacitor. *Journal of Materials Chemistry A* 5 (9): 4397–4403.

6 Deng, J., Zhang, Y., Zhao, Y. et al. (2015). A shape-memory supercapacitor fiber. *Angewandte Chemie International Edition* 54 (51): 15419–15423.

7 Hu, Y., Cheng, H., Zhao, F. et al. (2014). All-in-one graphene fiber supercapacitor. *Nanoscale* 6 (12): 6448–6451.

8 Sun, H., You, X., Deng, J. et al. (2014). Novel graphene/carbon nanotube composite fibers for efficient wire-shaped miniature energy devices. *Advanced Materials* 26 (18): 2868–2873.

9 Le, V.T., Kim, H., Ghosh, A. et al. (2013). Coaxial fiber supercapacitor using all-carbon material electrodes. *ACS Nano* 7 (7): 5940–5947.

10 Wang, K., Meng, Q., Zhang, Y. et al. (2013). High-performance two-ply yarn supercapacitors based on carbon nanotubes and polyaniline nanowire arrays. *Advanced Materials* 25 (10): 1494–1498.

11 Shang, Y., Wang, C., He, X. et al. (2015). Self-stretchable, helical carbon nanotube yarn supercapacitors with stable performance under extreme deformation conditions. *Nano Energy* 12: 401–409.

12 Su, F. and Miao, M. (2014). Flexible, high performance two-ply yarn supercapacitors based on irradiated carbon nanotube yarn and PEDOT/PSS. *Electrochimica Acta* 127: 433–438.

13 Choi, C., Lee, J.A., Choi, A.Y. et al. (2014). Flexible supercapacitor made of carbon nanotube yarn with internal pores. *Advanced Materials* 26 (13): 2059–2065.

14 Lu, X., Bai, Y., Wang, R., and Sun, J. (2016). A high-performance flexible and weavable asymmetric fiber-shaped solid-state supercapacitor enhanced by surface modifications of carbon fibers with carbon nanotubes. *Journal of Materials Chemistry A* 4 (46): 18164–18173.

15 Yu, X., Fu, Y., Cai, X. et al. (2013). Flexible fiber-type zinc-carbon battery based on carbon fiber electrodes. *Nano Energy* 2 (6): 1242–1248.

16 Goodenough, J.B. and Park, K.S. (2013). The Li-ion rechargeable battery: a perspective. *Journal of the American Chemical Society* 135 (4): 1167–1176.

17 Kwon, Y.H., Woo, S.W., Jung, H.R. et al. (2012). Cable-type flexible lithium ion battery based on hollow multi-helix electrodes. *Advanced Materials* 24 (38): 5192–5197.

18 Ren, J., Li, L., Chen, C. et al. (2013). Twisting carbon nanotube fibers for both wire-shaped micro-supercapacitor and micro-battery. *Advanced Materials* 25 (8): 1155–1159.

19 Ren, J., Zhang, Y., Bai, W. et al. (2014). Elastic and wearable wire-shaped lithium-ion battery with high electrochemical performance. *Angewandte Chemie International Edition* 53 (30): 7864–7869.

20 Luo, Y., Zhang, Y., Zhao, Y. et al. (2015). Aligned carbon nanotube/molybdenum disulfide hybrids for effective fibrous supercapacitors and lithium ion batteries. *Journal of Materials Chemistry A* 3 (34): 17553–17557.

21 Zhang, Y., Wang, Y., Wang, L. et al. (2016). A fiber-shaped aqueous lithium ion battery with high power density. *Journal of Materials Chemistry A* 4 (23): 9002–9008.

22 Kaiser, M.R., Han, Z.J., Liang, J. et al. (2019). Lithium sulfide-based cathode for lithium-ion/sulfur battery: recent progress and challenges. *Energy Storage Materials* 19: 1–15.

23 Fang, X., Weng, W., Ren, J., and Peng, H. (2016). A cable-shaped lithium sulfur battery. *Advanced Materials* 28 (3): 491–496.

24 Chong, W.G., Huang, J., Xu, Z. et al. (2017). Lithium-sulfur battery cable made from ultralight, flexible graphene/carbon nanotube/sulfur composite fibers. *Advanced Functional Materials* 27 (4): 1604815.

25 Liu, R., Liu, Y., Chen, J. et al. (2017). Flexible wire-shaped lithium-sulfur batteries with fibrous cathodes assembled via capillary action. *Nano Energy* 33: 325–333.

26 Li, L., Chang, Z., and Zhang, X. (2017). Recent progress on the development of metal-air batteries. *Advanced Sustainable Systems* 1 (10): 1700036.

27 Yin, Y., Yang, X., Chang, Z. et al. (2018). A water-/fireproof flexible lithium-oxygen battery achieved by synergy of novel architecture and multifunctional separator. *Advanced Materials* 30 (1): 1703791.

28 Xu, Y., Zhang, Y., Guo, Z. et al. (2015). Flexible, stretchable, and rechargeable fiber-shaped zinc-air battery based on cross-stacked carbon nanotube sheets. *Angewandte Chemie International Edition* 54 (51): 15390–15394.

29 Xu, Y., Zhao, Y., Ren, J. et al. (2016). An all-solid-state fiber-shaped aluminum-air battery with flexibility, stretchability, and high electrochemical performance. *Angewandte Chemie International Edition* 55 (28): 7979–7982.

30 Zhang, Y., Jiao, Y., Lu, L. et al. (2017). An ultraflexible silicon–oxygen battery fiber with high energy density. *Angewandte Chemie International Edition* 56 (44): 13741–13746.

31 Konarov, A., Voronina, N., Jo, J.H. et al. (2018). Present and future perspective on electrode materials for rechargeable zinc-ion batteries. *ACS Energy Letters* 3: 2620–2640.

32 Li, H., Liu, Z., Liang, G. et al. (2018). Waterproof and tailorable elastic rechargeable yarn zinc ion batteries by a cross-linked polyacrylamide electrolyte. *ACS Nano* 12 (4): 3140–3148.

33 Zhu, Y., Yuan, S., Bao, D. et al. (2017). Decorating waste cloth via industrial wastewater for tube-type flexible and wearable sodium-ion batteries. *Advanced Materials* 29 (16): 1603719.

34 Song, C., Li, Y., Li, H. et al. (2019). A novel flexible fiber-shaped dual-ion battery with high energy density based on omnidirectional porous Al wire anode. *Nano Energy* 60: 285–293.

35 Zhang, Q., Zhou, Z., Pan, Z. et al. (2018). All-metal-organic framework-derived battery materials on carbon nanotube fibers for wearable energy-storage device. *Advanced Science* 5 (12): 1801462.

36 Jin, Z., Li, P., Jin, Y., and Xiao, D. (2018). Superficial-defect engineered nickel/iron oxide nanocrystals enable high-efficient flexible fiber battery. *Energy Storage Materials* 13: 160–167.

37 Liu, G., Kim, J.Y., Wang, M. et al. (2018). Soft, highly elastic, and discharge-current-controllable eutectic gallium-indium liquid metal-air battery operated at room temperature. *Advanced Energy Materials* 8 (16): 1703652.

38 Qiu, J., Huo, D., and Xia, Y. (2020). Phase-change materials for controlled release and related applications. *Advanced Materials* 32 (25): 2000660.

39 McCann, J.T., Marquez, M., and Xia, Y. (2006). Melt coaxial electrospinning: a versatile method for the encapsulation of solid materials and fabrication of phase change nanofibers. *Nano Letters* 6 (12): 2868–2872.

40 Sun, S., Xie, R., Wang, X. et al. (2015). Fabrication of nanofibers with phase-change core and hydrophobic shell, via coaxial electrospinning using nontoxic solvent. *Journal of Materials Science* 50 (17): 5729–5738.

41 Cai, Y., Zong, X., Zhang, J. et al. (2013). Electrospun nanofibrous mats absorbed with fatty acid eutectics as an innovative type of form-stable phase change materials for storage and retrieval of thermal energy. *Solar Energy Materials and Solar Cells* 109: 160–168.

42 Ke, H., Pang, Z., Peng, B. et al. (2016). Thermal energy storage and retrieval properties of form-stable phase change nanofibrous mats based on ternary fatty acid eutectics/polyacrylonitrile composite by magnetron sputtering of silver. *Journal of Thermal Analysis and Calorimetry* 123 (2): 1293–1307.

43 Ke, H. (2017). Morphology and thermal performance of quaternary fatty acid eutectics/polyurethane/Ag form-stable phase change composite fibrous membranes. *Journal of Thermal Analysis and Calorimetry* 129 (3): 1533–1545.

44 Chen, C., Wang, L., and Huang, Y. (2009). Crosslinking of the electrospun polyethylene glycol/cellulose acetate composite fibers as shape-stabilized phase change materials. *Materials Letters* 63 (5): 569–571.

45 Chen, C., Wang, L., and Huang, Y. (2011). Electrospun phase change fibers based on polyethylene glycol/cellulose acetate blends. *Applied Energy* 88 (9): 3133–3139.

46 Chen, C., Liu, K., Wang, H. et al. (2013). Morphology and performances of electrospun polyethylene glycol/poly (dl-lactide) phase change ultrafine fibers for thermal energy storage. *Solar Energy Materials and Solar Cells* 117: 372–381.

47 Dang, T.T., Nguyen, T.T.T., Chung, O.H., and Park, J.S. (2015). Fabrication of form-stable poly (ethylene glycol)-loaded poly (vinylidene fluoride) nanofibers via single and coaxial electrospinning. *Macromolecular Research* 23 (9): 819–829.

48 Babapoor, A., Karimi, G., Golestaneh, S.I., and Mezjin, M.A. (2017). Coaxial electro-spun PEG/PA6 composite fibers: fabrication and characterization. *Applied Thermal Engineering* 118: 398–407.

49 Chen, C., Wang, L., and Huang, Y. (2009). Ultrafine electrospun fibers based on stearyl stearate/polyethylene terephthalate composite as form stable phase change materials. *Chemical Engineering Journal* 150 (1): 269–274.

50 Chen, C., Liu, S., Liu, W. et al. (2012). Synthesis of novel solid-liquid phase change materials and electrospinning of ultrafine phase change fibers. *Solar Energy Materials and Solar Cells* 96: 202–209.

51 Zdraveva, E., Fang, J., Mijovic, B., and Lin, T. (2015). Electrospun poly (vinyl alcohol)/phase change material fibers: morphology, heat properties, and stability. *Industrial & Engineering Chemistry Research* 54 (35): 8706–8712.

52 Ke, H., Pang, Z., Xu, Y. et al. (2014). Graphene oxide improved thermal and mechanical properties of electrospun methyl stearate/polyacrylonitrile form-stable phase change composite nanofibers. *Journal of Thermal Analysis and Calorimetry* 117 (1): 109–122.

6

Smart Energy Textiles

With wearable technology development, power sources are gradually developing in miniaturization, portability, and efficiency. The emerging fibrous energy devices with one-dimensional structures are thinner in diameter, lighter in weight, and higher flexibility in three-dimension. Textiles are the final form for human wearing. By weaving fibrous energy devices into textiles, the overall power output can be greatly increased. With the characteristics of general textiles and energy devices, smart energy textiles will better meet large-scale wearable applications' needs.

There are usually four weaving methods to fabricate a smart energy textile.

The first method is the stacking of two or more textile-type electrodes, like a sandwich. The fabric-type electrode was manufactured by directly coating the conductor and semiconductor materials on a textile substrate, like the method to fabricate a conventional plate-type device. The coating process can be accomplished by dip coating, spray coating, and in situ chemical growth.

The second method is layer-by-layer assembling on textile electrodes. During which, various conductor and semiconductor materials were successively coated on a textile substrate layer-by-layer. Unlike the first method, a layer of the conductor was finally coated on the semiconductor layer as the counter electrode, rather than stacking another textile-type electrode.

The third method is the interweaving of fiber electrodes. During which, different types of fiber electrodes were fabricated. Two sets of fiber electrodes were fixed on a weaving machine as the weft and warp and were interwoven together by a so-called shuttle-flying process to form the smart energy textiles.

The fourth method is the weaving of as-fabricated fiber devices. During which, fiber devices were assembled from fiber electrodes, as discussed in the previous chapter. As a whole, the device can be fed into the weaving machine as the weft or warp and be woven together with cotton or other polymer wires to obtain the smart energy textiles.

According to the type of energy it harvested, smart energy textiles can be divided into different types: energy harvesting textiles, energy storage textiles, and hybrid energy textiles. This chapter mainly describes the state-of-the-art of these smart energy textiles.

Textile-Based Energy Harvesting and Storage Devices for Wearable Electronics, First Edition.
Xing Fan, Nannan Zhang, and Yi Wang.

6.1 Energy Harvesting Textiles

All electronic devices in our life need electricity. Nevertheless, power cable or charging, would greatly limit wearable or distributed microelectronic devices, such as the ubiquitous sensors and portable communication equipment. We need special energy harvesting technology as a suitable power source for these devices.

There is much energy around the human body, such as solar energy, thermal energy, and mechanical energy. Although these energy sources are scattered and weak, they can be continuously collected to facilitate electricity in special environments. In recent years, most works on energy harvesting technology for weak energy around the human body are focused on increasing the energy density of outputs and improving the wearing comfort. However, the problem of "wear" remains a challenge for many energy harvesting technologies, far more than flexibility and breathability. If energy harvesting devices can be worn on the human body like a common textile, its wearing comfort can be greatly improved.

This section mainly discussed these three energy harvesting textiles: photovoltaic energy harvesting textiles, thermoelectric energy harvesting textiles, and mechanical energy harvesting textiles.

6.1.1 Photovoltaic Energy Harvesting Textiles

A solar cell is one of the most common devices to achieve photoelectric conversion for electrical energy. It is usually formed by joining P-type and N-type semiconductors to form a P-N junction on the interface. When irradiated by sunlight, the electrons at the interfacial P-N junction are excited to produce pairs of electrons and holes. These electrons and holes are affected by the electric field resulting from polarity difference between the materials, which then separate and inject into different layers of N-type or P-type semiconductors. Finally, they were collected by two electrodes, respectively. It will form a circuit loop for the electric current to run through by connecting the positive electrode and the negative electrode.

As discussed in previous chapters, a conventional solar cell with a rigid panel structure or softer film structure have disadvantages of poor flexibility and low deformability. Thus, emerging fiber-type solar cells have gained wide attention, owing to the advantages of easy adapting to complex deformations caused by human movement. However, it is not that easy to merge fiber-type solar cells into textiles because it remains a great challenge to realize comfortable wearing while maintaining the sustainability and stability of a highly efficient power supply.

Unlike conventional solar cells, which are based on crystalline silicon, cadmium telluride, copper selenide, or other materials, they are not suitable for textile-type solar cells because of their inherent rigidity. Thus, previously-reported textile-type solar cells included polymer solar cells, dye-sensitized solar cells, and perovskite solar cells, fabricated on flexible photo-electrodes, low-cost metal wire, optical fibers, and carbon fibers.

We introduced polymer solar cell textiles. Its fabrication process always includes three steps:

1. Fiber-type or textile-type electrodes are fabricated by depositing indium tin oxide (ITO) or metal materials on fiber or textile substrate;
2. A blend film of conjugated polymer donor and fullerene derivative acceptor is coated on fiber-type or textile-type electrode, as photo-electrode;
3. Another kind of conductive material is employed as the counter electrode.

As discussed above, there are four methods for pairing photo-electrodes and counter electrodes to fabricate a smart energy textile. The working principle of the polymer solar cell textiles is similar to that of the flat-type polymer solar cells. It has the advantages of simple structure, easy preparation, low cost, and lightweight, which has become a research hotspot in this area.

For example, Peng and coworkers [1] reported a polymer solar cell textile, which makes the circuit connection easier by designing the interlaced structure of cathode and anode fibers. As shown in the illustration of a polymer solar cell textile, it can be designed as a module connected in series or parallel (Figure 6.1a). Each module contains a staggered structure of warp and weft in organic photovoltaic textiles. The warp and weft yarns described above serve as the anode and modified cathode of the polymer solar cell fabric, respectively. Therefore, they can be effectively connected to power external electronic products. The polymer solar cell textiles were thin, lightweight, and flexible with high working stability and excellent photovoltaic performance (Figure 6.1b). They were close to materials used for daily clothes, affording them potential in various emerging new application fields such as

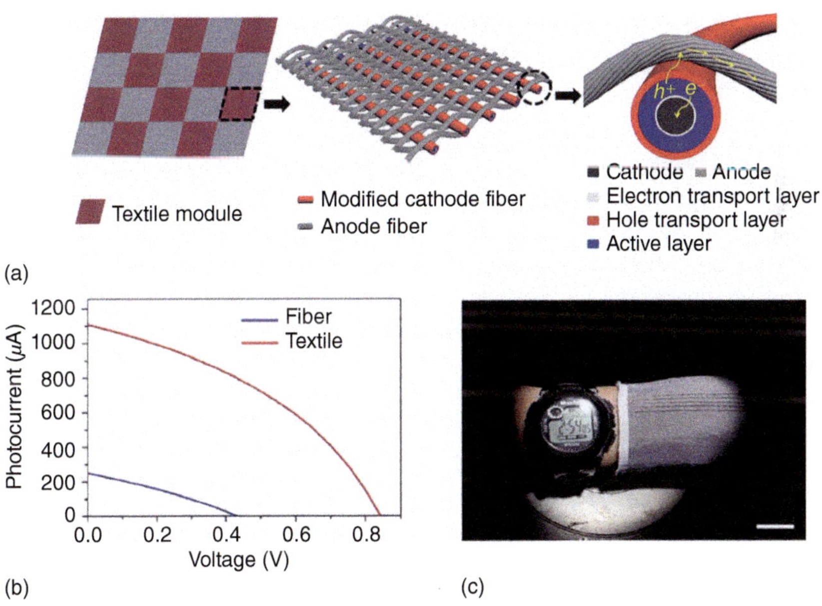

Figure 6.1 Polymer solar cell textile: (a) Schematic illustration of the polymer solar cell textile. (b) Photovoltaic performance of the textile device and one modified cathode fiber in the photovoltaic textile. (c) Demonstration of the photovoltaic textiles used to power an electronic watch under solar irradiation. Scale bar, 2 cm. Source: (a, b) Liu et al. [1]. © 2018, Royal Society of Chemistry, (c) Adapted with permission from Liu et al. [1]. Copyright (2018) by WILEY-VCH Verlag GmbH & Co. KGaA, Weinheim.

wearable electronics, biomedical electronics, and artificial intelligence (Figure 6.1c). This novel structure has successfully integrated the device fabrication, textile weaving, and circuit connection all in one process, which significantly enhances the performance of the organic photovoltaic textiles.

We introduced the dye-sensitized solar cell textiles. Dye-sensitized solar cells use low-cost nanopores metal oxides and dyes as the photosensitized functional materials. There are four steps to prepare dye-sensitized solar cell textiles:

1. A layer of conductive material is loaded on fiber or textile substrate;
2. A nanoporous semiconductor film made of metal oxide is gathered on the conductive fiber or textile;
3. The photosensitive dye is adsorbed on the nanoporous semiconductor film;
4. After coating an electrolyte layer, another kind of conductive material is employed as the counter electrode.

As discussed above, four methods for the pairing of photo-electrode and counter electrode are used to fabricate a smart energy textile. The working principle of the dye-sensitized solar cell textiles is the same as that of the flat-type dye-sensitized solar cells. It has some advantages that the preparation process is relatively green, and large-area manufacturing is easy. However, most dye-sensitized solar cells (DSSCs) fiber devices typically used liquid electrolytes, which may leak and lead to a decrease in efficiency and stability. Therefore, to solve this problem, much work in recent years has been researched.

For example, Fan and coworkers [2] presented a solution for the first time to power portable electronics in a wearable manner by fabricating an all-solid dye-sensitized photovoltaic textile. Unlike traditional mosaic-like solar modules composed of small solar cell units, strings of the wire-shaped photo-anodes and counter electrodes (CE) were woven in an interlaced manner to form a single-layered textile (Figure 6.2a). Under the standard light intensity of 100 mW/cm^2, the fabric cell's open-circuit voltage is 4.6 V. The short-circuit current density is 7.8 mA/cm^2, and the efficiency of the corresponding single cell is 1.3%. Peng and coworkers [3] reported the discovery of a new type of dye-sensitized solar cell with a power conversion efficiency (PCE) of 10.00% (Figure 6.2b). The design of new fiber electrodes creates a hydrophobic aligned carbon nanotube (CNT) core that provides high electrical conductivity and mechanical strength and a hydrophilic aligned CNT sheath that can effectively incorporate other active phases. The fiber-shaped solar cells show high flexibility and only lost 18% of the initial PCE after bending at 90° for 2000 cycles, and five 3 cm fibrous DSSCs are attached to a pedometer and woven into a T-shirt. Peng and coworkers [4] improved their work using flexible self-powered textiles on a large-scale industrial loom by bridging photoactive and electrochemically active fibers via warp yarns (Figure 6.2c). This novel weaving method significantly enhanced the integration efficiency and enabled the production of self-powered textiles without any external electrical connection. The textiles showed high powering performances of 14% PCE.

Thirdly, we introduced the perovskite solar cell textiles. Perovskite solar cells are solar cells that use perovskite-type organic metal halide semiconductors as

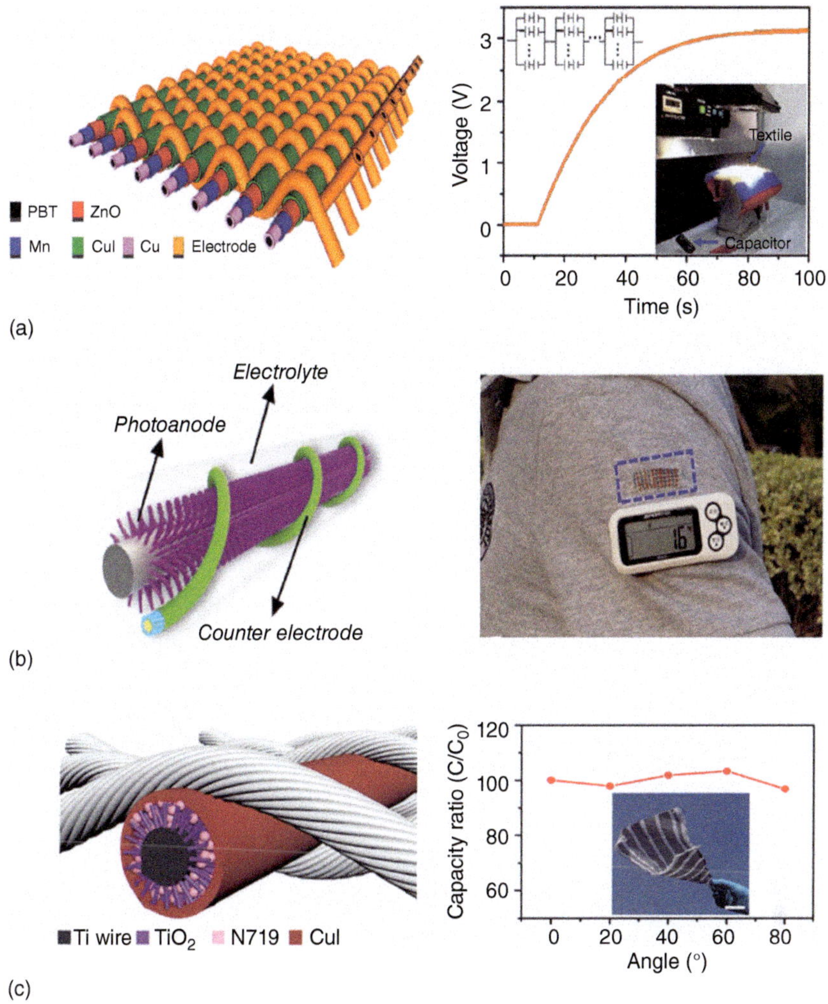

Figure 6.2 Dye-sensitized solar cell textiles: (a) Schematic illustrations of the photovoltaic textile and demonstration of the photovoltaic textile to charge a 2 mF capacitor under natural solar irradiation. Source: Adapted with permission from Zhang et al. [2]. Copyright (2016) by WILEY-VCH Verlag GmbH & Co. KGaA, Weinheim. (b) Schematic illustration of a DSSC with Pt-composite fiber as the counter electrode and step number displayed on the pedometer after 16 times of arm swing under outdoor sunlight. Source: Adapted with permission from Fu et al. [3]. Copyright (2018) by Royal Society of Chemistry. (c) Schematic illustration of the structure of the PC part and dependence of capacity on the bending angle for the ES part; the inset indicates the high flexibility of the self-powered textile. Scale bar, 5 cm. Source: Adapted with permission from Gao et al. [4]. Copyright (2019) by Royal Society of Chemistry.

light-absorbing materials. The steps for assembling perovskite solar cell textiles are as follows: First, a compact layer, a mesoporous TiO_2 layer, and a perovskite $CH_3NH_3PbI_3$ sensitizer are sequentially coated onto a fiber or textile wire electrode. Second, the hole-transport material is subsequently coated as the photo-electrode. Third, another kind of conductive material is employed as the counter electrode. As discussed above, there are four methods for pairing photo-electrode and counter electrode to fabricate a smart energy textile. The working principle of the perovskite solar cell textiles is similar to that of the flat-type perovskite solar cells. The research on perovskite solar cells has continuously updated photoelectric conversion efficiency [5, 6]. Although the perovskite solar cells exhibit high PCE, it is difficult to control the morphology of these functional layers on a non-planar substrate, so their performance in textiles is not ideal.

To solve this problem of perovskite solar cells textiles, Li et al. [7] suggested a CNT fiber-supported double-twisted perovskite solar cell featuring high stability perovskite nano-crystals on the CNT fiber sustain the interfacial adhesion under stress (Figure 6.3a). The P_3HT/SWNT was used as a hole transport layer, and the Ag nanowire was added to increase the surface area between electrodes. The fabricated perovskite solar cell fiber can be seen in Figure 6.3a, then weaved into textiles. The optimized maximum PCE was 3.03%. Peng and coworkers [8] have synthesized a new type of fibrous perovskite solar cell through a simple dip-coating process (Figure 6.3b). Its energy conversion efficiency is 3.3%, which is the highest reported coaxial all-solid fibrous solar cells. The prepared fibrous solar cells are flexible and can be woven into textiles. The abovementioned textile-based

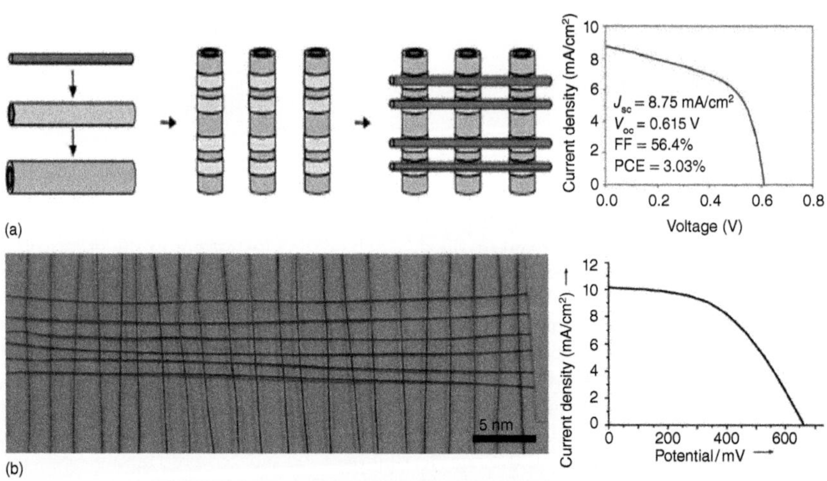

Figure 6.3 Perovskite solar cell textile: (a) Process flow for weave-patterned transistors on fiber(left). The current J-V curve is shown for electrical performance (right). Source: Li et al. [7]. © 2015, John Wiley & Sons. (b) Photograph of a perovskite solar cell textile (left). Current density–voltage curve of a coaxial fiber-shaped perovskite solar cell with an energy conversion efficiency of 3.3% (V_{OC} = 0.664 V, I_{SC} = 10.2 mA/cm^2, FF = 0.487) (right). Source: Adapted with permission from Qui et al. [8]. Copyright (2014) by WILEY-VCH Verlag GmbH & Co. KGaA, Weinheim.

photovoltaic harvesters have limitations such as mass production, high cost, and low energy conversion efficiency that remained obstacles to real fabric utilization. Textiles-based solar cells look promising for future wearable solar cells.

6.1.2 Thermoelectric Energy Harvesting Textiles

Thermoelectric energy harvesting technology refers to the technology that converts thermal energy into electricity through the thermoelectric conversion mechanisms. There are two thermoelectric conversion mechanisms. One is converting temperature gradient (dT/dx) into electrical energy through the Seebeck effect. Another is converting temperature fluctuations (dT/dt) into electrical energy through thermal diffusion [9, 10]. Because temperature fluctuations are not easily captured, most thermoelectric devices use the former conversion mechanism. The Seebeck effect refers to the two ends of two conductors (or semiconductors) contact to form a loop. An electric potential will be generated in the loop when the temperatures of the two contact points are different.

There is a natural temperature difference between the human body and the environment, which can be used as an energy source for thermoelectric energy harvesting devices. So thermoelectric harvesting devices are very suitable to be combined with textiles to power wearable devices. There are two methods to fabricate thermoelectric textiles, either depositing small pieces of brittle thermoelectric materials on flexible substrates or coating flexible thermoelectric materials on textile substrates. This chapter mainly introduced thermoelectric energy harvesting textiles using the Seebeck effect as the thermoelectric conversion mechanism, for which both fabricating methods have been reported.

For example, Kim et al. [11] described a wearable thermoelectric generator (TEG) that uses a combination of fabric and printed thermocouples that can be easily embedded in clothing to capture low-temperature energy (Figure 6.4a). When the thermoelectric generator is integrated into the shirt, the skin temperature is maintained at 32 °C at low (5 °C) or high (25 °C). Interestingly, the output power of ~147 nw in a cold environment is significantly higher than that of ~8.1 nW in a hot environment. The device did not deform after it was bent and stretched several times. The prospect of using the TEGs in clothing applications was confirmed under realistic conditions.

The TEG textile application was explored by integrating with advanced wearable organic electronics, such as flexible textile-based strain sensors. Jiang and coworkers [12] proposed an efficient strategy for the fabrication of a highly conductive commercial textile coated with poly(3,4-ethylenedioxythiophene) (PEDOT) via vapor phase polymerization (VPP) as a wearable thermoelectricity (TE) devices to a drive strain sensor (Figure 6.4b). An output voltage of 5.0 mV of the TE device was achieved at a temperature gradient of 25 K. An optimized gauge factor of the strain sensor reached 54 at a strain of 1.5%, which can fully satisfy the demands of wearable electronic sensor devices. Baughman and coworkers [13] described the fabrication of flexible, woven, and knitted textiles based on electrospun polymer nanofiber cores that are coated with N- and P-type semiconductor sheaths (Bi_2Te_3 and Sb_2Te_3, respectively)

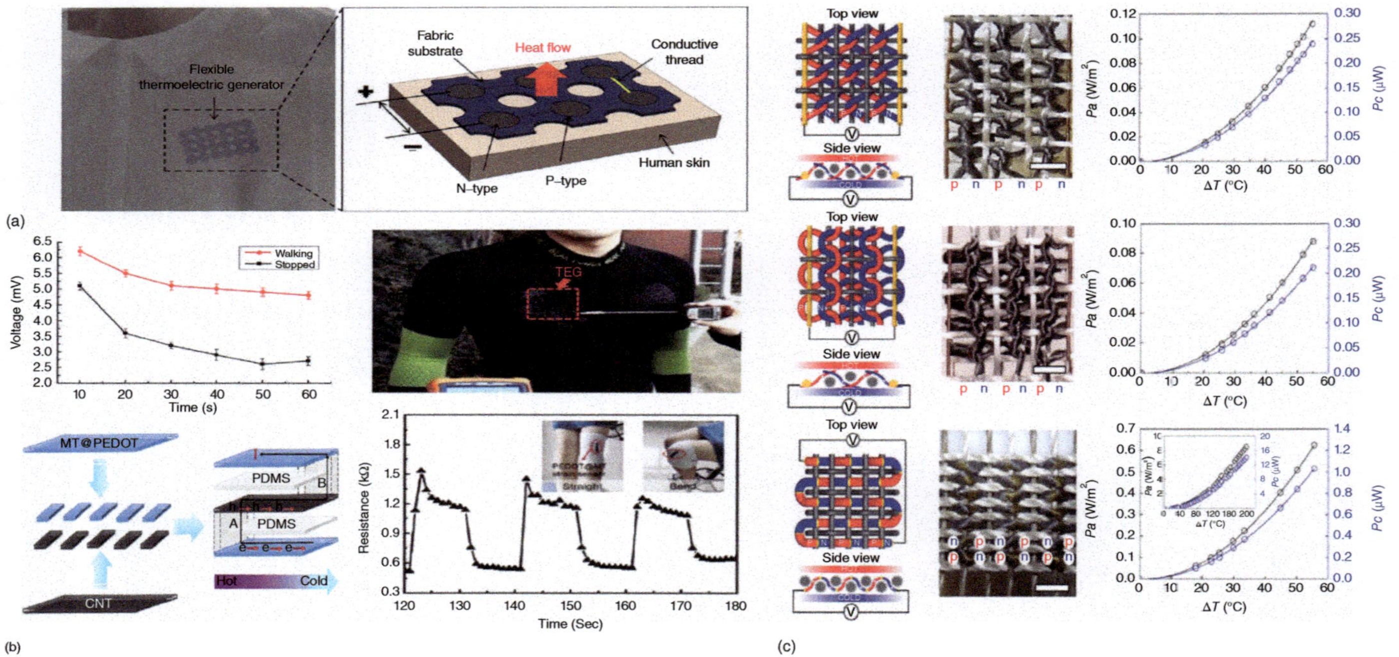

Figure 6.4 Thermoelectric generator textiles: (a) A shirt with the wearable TEGs, voltage output on the human body when moving at an ambient temperature of 25 °C, and photographs of the shirt with the TEG embedded for clothing applications. Source: Adapted with permission from Kim et al. [11]. Copyright (2014) by Iop science. (b) Schematic illustration for assembling an MT@PEDOT-based TE generator device and time dependence of resistance of the MT@PEDOT strain sensor with enlarged test data for 60 seconds (bent to straight about 90°). Source: Adapted with permission from Jia et al. [12]. Copyright (2019) by Royal Society of Chemistry. (c) The fabrication of flexible, woven, and knitted textiles. Source: Adapted with permission from Lee et al. [13]. Copyright (2016) by WILEY-VCH Verlag GmbH & Co. KGaA, Weinheim.

and twisted into flexible yarns (Figure 6.4c). They have used three different textile designs to demonstrate the more practically useful thermal-energy harvesting in the textile thickness direction. The output power of up to 8.56 W/m^2 was obtained for a temperature difference of 200 °C in the textile thickness direction.

6.1.3 Mechanical Energy Harvesting Textiles

The mechanical energy harvesting technologies applied to textile devices include piezoelectric nanogenerator (PENG) and triboelectric nanogenerator (TENG). The electricity generation principle of the PENGs is based on the piezoelectric effect, for which the dipole of the special nano-materials (such as zinc oxide) will change during bending and compression, thus lead to a variety of internal electric field to convert the mechanical energy into electrical energy. The electricity generation principle of the TENGs is based on the triboelectric effect. Two different materials with different electron-binding capabilities will gain and lose electrons when they contact and separate from each other, leading to various interfacial potential differences to generate current in the external circuit.

Human movement is an inexhaustible source of mechanical energy used for charging portable electronic devices and biomedical applications [16–20]. Traditional flat structure mechanical energy harvesting devices, with large volume and rigidity disadvantages, are not suitable for wearing. It should be noted that the human body is accustomed to wearing textiles. Therefore, making the mechanical energy harvesting device into a textile should have extensive application prospects, especially for on-body applications [21–26]. This section mainly introduced the development of TENG textiles and PENG textiles to apply wearable products.

For the piezoelectric textiles, there are two preparing methods. The first method is to weave fiber-type devices into textile-type devices. The fiber substrate is coated with special semiconductor materials (such as ZnO nanowires), which have piezoelectric properties. A fiber substrate is coated with conductive material to form a counter electrode, and the two fiber electrodes were integrated as the weft and warp into a conventional textile. The second method is to directly coat a special semiconductor material with piezoelectric properties on a textile substrate and then stack another conductive textile, coated with a conductive material as the counter electrode.

For example, Magniez et al. [14] reported a method of developing a piezoelectric power sensor by using a melt-spun piezoelectric polyvinylidene fluoride (PVDF) fiber of approximately 80% β-phase (Figure 6.5a). After melt crystallization, the fibers are stretched at 120 °C to between 25% and 75% of their original length, resulting in molecular orientation, significant changes in polymorphism and tensile properties, and molecular alignment during stretching. To further assemble into a piezoelectric textile structure, PVDF fibers and conductive fibers were integrated into a conventional polyester braid and 2/2 twill weave. Fabrics tested at a frequency of 1 Hz under an impact load of 70 N provided an output of 6 V with a sensitivity value of 55 mV/N. Zhang et al. [15] reported a flexible two-dimensional fabric nanogenerator based on hybrid piezoelectric fibers. It is promising for easily integrating with clothing and converting the mechanical energy of human body motion

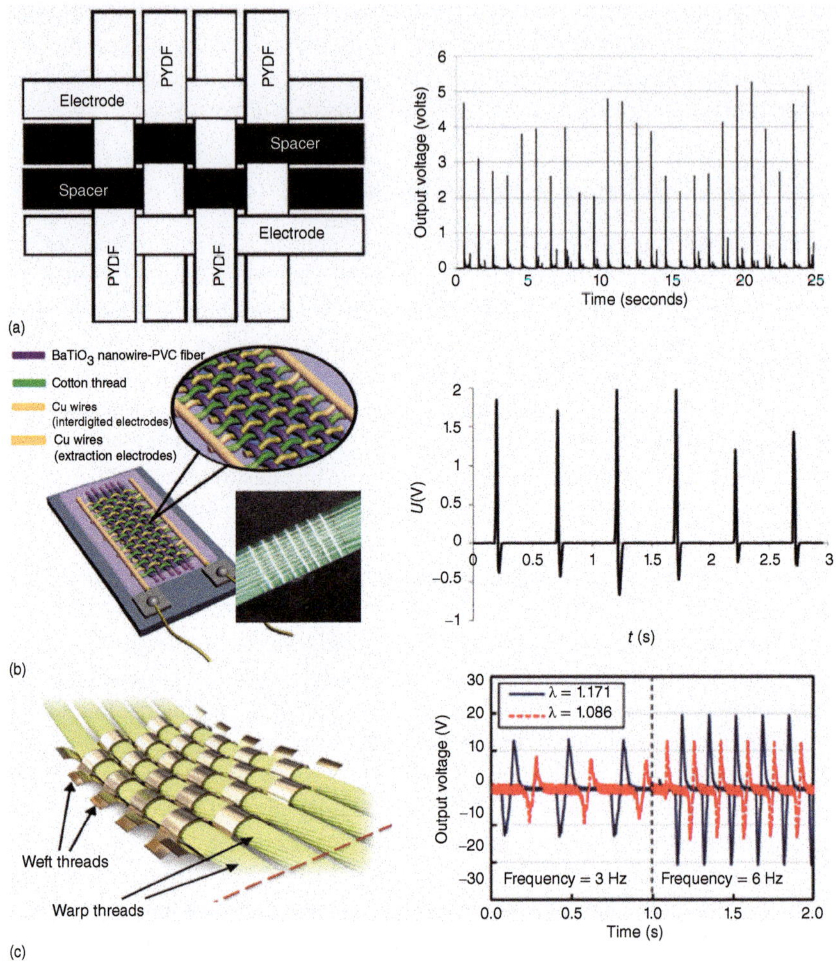

Figure 6.5 Nanogenerator textiles based on piezoelectric effect: (a) Configuration of the 2D flexible piezoelectric woven sensor and the output voltage measured on the 2D sensors upon a 1 Hz frequency, 70 Newton spike-shaped stimulus (bottom curve). Source: Magniez et al. [14]. © 2013, John Wiley & Sons. (b) The structure of the fabric nanogenerator and the voltage output of the FNG. Source: Adapted with permission from Zhang et al. [15]. Copyright (2015) by Elsevier. (c) A scalable woven PENG composed of warp and weft threads made of yarn strings and piezoelectric film straps, and an output profile of the open-circuit output voltage from the device composed of seven warp and five weft threads at two different stretch ratios. Source: Song et al. [35]. © 2015, Royal Society of Chemistry.

into electric energy (Figure 6.5b). The hybrid piezoelectric fiber is comprised of aligned $BaTiO_3$ nanowires and polyvinyl chloride (PVC) polymer. The metal copper wires and cotton threads were woven into the textile to construct the nanogenerator with interdigitated electrodes. By attaching the textile nanogenerator on an elbow pad bent by a human arm, the nanogenerator can generate 1.9 V output voltage and 24 nA output current. The output is large enough to power the liquid crystal display (LCD). Comparing with the above complicated fiber-shaped piezoelectric components, it seems more convenient and faster to prepare 2D fabric-based PENGs directly from textiles or film straps. For example, Song et al. [35] designed a scalable woven fabric-based PENG by interlacing piezoelectric film straps with yarn strings in an orthogonal direction (Figure 6.5c). The piezoelectric straps were made of PVDF films with metal electrodes on both surfaces. The fabric-based PENG with piezoelectric film straps can further be optimized into tactile sensor arrays.

For the triboelectric textiles, there are two preparing methods. The first method is to weave fiber-type electrodes into textile-type devices directly. The materials with different electrophilic properties are coated on the conductive fiber substrate to prepare the positive electrode and the negative electrode. These two electrodes are woven as warp and weft together with yarns. The second method is to directly coat functional materials with different electrophilic properties on different textile substrates, stacked together to make a multiple-layered textile.

For instance, Fan and coworkers [27] interlaced copper-coated polytetrafluoroethylene (PTFE) tape and copper wires on an industrial loom through a shuttle flight process to prepare a power textile with a micro-cable structure (Figure 6.6a). The power textile had an ultrathin single-layered interlaced structure with a thickness of 320 μm. Under ambient sunlight with mechanical excitation, the as-woven textile could generate sufficient power for various practical applications. They include charging a 2 mF commercial capacitor up to 2 V in 1 minute, continuously driving an electronic watch, directly charging a cell phone, and driving the water-splitting reactions. Due to thin films and metal foils' low flexibility, common cloth strips are gradually used to prepare fabric band-based TENGs. Wang and coworkers [28] fabricated a woven structured triboelectric nanogenerator (W-TENG) by interlacing Polyethylene terephthalate (PET) fabric wrapped in Ag-coated cloth strips and nylon fabric wrapped Ag-coated cloth strips into a plain pattern (Figure 6.6b). The W-TENG is integrated into shoes, coats, and trousers to harvest different kinds of mechanical energy from human motion to demonstrate the potential applications of the W-TENG. This work presented a new approach in applying triboelectric nanogenerators to wearable devices. Conductive or dielectric fibers can first be woven into corresponding fabrics, and then electrical outputs are realized through the relative contact and separation motion between fabrics. Kim and coworkers [29] produced a fully stretchable knitted TENG structured in a double-curved structure by sandwiching a layer of silver-coated conductive fabric between two silver-coated PTFE fiber fabrics (Figure 6.6c). The looping structure and microporous structure endow the knitted TENG with good stretchability and large contact area. They investigated plain-, double-, and rib-fabric structures and analyzed their potentials for textile-based energy harvesting. The results of this work show the feasibility of a cloth-integrated and industrial-ready TENG for the harvesting of energy from human biomechanical movements in cloth and garments.

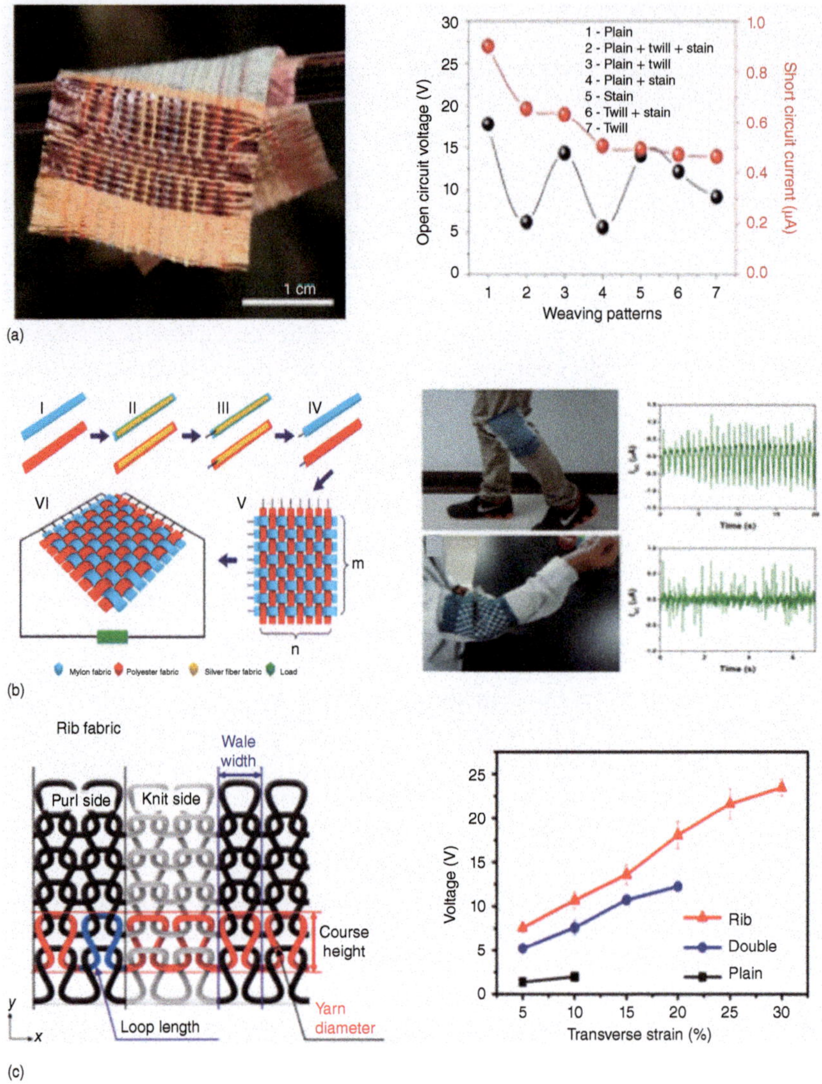

Figure 6.6 Triboelectric nanogenerators textiles: (a) A photograph of the textile mixed with colored wool wires and weaving-pattern-dependent electrical output. Source: Adapted with permission from Chen et al. [27]. Copyright (2016) by Springer Nature Limited. (b) Schematic diagram illustrating the fabricating process of a W-TENG and the W-TENG harvesting energy from leg joints bend and arm joints. Source: Adapted with permission from Zhou et al. [28]. Copyright (2014) by American Chemical Society. (c) Nomenclature of the knitted fabrics and experiment results of the electrode-potential difference (40 MΩ load) generated by the plain, double, and rib S-TENGs under various transversal strains from 5% to 30%. Source: Kwak et al. [29]. © 2017, American Chemical Society.

6.2 Energy Storage Textiles

Textiles with only energy harvesting function cannot meet the continuous working requirements of the device for daily applications, so efficient and flexible energy storage textiles are essential for constructing wearable uninterrupted power supply.

Traditional energy storage devices possess bulky planar structures or cylinder structures, and they cannot fit with the curved human body surface very well. To meet people's long-term wearing habits, fiber-type, and textile-type energy storage devices, such as supercapacitor and battery, have broadened attention.

Supercapacitor textiles and battery textiles have become widely used energy storage textiles due to their high output power. This section described the research progress, the challenges, and future development directions of supercapacitor textiles, primary batteries textiles, and secondary battery textiles, to put forward some suggestions from practical applications.

6.2.1 Supercapacitor Textiles

Supercapacitors refer to a new type of energy storage device between traditional capacitors and rechargeable batteries that store energy through the double layer of the interface formed between the electrode and the electrolyte. Comparing with batteries, they have a higher power density for fast charging and discharging. According to the energy storage mechanism, supercapacitors can be divided into electrostatic double-layer capacitors and pseudo-capacitors. Electrostatic double-layer capacitors are usually made of carbon electrodes, and charge separation is achieved at the interface between the electrodes and the electrolyte. Pseudo capacitors are usually made of metal oxide or conductive polymer electrodes and use redox reactions to store energy.

At present, both types of capacitors have been made into textiles, which are expected to be the development direction of the supercapacitors. There are two methods to prepare supercapacitor textiles. The first is to weave fiber-type devices into textile-type devices. One fiber substrate is usually coated with metallic oxide, and another fiber substrate is usually coated with active carbon materials. The two fiber electrodes are twisted or placed side by side to form a fiber-shaped supercapacitor or directly integrated as the weft and warp into a conventional textile.

Furthermore, a layer of electrolyte is brushed on the textile or fiber-type devices and then are encapsulated. The second method is to directly coat metallic oxide or active carbon materials on two different textile substrates as the positive electrode and the negative electrode. The two textile-type electrodes are stacked together to form the multi-layered textile-type supercapacitors.

At the early stage, supercapacitor mainly employed liquid-type electrolytes for higher performance. However, they have serious issues of encapsulation, so solid-type electrolyte has been pursued. This chapter mainly introduced the research progress of supercapacitor textiles from the aspects of liquid electrolyte and solid electrolyte.

There are supercapacitor textile based on liquid-type electrolytes, such as sulfuric acid, potassium hydroxide aqueous solutions, ionic liquid (BMIPF6, EMIPF6). In comparison, such a piece of supercapacitors' textile is fabricated by stitching supercapacitor fibers/yarns encapsulated in polymer tubes into a piece of the existing fabric. Jost et al. [30] developed a natural fiber welding (NFW) coating technique that can produce hundreds of feet of coated yarns at a time, realizing consistent mass loading of active materials (Figure 6.7a). The resultant fabric-based supercapacitor can be stretched like normal textile. However, the main drawback of this technique is that the electrolyte needs to be coated onto the fabric after knitting, which requires further encapsulation that ultimately leads to less flexibility. To address this concern, Cheng and coworkers [31] directly sewed three yarn-shaped supercapacitors in series into a textile (Figure 6.7b). These yarn-shaped supercapacitors were composed of Redox graphene (rGO) fiber electrodes fabricated by a scalable nonliquid-crystal spinning method. Owing to the willow-bark-like fine micro-structure of rGO that facilitated the access to and fast diffusion of electrolyte ions, the as-assembled solid-state yarn-shaped supercapacitor showed a specific

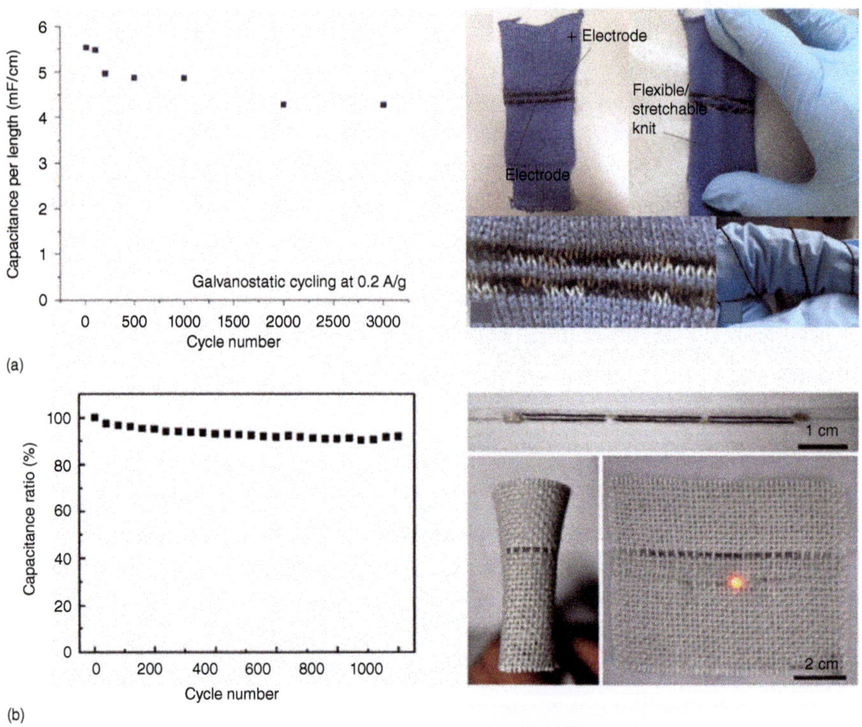

Figure 6.7 Supercapacitor fabrics using liquid electrolyte: (a) Long term cycling at 0.2 A/g over 3000 galvanostatic cycles and photographs of the knitted sample. Source: Adapted with permission from Jost et al. [30]. Copyright (2015) by WILEY-VCH Verlag GmbH & Co. KGaA, Weinheim. (b) The cycling performance at a current density of 1 A/g and the supercapacitors light LED lamps. Source: Adapted with permission from Chen et al. [31]. Copyright (2015) by Elsevier.

capacitance of 226 F/cm^3 and an energy density of 7.03 mWh/cm^3 at 0.244 A/cm^3. The sewed textile supercapacitor could be immediately charged to 3 V to power a red LED for 5 minutes, with a device capacitance of 1.12 mF at a current of 7.45 μA. This non-liquid-crystal spinning strategy could be extended to installing other two-dimensional nano-materials into macroscopic fibers for applications in micro-devices, wearable electronics, and smart textile.

For supercapacitor textile based on solid-type electrolyte, the electrolyte is composed of gel electrolytes, such as potassium hydroxide solution of potassium polyacrylate and sulfuric acid solution of ammonium polyacrylate. A similar device structure as a supercapacitor textile based on liquid-type electrolyte shown better stability and easier processability for encapsulation. Chen and coworkers [32] demonstrated all-carbon solid-state yarn supercapacitors using commercially available activated carbon and carbon fiber yarns for smart textiles (Figure 6.8a). Two hybrid carbon yarn electrodes were twisted together in polyvinyl alcohol/H_3PO_4 polymer gel, which is used as both an electrolyte and a separator. A 10 cm long yarn supercapacitor, with the optimum composition of 2.2 mg/cm activated carbon and 1 mg/cm carbon fiber, shows a specific length capacitance of 45.2 mF/cm at 2 mV/s, an energy density of 6.5 μWh/cm, and a power density of 27.5 μW/cm. Zheng and coworkers [33] reported a fabric-type all-solid SCs (Figure 6.8b), and the high-performance composite fabric electrode was synthesized by direct selective filtration of multi-walled carbon nanotubes (MWCNT) and rGO using Ni-coated cotton as a collector. The mass ratio of the capacitor remains at 250 F/g regardless of the thickness of the electrode. This enhances the conductance and flexibility of the electrode so that the all-solid SCs device can achieve discharge currents of

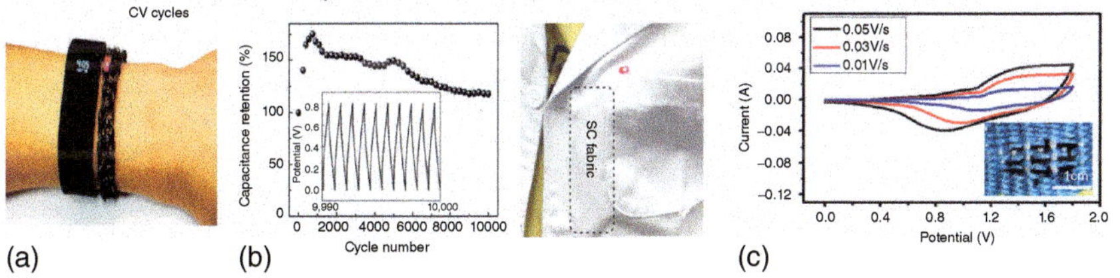

Figure 6.8 All solid supercapacitor textiles: (a) Long term cycling at the scan rate of 20 m/V for 100 00 cycles and the photo of our wristband worn together using a Fitbits wristband, the wristband can light an LED. Source: Adapted with permission from Zhai et al. [32]. Copyright (2015) by Royal Society of Chemistry. (b) Schematic diagram of the composite fabric electrodes, the stability tests of the all-solid-state fabric-type SCs, and the supercapacitor fabric photo. Source: Adapted with permission from Yang et al. [33]. Copyright (2017) by WILEY-VCH Verlag GmbH & Co. KGaA, Weinheim. (c) Fabricating a flexible textile-type or fiberlike supercapacitor from the dendritic cable electrode and its electrochemical performance. Source: Adapted with permission from Yang et al. [34]. Copyright (2017) by American Chemical Society.

0.2–1 A, and the specific capacitance increases to 3.2 F/cm^2 after 10 000 charge and discharge cycles, with no significant performance degradation after more than 10 000 bending tests. Fan and coworkers [34] fabricated solid supercapacitor fabrics by weaving the positive and negative electrodes of a dendritic cable with cotton threads (Figure 6.8c). Compared with assembling a thick layer of dendritic capacitive oxide on a smooth current collector, assembling a thin layer of capacitive oxide on a dendritic cable current collector would further shorten the average charge transfer distance while reserving a high usable area. With good mechanical and charge–discharge stability, it could be charged up to 1.8 V, outputting an energy density of 0.1408 mWh/cm^2 and power density of 3.01 mW/cm^2. The idea of the micro dendritic electrode structure provides an inspiring strategy to promote nano-materials in final devices, which is suitable for future mass production and could be extensively applied to other types of energy harvesting and storage devices.

6.2.2 Primary Battery Textiles

Primary batteries, such as primary alkaline zinc manganese batteries, that cannot be recharged and restored after being discharged, have been used in many commercial applications. In comparison with supercapacitors, batteries possess lower power density but higher energy density.

The current preparing method for primary battery textiles is to weave fiber-type devices into textile-type devices. Fiber-shaped conductive substrates are coated with different active materials to form negative and positive electrodes, respectively. After the two electrodes are twisted or placed side by side, they are coated with electrolyte and then encapsulated in plastic tubes to form the fiber-type battery. These fiber-type batteries can be woven with other polymer yarns to form a common-looking textile.

For example, Zou and coworkers [36] have assembled a flexible fiber-shaped zinc ion battery (ZIB) (Figure 6.9). With commercial carbon fiber as electrode base material, nanometer manganese dioxide powder and zinc powder were coated and bonded to the two electrodes and then encapsulated in a plastic tube after electrolyte injection. The battery has a capacity of 143 mAh/g at a discharge current density of 70 mA/g. Wrapping the battery around a finger for a physical display can

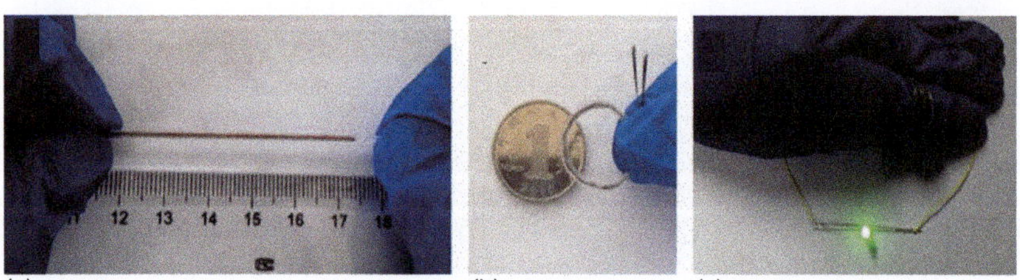

(a) (b) (c)

Figure 6.9 Primary zinc ion battery textile: (a) optical picture of fiber battery. (b) Optical picture of flexible fiber battery. (c) Two fiber batteries connected in series continuously driven a commercial green LED when bent around a human finger. Source: Adapted with permission from Yu et al. [36]. Copyright (2013) by Elsevier.

light up a small light-emitting diode (LED) lamp. In the form of fiber, this battery can be post-processed into the form of textile.

6.2.3 Secondary Battery Textiles

Secondary batteries, also known as rechargeable batteries or accumulators, refer to the batteries to activate active materials through charging after the batteries are discharged. Compared with primary batteries, secondary batteries have been researched extensively in energy storage textiles. By weaving secondary batteries fibers into textiles, the overall power output can be greatly increased.

Compared with the primary battery, there are more preparing methods reported for the secondary battery. The first preparation method is similar to that of the primary battery. For the second preparation method, layers of different active materials can be directly coated on a textile-shaped conducted substrate to form different textile-shaped electrodes. Different textile-shaped electrodes can be stacked together to form a multi-layered textile-type battery. For the third preparation method, layers of different active materials can be assembled layer-by-layer on a fiber conductive substrate to form a fiber battery with a coaxial structure. Then the fiber battery is woven with other polymer yarns into a textile.

Various secondary battery textiles have been reported, including lithium-ion batteries, zinc ion batteries, metal-air batteries, and other metal/metal oxide energy storage batteries.

Firstly, for the lithium-ion battery textiles, the preparing method is to weave fiber-type devices into textile-type devices. One fiber substrate is coated with oxides containing lithium salts, binders, and conductive agents, to form the positive electrode. Another fiber substrate is coated with graphite material to form the negative electrode. The two electrodes are twisted or placed side by side, then encapsulated in a plastic tube after electrolyte injection, such as oxides containing lithium. Lithium-ion batteries work rely on lithium ions to move between the positive and negative electrodes. Because of the advantages of green, safety, and high energy density, lithium-ion batteries are the mainstream in the battery field. With the new features such as flexibility and stretchability, lithium-ion batteries textiles have received extensive attention. For example, Peng and coworkers [37] reported a wearable fiber-shaped aqueous lithium with high power density, which is developed using a polyimide/CNT hybrid fiber as the anode and $LiMn_2O_4$/CNT hybrid fiber as the cathode (Figure 6.10a). This battery outputs a power density of 10 217.74 W/kg. These fiber-shaped Li-ion batteries (LIBs) can be further woven into textiles that can be bent, folded, and twisted into various architectures and connected in series, with maintained electrochemical performances under various deformations. Because batteries using aqueous electrolytes are prone to electrolyte leakage and low flexibility, people are increasingly paying attention to batteries using solid electrolytes. Hu and coworkers [38] studied the preparation of fiber-type solid-state lithium-ion flexible batteries using 3D printing technology (Figure 6.10b). Among them, ordinary binder PVDF is used as a viscosity modifier. CNTs are used as a conductive agent, and lithium iron phosphate (LFP) and lithium

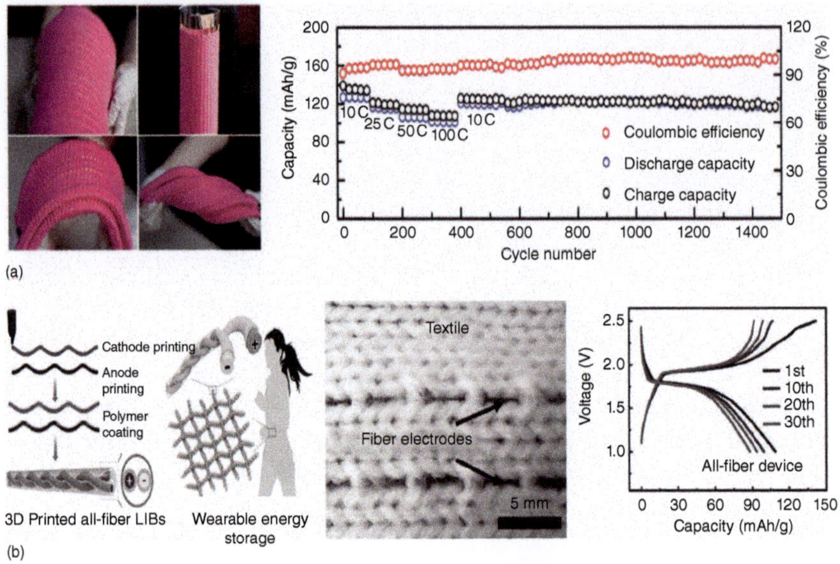

Figure 6.10 Flexible lithium-ion battery textiles: (a) The photo of high energy density lithium-ion battery textile and its electrochemical performance. Source: Adapted with permission from Zhang et al. [37]. Copyright (2016) by Royal Society of Chemistry. (b) Schematic of the design concept and fabrication process of 3D printed all solid-state flexible LIBs, its photo, and the charge and discharge profiles of all solid-state LIB device. Source: Adapted with permission from Wang et al. [38]. Copyright (2017) by WILEY-VCH Verlag GmbH & Co. KGaA, Weinheim.

titanate (LTO) are used as electrode activities of the positive electrode and the negative electrode, respectively. A flexible full-cell battery that maintains stable electrochemical performance under bending conditions and can be combined with conventional fabrics in the future as an important energy storage device for wearable electronic devices. This simple and fast method provides a new idea for the large-scale production of flexible lithium-ion batteries. It can also be effectively applied to flexible one-dimensional batteries of other active material systems.

Secondly, for the ZIB textiles, there are three types of preparing methods. The first method is similar to that of lithium-ion batteries. However, manganese or other oxide-based material is used as the active material for the positive electrode. Metallic zinc is used as the active material for the negative electrode, and an oxide containing zinc salts is used as the electrolyte. The second method of preparing ZIB textiles is directly coating a layer of active materials on a textile-shaped conducted substrate, which is then stacked with another textile-shaped metallic zinc electrodes. The third method for preparing ZIB textiles is to assemble active materials layer-by-layer on the fiber conductive substrate to obtain a fiber battery with a coaxial structure. The electrode can only employ all-solid electrolytes. And then, the fiber battery is woven into a textile with yarn.

Zinc ion batteries rely on zinc ions to move between the positive and negative electrodes [42]. They have been widely used because of the advantages of high

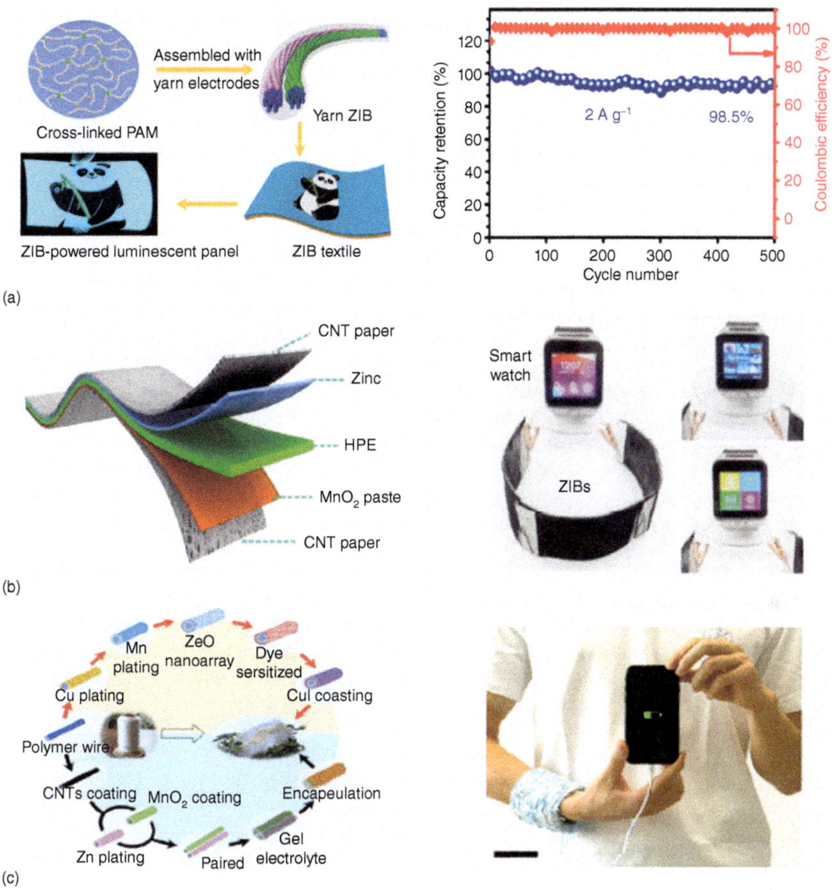

Figure 6.11 Zinc ion battery textiles: (a) Schematic diagram of fabrication and encapsulation of the yarn ZIB, cycling performance, and coulombic efficiency at 2 A/g. Source: Li et al. [40]. © 2018, American Chemical Society. (b) Schematic illustration of the structure of the solid-state ZIB and three ZIBs were connected in series to power a commercial smartwatch. Source: Adapted with permission from Li et al. [41]. Copyright (2018) by Royal Society of Chemistry. (c) Fabrication process of the functional fibers as the building blocks and demonstration of the photo-rechargeable fabric used to power a cell phone. Source: Adapted with permission from Zhang et al. [43]. Copyright (2020) by Elsevier.

energy density, low cost, high safety, and easy fabrication [39]. Many researchers have begun to prepare flexible zinc ion batteries textiles to make them more suitable for wearable devices. Li et al. [40] developed a high-performance, waterproof, tailorable, and stretchable yarn ZIB using double-helix yarn electrodes and a cross-linked polyacrylamide (PAM) electrolyte (Figure 6.11a). The yarn ZIB delivers a high specific capacity and volumetric energy density (302.1 mAh/g and 53.8 mWh/cm^3, respectively) and excellent cycling stability (98.5% capacity retention after 500 cycles). More importantly, the quasi-solid-state yarn ZIB also demonstrates superior knittability, good stretchability (up to 300%strain), and superior waterproof capability (high capacity retention of 96.5% after 12 hours underwater operation).

Besides, Zhi et al. [41] introduced an extremely safe and wearable solid-state ZIB comprising novel gelatin and PAM-based hierarchical polymer electrolyte (HPE) and an α-MnO_2 nano-rod/CNT cathode (Figure 6.11b). The flexible solid-state ZIB delivers a high areal energy density and power density (6.18 mWh/cm^2 and 148.2 mW/cm^2, respectively), high specific capacity (306 mAh/g), and excellent cycling stability (97% capacity retention after 1000 cycles at 2772 mA/g). Zhang et al. [43] demonstrated a flexible energy textile based on fiber Zn-MnO_2 battery cells and photovoltaic cells (Figure 6.11c). In the battery cells, both MnO_2 electrodes and Zn electrodes are made on extremely light wires. These wires are formed by coating a layer of MWCNT on multiple strands of PET wires. This kind of energy fabric can be made into clothing in various wearable forms and used as a portable power source for wearable electronic devices.

Thirdly, for various air battery textiles, there are two kinds of preparation methods. The first method includes three steps:

1. A fiber substrate is coated with a metallic material to form the positive electrode;
2. A fiber substrate is coated with a porous material to form the negative substrate;
3. These two electrodes are twisted or placed side-by-side then encapsulated in a plastic tube after electrolyte injection.

The second method is to assemble functional materials layer-by-layer on the fiber conductive substrate to obtain a fiber battery with a coaxial structure. For both fabrication methods, the as-fabricated fiber batteries can be woven into a textile with yarns. The air battery is working as an opened system, and its theoretical capacity is much larger than the theoretical capacity of other closed batteries. Therefore, various kinds of metal-air batteries are prospering, such as lithium-air batteries, aluminum-air batteries, and zinc-air batteries.

The lithium-air battery uses air as the positive electrode and metal lithium as the negative electrode. Peng and coworkers [44] developed a coaxial fiber-shaped Li-air battery, of which the aligned CNTs in the cathode could facilitate air diffusion and electron conduction. Meanwhile, the polymer gel electrolyte could prevent air diffusion and alleviate lithium corrosion (Figure 6.12a). It exhibited a discharge capacity of 12 470 mAh/g at a current density of 1400 mA/g and could effectively work for 100 cycles in the air with a cutoff capacity of 500 mAh/g. Furthermore, three fiber-shaped Li-air batteries were woven into a flexible textile with a discharging voltage up to 8 V. It was woven into power textiles to support electronic devices. To improve the cyclability of the Li-air battery in ambient air, Peng and coworkers [45] further combined a low-density polyethylene film and a gel electrolyte (Figure 6.12b). The low-density polyethylene film can restrain the side reactions of the discharge product of Li_2O_2 to Li_2CO_3 in ambient air, while the LiI can facilitate the electrochemical decomposition of Li_2O_2 during charging. The as-fabricated Li-air battery achieved an impressive life of over 600 cycles in ambient air. Six fiber-shaped batteries were integrated into clothes, which could effectively charge a smartphone.

Aluminum-air battery possesses high energy density, but lower cost compared to lithium. Peng and coworkers [46] described creating a new family of all-solid-state

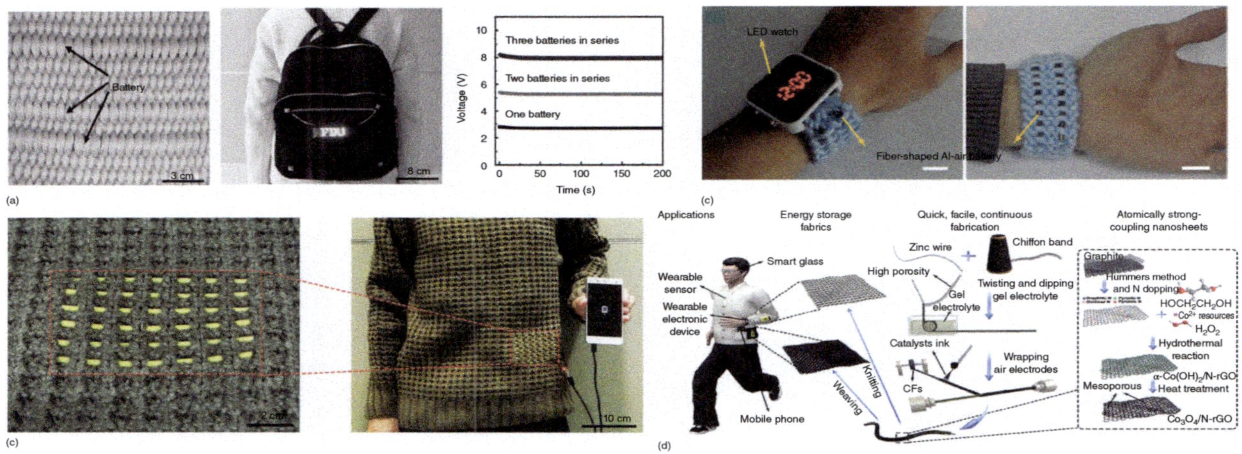

Figure 6.12 Air battery textiles: (a) Photograph of a Li-air battery textile, photograph of fiber-shaped Li-air batteries woven into a knapsack, and discharge curves. Source: Adapted with permission from Peng et al. [44]. Copyright (2016) by WILEY-VCH Verlag GmbH & Co. KGaA, Weinheim. (b) Photographs of flexible fiber-shaped Li-air batteries. Source: Adapted with permission from Wang et al. [45]. Copyright (2018) by WILEY-VCH Verlag GmbH & Co. KGaA, Weinheim. (c) Photographs of a commercial LED watch powered by fiber-shaped Al-air batteries. Source: Adapted with permission from Xu et al. [46]. Copyright (2016) by WILEY-VCH Verlag GmbH & Co. KGaA, Weinheim. (d) Schematic illustration of the fabrication process of flexible fiber-shaped Zn-air batteries. Source: Adapted with the permission from Li et al. [47]. Copyright (2018) by WILEY-VCH Verlag GmbH & Co. KGaA, Weinheim.

fiber-shaped aluminum-air batteries with a specific capacity of 935 mAh/g and an energy density of 1168 Wh/kg (Figure 6.12c). The synthesis of an electrode composed of cross-stacked aligned carbon-nanotube/silver-nanoparticle sheets contributes to the remarkable electrochemical performance. As a demonstration of wearable applications, two such batteries connected in series were woven into a wrist textile that successfully powered an LED watch.

Zinc-air battery uses metallic zinc as the negative electrode and oxygen in the air as the positive electrode. They have been extensively studied because of their theoretical energy density as high as 1084 Wh/kg. Zhong and coworkers [47] prepared a braided, fibrous Zn-air battery using a continuous, simple process path using low-cost electrode materials (Figure 6.12d). To improve the performance of braided fiber-like zinc-empty batteries, the team used in situ recombination to coat nitrogen-doped REDOX graphene (NrGO) with a layer of atomic-thick mesoporous Co_3O_4 nanosheets. The zinc-air battery showed superior charge-discharge performance. Under the charge and discharge current density of 3 mA/cm^3, 1.2 V discharge voltage, and 2 V charging voltage, energy efficiency reached 60%. The mass and energy density up to 649 Wh/kg and the ultra-high-volume energy density of 36.1 mWh/cm^3 also meet the flexible energy storage system requirements with high energy density. Flexible fiber-like zinc-empty batteries woven into the fabric can be connected in series and parallel to power an LED watch, charge an LED display, and charge an iPhone.

Fourthly, for other metals/metal oxides battery textiles, there are three kinds of preparation methods, all of which are like the preparation method of ZIB, only for the composition of electrode materials and electrolyte are different. For example, Guan et al. [48] rationally designed a flexible Ni/Fe battery with high energy and power densities by directly growing needle-like Fe_3O_4 and flake-like NiO anode and cathode, respectively, onto a carbon-cloth/carbon-nanofiber matrix (Figure 6.13a). It demonstrates optimal electrochemical performance, such as high energy density (5.2 mWh/cm^3 and 94.5 Wh/kg), high power density (0.64 W/cm^3 and 11.8 kW/kg), together with excellent cycling ability. Peng and coworkers [49] developed a silicon–oxygen battery fiber with high energy density and ultra-high flexibility by designing a coaxial architecture with a lithiated silicon/CNT hybrid fiber as an inner anode, a polymer gel as a middle electrolyte, and a bare CNT sheet as the outer cathode (Figure 6.13b). The fiber showed a high energy density of 512 Wh/kg and could effectively work after bending for 20 000 cycles. These battery fibers have been further woven into flexible textiles for a large-scale application.

6.3 Hybrid Energy Textiles

Hybrid energy harvesting and storage is an effective strategy to supply energy continuously and stably as an uninterrupted power supply. For one thing, solar energy, wind energy, and mechanical energy cannot be collected by one single energy harvesting device. Thus, different types of fiber-shaped or textile-shaped energy harvesting devices should be woven together to achieve multi-sourced energy harvesting.

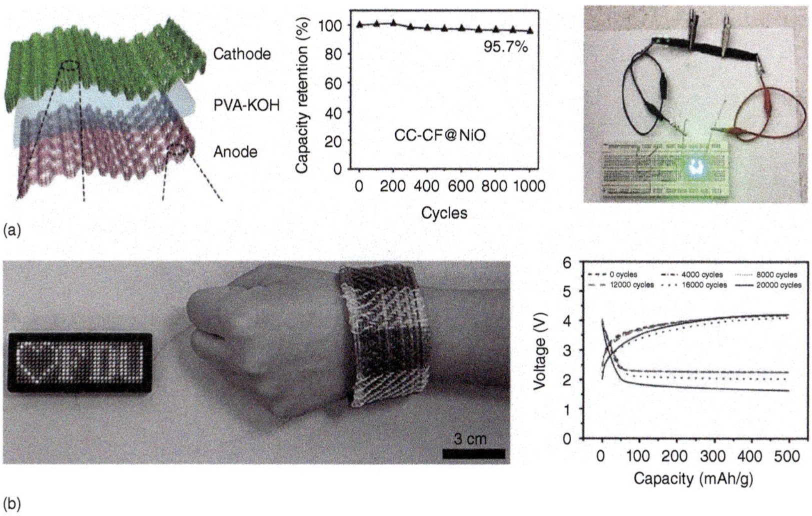

Figure 6.13 Other metal battery textiles: (a) Schematic illustration of the flexible Ni/Fe cell, cycling test, and Ni/Fe full cells lighting up LEDs. Source: Adapted with permission from Guan et al. [48]. Copyright (2017) by WILEY-VCH Verlag GmbH & Co. KGaA, Weinheim. (b) The SOB textile lighting up an LED screen and charge and discharge curves. Source: Adapted with permission from Zhang et al. [49]. Copyright (2016) by WILEY-VCH Verlag GmbH & Co. KGaA, Weinheim.

For another thing, solar light, wind, and human movement are not stable for time and space. Thus, people need to hybrid energy harvesting devices and energy storage devices to achieve a stable power output.

Compared with traditional block-type and flat-type hybrid energy devices, fabric-type hybrid energy devices, namely hybrid energy textiles, which have many advantages such as lightweight, excellent portability, and good wearability, were widely researched. There are two main methods to prepare hybrid energy textiles. The first method is to weave energy harvest fibers and the energy storage fibers into one single-layered textile. The second method is to directly connect more than one separated piece of energy textiles through wires and stack or mosaic them together. Therefore, these fabrication processes are still facing key issues such as structural design issues, packaging issues, and washing issues.

This section mainly introduces the development of the multiple energy harvesting hybrid textiles and the harvest-storage hybrid energy textiles.

6.3.1 Multiple Energy Harvesting Hybrid Textiles

Multiple energy harvest hybrid textiles are the textiles that can harvest multiple energy simultaneously, which was fabricated by integrating different energy-harvesting fibers or textiles. This section introduced various combinations of energy harvesting units to harvest different types of energies, such as hybrid textiles to harvest mechanical energy and solar energy, hybrid textiles to harvest mechanical energy and thermal energy, and hybrid textiles to harvest thermal, solar, and mechanical energy.

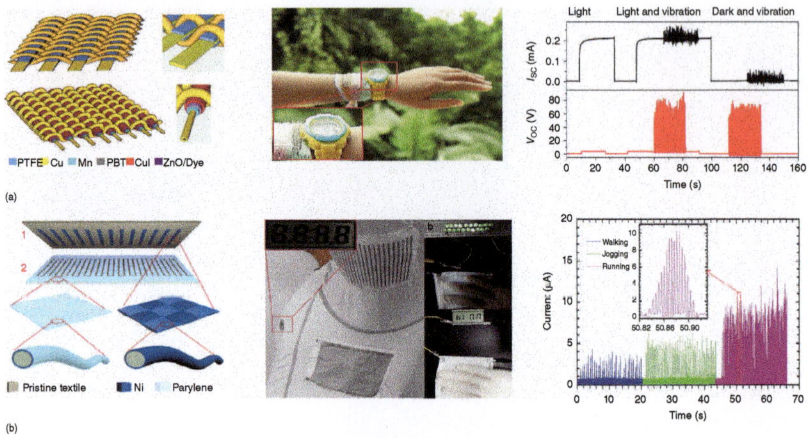

Figure 6.14 Simultaneously harvest energy from irradiated solar energy and mechanical motion: (a) Schematic illustration, electrical output of the wearable all-solid hybrid power textile and the hybrid power textile continuously powering an electronic watch. Source: Adapted with permission from Chen et al. [27]. Copyright (2016) by Springer Nature Limited. (b) The scheme, the demonstration, and the rectified current of the TENG fabrics. Source: Adapted with permission from Pu et al. [50]. Copyright (2016) by WILEY-VCH Verlag GmbH & Co. KGaA, Weinheim.

In terms of hybrid energy textiles to harvest solar energy and mechanical energy, Fan and coworkers [27] have developed a fabric-based, all-solid-state hybrid energy device that can simultaneously harvest energy from irradiated solar energy and mechanical motion using economically viable materials and scalable manufacturing techniques (Figure 6.14a). A fabric-based hybrid energy device at 4×5 cm in size can harvest both photovoltaic energy from the sunlight with an intensity of 80 mW/cm^2 and the mechanical energy from walking that generates an average output power of 0.5 mW over a wider range of load resistances from 10 kΩ to 10 MΩ. Wang and coworkers [50] developed a grating-structured TENG fabric and its integration with fiber dye-sensitized solar cells (F-DSSCs) to achieve a whole textile-based energy harvesting system (Figure 6.14b). A route of laser-scribing masking and electroluminescent display (ELD) Ni plating was first proposed for the synthesis of conductive circuits/patterns on the textile. Interdigitated grating-structured TENG fabrics were then fabricated to convert low-frequency human motion energy into high-frequency current outputs. By reducing the grating size, large improvements were achieved in the current amplitude and output power.

As hybrid energy textiles harvest human body heat and biomechanical energy, Lee et al. [51] developed a flexible hybrid generator by vertically stacking a thermoelectric generator, which consisted of six pairs of TE legs with a size of 2 mm by 2 mm by 1 mm, on a ZnO-based PENG (Figure 6.15a). While attached to a human forearm, this flexible hybrid generator produced an enhanced output voltage of up to

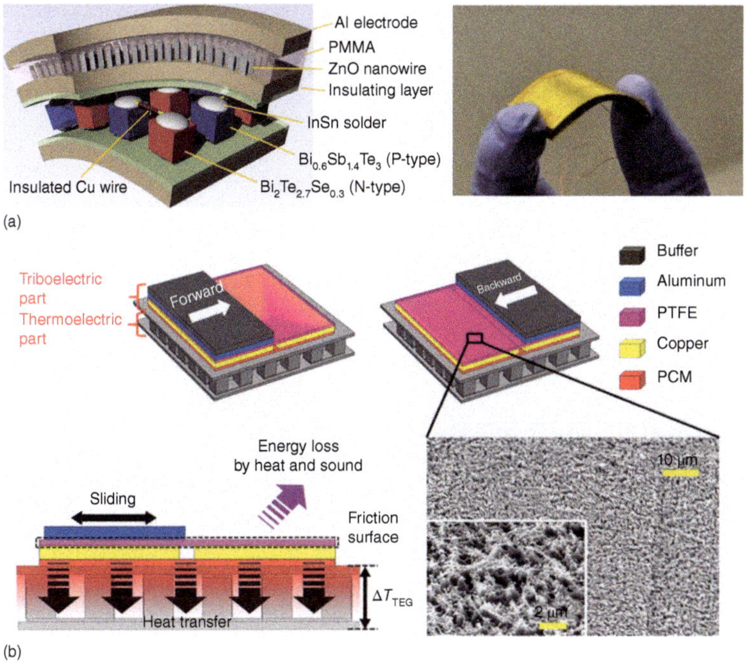

Figure 6.15 Simultaneously harvest energy from body heat and biomechanical energy: (a) Scheme and photographic of the vertically stacked hybrid device for simultaneously harvesting thermal and mechanical energies. Source: Adapted with permission from Lee et al. [51]. Copyright (2013) by 2013 Elsevier. (b) Proposed hybrid nanogenerator, detailed illustration of the working mechanism of the hybrid nanogenerator, and FESEM image of the nanostructured PTFE surface. Source: Adapted with permission from Kim et al. [52]. Copyright (2016) by IOP Publishing.

0.12 V and a current of 90 nA. Similarly, Kim et al. [52] demonstrated a hybrid generator by mounting a PTFE-based triboelectric generator on a thermoelectric generator consisting of Bi-Te pellet arrays (Figure 6.15b). Under a sliding rate of 12 cm/s and the ambient temperature of 24 °C, this hybrid nanogenerator (HG) delivered an output power density of 14.98 mW/cm^2, which could light up 100 commercial LED bulbs. It is worth mentioning that all reported wearable hybrid generators are built on a thin film platform and seldomly merged with textile.

In terms of hybrid energy textiles to harvest three forms of energy, Wang et al. [53] developed a hybrid nanogenerator based on multifunctional lead zirconate titanate piezoelectric ceramic (PZT) for simultaneously harvesting biomechanical, thermal, and solar energy (Figure 6.16). By harnessing three kinds of energy simultaneously, this hybrid generator with an active area of 70 by 70 mm could easily reach a voltage of up to 80 V, charging a 10 µF capacitor to 5.1 V in 90 seconds. However, developing smart textiles capable of simultaneously harvesting more than three energy forms has yet to be explored because of complicated structural requirements. Integrating these hybrid generators with textiles will be a promising research direction to provide a sustainable power source for on-body electronics in the future.

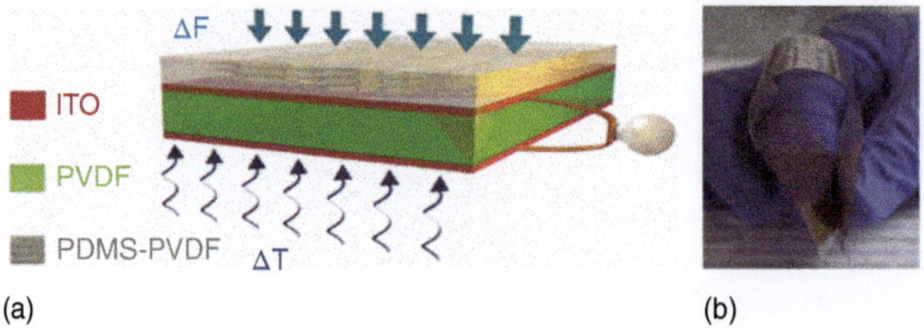

Figure 6.16 Simultaneously harvest energy from biomechanical, thermal, and solar energy: (a) Schematic diagram of the hybridized nanogenerator. (b) Photograph of the flexible hybridized nanogenerator. Source: (a) Wang et al. [53]. © 2016, John Wiley & Sons, (b) Adapted with permission from Wang et al. [53]. Copyright (2016) by WILEY-VCH Verlag GmbH & Co. KGaA, Weinheim.

6.3.2 Harvesting-Storage Hybrid Energy Textiles

Harvest-storage hybrid energy textiles are the textiles that simultaneously have the dual functions of both energy harvesting and energy storage, which is fabricated by combining different energy harvest units (fibers or textiles) and energy storage units. This section introduced the combinations of various energy harvesting units with different energy storage units, such as hybrid energy textile of solar cells and supercapacitors, hybrid energy textiles of TENGs and supercapacitors, hybrid energy textiles of solar cells, triboelectric NGs and supercapacitors.

In terms of energy textiles of solar cells and supercapacitors, Fan and coworkers [54] presented an all-solid tailorable energy textile possessing integrated function of simultaneous solar energy harvesting and storage, the fiber supercapacitor with merits of tailorability, ultrafast charging capability. Ultrahigh bending-resistance is used as the energy storage module. In contrast, an all-solid dye-sensitized solar cell textile is used as the solar energy harvesting module, and they call it a tailorable textile device (Figure 6.17a). The textile sample can be fully charged to 1.2 V in 17 seconds by self-harvesting solar energy and fully discharged in 78 seconds at a discharge current density of 0.1 mA.

In terms of energy textiles of TENGs and supercapacitors, Hu and coworkers [55] proposed a novel self-charging power textile (SCPT) consisting of a fabric triboelectric nanogenerator (FTENG) and a woven supercapacitor (W-SC) for simultaneously harvesting and storing human motion energy (Figure 6.17b). Utilizing traditional woven craft, this one-piece self-power/SCPT could be easily fabricated by alternating the woven wires/threads. This SCPT is demonstrated to have harvested energy through common daily activities such as running and walking, and simultaneously to have powered wearable electronics, such as an electric watch.

In terms of energy textiles of solar cells, TENGs, and supercapacitors, Wang and coworkers [56] proposed a hybridized SCPT system (Figure 6.17c). Both harvested energies can be easily converted into electricity by using fiber-shaped dye-sensitized solar cells (for solar energy) and fiber-shaped FTENGs (for random body motion energy) and then further stored as chemical energy in fiber-shaped supercapacitors. A single F-DSSC unit, a V_{OC} of 0.74 V, and a J_{SC} of 11.92 mA/cm^2 were achieved, corresponding to an overall PCE of 5.64%. The F-TENG can take advantage of human

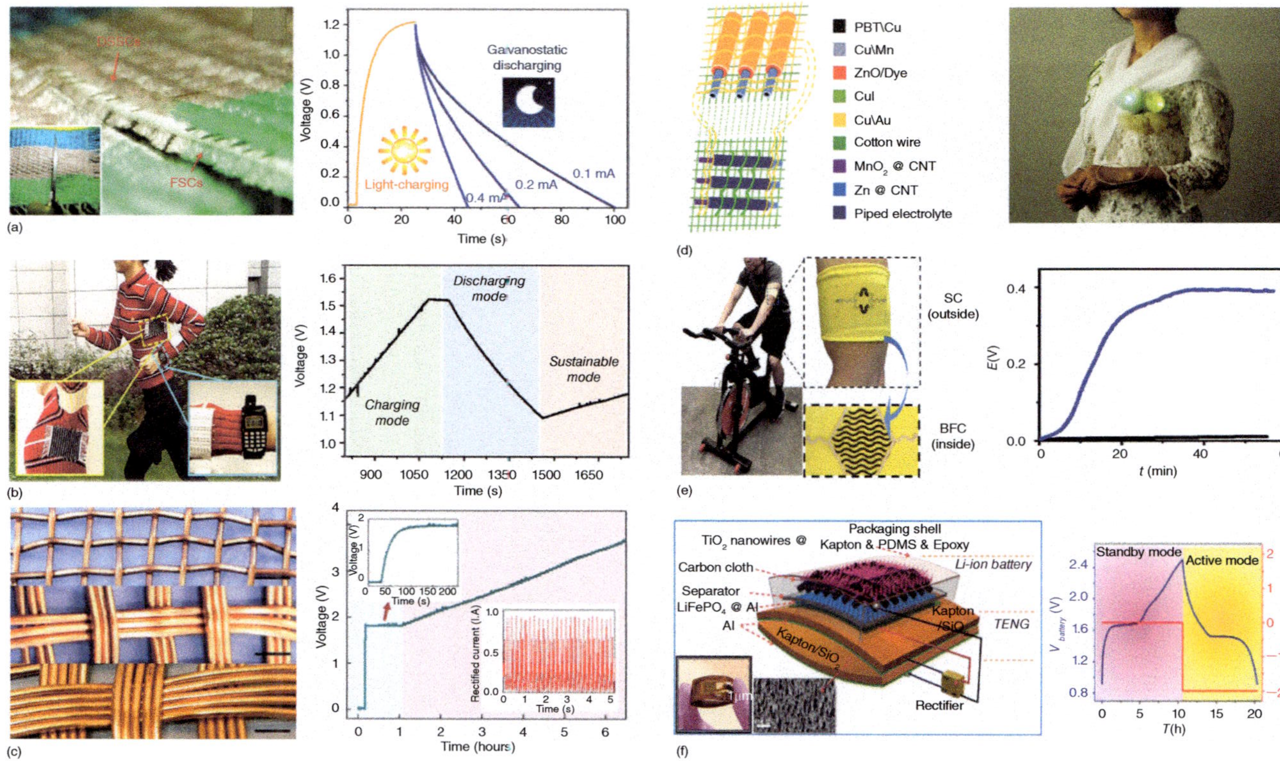

Figure 6.17 Harvest-storage hybrid energy textiles: (a) Light-charge and galvanostatic discharge performance and the tailorability of the textile cloth. Source: Adapted with permission from Fan et al. [54]. Copyright (2020) by American Chemical Society. (b) demonstration and Enlarged V–t curve of continuously driving an electric watch. Source: Adapted with permission from Chen et al. [55]. Copyright (2018) by Elsevier. (c) Physical map and V–t curve of the self-charging power textile. Source: Adapted with permission from Fan et al. [56]. Copyright (2018) by Elsevier. (d) Electrical integration of the battery units and the photovoltaic units and demonstration of the energy fabric to serve as the power source for light-emitting diode (LED) lights. Source: Adapted with permission from Huang et al. [57]. Copyright (2020) by American Chemical Society. (e) The integrated chemical self-powered system on one piece of textile and the real-time voltage of the printed SC charged from the on-body BFC during a constant cycling exercise. Source: Adapted with permission from Lv et al. [58]. Copyright (2018) by Energy & Environmental Science. (f) Schematic diagram showing the detailed structure of the SCPU and the voltage profile showing the LIB part fully charged by the TENG part (the "standby mode") and then discharging under a current of 2 μA (the "active mode"). Source: Adapted with permission from Wang et al. [59]. Copyright (2013) by American Chemical Society.

motions, such as jogging, to deliver an output current of up to 0.91 mA. With the excellent pseudo-capacitance of the as-synthesized $RuO_2 \cdot xH_2O$, the fiber supercapacitor (F-SC) unit exhibits a promising specific capacitance (1.9 mF/cm) makes it an effective and flexible electronic energy storage device.

In terms of energy textiles of solar cells and batteries, Fan and coworkers [57] proposed a reliable and industrially scalable electrode assembling strategy of couching embroidery for sustainable energy devices (Figure 6.17d). Simultaneously assembling different cable electrodes with huge length difference into one fabric-type device of even butterfly-shaped pattern appearance to form the tulle-type photovoltaic cell can assemble different cable electrodes Zn-MnO_2 battery and integrated power sources. The approaches simplified the fabrication of integrated fabric-type energy devices and improved the structural and functional design flexibility of a portable electronic power source.

In terms of energy textiles of biofuel cell (BFC) and supercapacitors, Lv et al. [58] demonstrated the first example of a stretchable and wearable textile-based hybrid supercapacitor–biofuel cell (SC-BFC) system (Figure 6.17e). The hybrid device, screen-printed on both sides of the fabric, is designed to scavenge biochemical energy from the wearer's sweat using the BFC module and store it in the SC module for subsequent use. Its attractive performance, demonstrated in successful on-body testing, along with the unique architecture and low-cost scalable fabrication, make the new garmented hybrid energy device useful for meeting the power and mechanical resiliency requirements of wearable electronics and smart textiles.

In terms of energy textiles of TENGs and batteries, Wang et al. [59] have developed a self-charging power unit (SCPU) that integrates the triboelectric-effect-based mechanical energy harvester and electrochemical energy storage into one device by building a flexible LIB on an arch-shaped structure (Figure 6.15f). They proposed a new mode for the integration, the "sustainable mode". The SCPU in this mode can provide a constant voltage with a direct current (DC) current of 2 μA for over 40 hours (compared to 3.5 hours for the LIB alone) and is capable of continuously powering the operation of a UV sensor at a constant voltage. Hence, the integrated power sources developed from the concept of the SCPU in the proposed "sustainable mode" can potentially serve as an independent, portable, sustainable, and stable power unit, which will meet the general requirement of almost any electronic device.

6.4 Commercialization Power Requirements of Smart Energy Textiles

As a summary of the energy harvesters previously mentioned, examples of the reported highest performances are listed in Table 6.1.

In terms of energy storage, supercapacitor textiles have higher power density but lower energy density, suitable for fast charge and discharge rate. The highest areal of the supercapacitor textiles was up to 3.42 F/cm^2 [63]. Battery textiles have higher energy density but lower power density, suitable for long-term power supply. The highest power density of the battery textiles was up to 3984.9 W/kg [64], and the

Table 6.1 The reported highest performances of wearable energy harvesting textiles.

Energy harvesting textiles	Performances	References
Photovoltaic energy harvesting textiles	PCE:14%	[4]
Thermoelectric energy harvesting textiles	15.79 $\mu W/cm^2$	[61]
Piezoelectric NGs textiles	64.9 $\mu W/cm^2$	[60]
Triboelectric NGs textiles	50 mW/cm^2	[62]

highest energy capacity of the battery textiles was up to 2200 mAh/g [65]. Currently, lithium-ion battery textiles have shown the highest capacity.

In the future, the development of smart energy textiles is developing towards the direction of higher capacity, longer lifetime, more beautiful design, and easier scalable fabrication technology for a wider application in the area of flexible energy source for microelectronics.

References

1 Liu, P., Gao, Z., Xu, L. et al. (2018). Polymer solar cell textiles with interlaced cathode and anode fibers. *Journal of Materials Chemistry A* 6 (41): 19947–19953.

2 Zhang, N., Chen, J., Huang, Y. et al. (2016). A wearable all-solid photovoltaic textile. *Advanced Materials* 28 (2): 263–269.

3 Fu, X., Sun, H., Xie, S. et al. (2018). A fiber-shaped solar cell showing a record power conversion efficiency of 10%. *Journal of Materials Chemistry A* 6 (1): 45–51.

4 Gao, Z., Liu, P., Fu, X. et al. (2019). Flexible self-powered textile formed by bridging photoactive and electrochemically active fiber electrodes. *Journal of Materials Chemistry A* 7 (10): 14447.

5 Zhou, H., Chen, Q., Li, G. et al. (2014). Interface engineering of highly efficient perovskite solar cells. *Science* 345 (6196): 542–546.

6 Jeon, N., Noh, J., Yang, W. et al. (2015). Compositional engineering of perovskite materials for high-performance solar cells. *Nature* 517 (7535): 476–480.

7 Li, R., Xiang, X., Tong, X. et al. (2015). Wearable double-twisted fibrous perovskite solar cell. *Advanced Materials* 27 (25): 3831–3835.

8 Qiu, L., Deng, D., Lu, X. et al. (2014). Integrating perovskite solar cells into a flexible fiber. *Agewandte Chemie International Edition* 53 (10): 10425–10428.

9 Liang, D., Yang, H., Finefrock, S.W., and Wu, Y. (2012). Flexible nanocrystal-coated glass fibers for high-performance thermoelectric energy harvesting. *Nano Letters* 12 (4): 2140–2145.

10 Kumar, B. and Kim, S. (2011). Recent advances in power generation through piezoelectric nanogenerators. *Journal of Materials Chemistry* 21 (47): 18946–18958.

11 Kim, M.K., Kim, M.S., Lee, S. et al. (2014). Wearable thermoelectric generator for harvesting human body heat energy. *Smart Materials and Structures* 23 (10): 105002.

12 Jia, Y., Shen, L., Liu, J. et al. (2019). An efficient PEDOT-coated textile for wearable thermoelectric generators and strain sensors. *Journal of Materials Chemistry C* 7 (12): 3496–3502.

13 Lee, J.A., Bykova, J.S., Kim, D. et al. (2016). Woven-yarn thermoelectric textiles. *Advanced Materials* 28 (25): 5038–5044.

14 Magniez, K., Krajewski, A., Neuenhofer, M., and Helmer, R. (2013). Effect of drawing on the molecular orientation and polymorphism of melt-spun polyvinylidene fluoride fibers: toward the development of piezoelectric force sensors. *Journal of Applied Polymer Science* 129 (5): 2699–2706.

15 Zhang, M., Gao, T., Wang, J. et al. (2015). A hybrid fibers based wearable fabric piezoelectric nanogenerator for energy harvesting application. *Nano Energy* 13: 298–305.

16 Wang, P., Pan, L., Wang, J. et al. (2018). An ultra-low-friction triboelectric-electromagnetic hybrid nanogenerator for rotation energy harvesting and self-powered wind speed sensor. *American Chemical Society Nano* 12 (9): 9433–9440.

17 Liu, R., Kuang, X., Deng, J. et al. (2018). Shape memory polymers for body motion energy harvesting and self-powered mechanosensing. *Advanced Materials* 30 (8): 1705195.

18 Wang, P., Liu, R., Ding, W. et al. (2018). Complementary electromagnetic-triboelectric active sensor for detecting multiple mechanical triggering. *Advanced Functional Materials* 28 (11): 1705808.

19 Wu, Z., Ding, W., Dai, Y. et al. (2018). Self-powered multifunctional motion sensor enabled by magnetic-regulated triboelectric nanogenerator. *American Chemical Society Nano* 12 (6): 5726–5733.

20 Chen, J. and Wang, Z. (2017). Reviving vibration energy harvesting and self-powered sensing by a triboelectric nanogenerator. *Joule* 1 (3): 480–521.

21 Xu, M., Wang, P., Wang, Y. et al. (2018). A soft and robust spring based triboelectric nanogenerator for harvesting arbitrary directional vibration energy and self-powered vibration sensing. *Advanced Energy Materials* 8 (9): 1702432.

22 Chen, B., Yang, Y., and Wang, Z. (2018). Scavenging wind energy by triboelectric nanogenerators. *Advanced Energy Materials* 8 (10): 1702649.

23 Xie, Y., Wang, S., Lin, L. et al. (2013). Rotary triboelectric nanogenerator based on a hybridized mechanism for harvesting wind energy. *American Chemical Society Nano* 7 (8): 7119–7125.

24 Zhang, L., Zhang, B., Chen, J. et al. (2016). Lawn structured triboelectric nanogenerators for scavenging sweeping wind energy on rooftops. *Advanced Materials* 28 (8): 1650–1656.

25 Xu, M., Zhao, T., Wang, C. et al. (2019). High power density tower-like triboelectric nanogenerator for harvesting arbitrary directional water wave energy. *American Chemical Society Nano* 13 (2): 1932–1939.

26 Wang, Z., Jiang, T., and Xu, L. (2017). Toward the blue energy dream by triboelectric nanogenerator networks. *Nano Energy* 39: 9–23.

27 Chen, J., Huang, Y., Zhang, N. et al. (2016). Micro-cable structured textile for simultaneously harvesting solar and mechanical energy. *Nature Energy* 1: 16138.

28 Zhou, T., Zhang, C., Han, C.B. et al. (2014). Woven structured triboelectric nanogenerator for wearable devices. *ACS Applied Materials & Interfaces* 4 (16): 14695–14701.

29 Kwak, S., Kim, H., Seung, W. et al. (2017). Fully stretchable textile triboelectric nanogenerator with knitted fabric structures. *American Chemical Society Nano* 11 (11): 10733–10741.

30 Jost, K., Durkin, D.P., Haverhals, L.M. et al. (2015). Natural fiber welded electrode yarns for knittable textile supercapacitors. *Advanced Energy Materials* 5 (4): 1401286.

31 Chen, S., Ma, W., Cheng, Y. et al. (2015). Scalable non-liquid-crystal spinning of locally aligned graphene fibers for high-performance wearable supercapacitors. *Nano Energy* 15: 642–653.

32 Zhai, S., Jiang, W., Wei, L. et al. (2015). All-carbon solid-state yarn supercapacitors from activated carbon and carbon fibers for smart textiles. *Materials Horizons* 2 (6): 598–605.

33 Yang, Y., Huang, Q., Niu, L. et al. (2017). Waterproof, ultrahigh areal-capacitance, wearable supercapacitor fabrics. *Advanced Materials* 29 (19): 1606679.

34 Yang, Y., Zhang, N., Zhang, B. et al. (2017). Highly-efficient dendritic cable electrodes for flexible supercapacitive fabric. *ACS Applied Materials & Interfaces* 9: 40207–40214.

35 Song, S. and Yun, K.S. (2015). Design and characterization of scalable woven piezoelectric energy harvester for wearable applications. *Smart Materials and Structures* 24: 045008.

36 Yu, X., Fu, Y., Cai, X. et al. (2013). Flexible fiber-type zinc-carbon battery based on carbon fiber electrodes. *Nano Energy* 2 (6): 1242–1248.

37 Zhang, Y., Wang, Y., Wang, L. et al. (2016). A fiber-shaped aqueous lithium ion battery with high power density. *Journal of Materials Chemistry A* 4 (23): 9002–9008.

38 Wang, Y., Chen, C., Xie, H. et al. (2017). 3D-printed all-Fiber li-ion battery toward wearable energy storage. *Advanced Functional Materials* 27 (43): 1703140.

39 Cheng, F., Chen, J., Gou, X.L., and Shen, P. (2005). High-power alkaline Zn-MnO_2 batteries using γ-MnO_2 nanowires/nanotubes and electrolytic zinc powder. *Advanced Materials* 17 (22): 2753–2756.

40 Li, H., Liu, Z., Liang, G. et al. (2018). Waterproof and tailorable elastic rechargeable yarn zinc ion batteries by a cross-linked polyacrylamide electrolyte. *American Chemical Society Nano* 12 (4): 3140–3148.

41 Li, H., Han, C., Huang, Y. et al. (2018). An extremely safe and wearable solid-state zinc ion battery based on a hierarchical structured polymer electrolyte. *Energy & Environmental Science* 11 (4): 941–951.

42 Huang, F., Yu, H., Xiang, S. et al. (2020). Embroidering a filmsy photorechargeable energy fabric with wide weather adaptability. *ACS Applied Materials & Interfaces* 12 (3): 3654–3660.

43 Zhang, N., Huang, F., Zhao, S. et al. (2020). Photo-rechargeable fabrics as sustainable and robust power sources for wearable bioelectronics. *Matter* 2 (5): 1260–1269.

44 Zhang, Y., Wang, L., Guo, Z. et al. (2016). High-performance lithium-air battery with a coaxial-fiber architecture. *Angewandte Chemie International Edition* 55 (14): 4487–4491.

45 Wang, L., Pan, J., Zhang, Y. et al. (2018). A Li-air battery with ultralong cycle life in ambient air. *Advanced Materials* 30 (3): 1704378.

46 Xu, Y., Zhao, Y., Ren, J. et al. (2016). An all-solid-state fiber-shaped aluminum-air battery with flexibility, stretchability, and high electrochemical performance. *Angewandte Chemie International Edition* 55 (28): 7979–7982.

47 Li, Y., Zhong, C., Liu, J. et al. (2018). Atomically thin mesoporous Co_3O_4 layers strongly coupled with N-rGO nanosheets as high-performance bifunctional catalysts for 1D knittable zinc-air batteries. *Advanced Materials* 30 (4): 1703657.

48 Guan, C., Zhao, W., Hu, Y. et al. (2016). High-performance flexible solid-state Ni/Fe battery consisting of metal oxides coated carbon cloth/carbon nanofiber electrodes. *Advanced Energy Materials* 6 (20): 1601034.

49 Zhang, Y., Jiao, Y., Lu, L. et al. (2017). An ultraflexible silicon-oxygen battery fiber with high energy density. *Angewandte Chemie International Edition* 56 (44): 13741–13746.

50 Pu, X., Song, W., Liu, M. et al. (2016). Wearable power-textiles by integrating fabric triboelectric nanogenerators and fiber-shaped dye-sensitized solar cells. *Advanced Energy Materials* 6 (20): 1601048.

51 Lee, S., Bae, S.H., Lin, L. et al. (2013). Flexible hybrid cell for simultaneously harvesting thermal and mechanical energies. *Nano Energy* 2: 817–825.

52 Kim, M.K., Kim, M.S., Jo, S.E., and Kim, Y.J. (2016). Triboelectric-thermoelectric hybrid nanogenerator for harvesting frictional energy. *Smart Materials and Structures* 25: 125007.

53 Wang, S., Wang, Z., and Yang, Y. (2016). A one-structure-based hybridized nanogenerator for scavenging mechanical and thermal energies by triboelectric-piezoelectric-pyroelectric effects. *Advanced Materials* 28: 2881–2887.

54 Chai, Z., Zhang, N., Sun, P. et al. (2016). Tailorable and wearable textile devices for solar energy harvesting and simultaneous storage. *American Chemical Society Nano* 10 (10): 9201–9207.

55 Chen, J., Guo, H., Pu, X. et al. (2018). Traditional weaving craft for one-piece self-charging power textile for wearable electronics. *Nano Energy* 50: 536–543.

56 Wen, Z., Yeh, M.H., Guo, H. et al. (2016). Self-powered textile for wearable electronics by hybridizing fiber-shaped nanogenerators, solar cells, and supercapacitors. *Science Advances* 2 (10): 1600097.

57 Huang, F., Yu, H., Xiang, S. et al. (2020). Embroidering a filmsy photo-rechargeable energy fabric with wide weather adaptability. *ACS Applied Materials & Interfaces* 12 (3): 3654–3660.

58 Lv, J., Jeerapan, I., Tehrani, F. et al. (2018). Sweat-based wearable energy harvesting-storage hybrid textile devices. *Energy & Environmental Science* 11 (12): 3431–3442.

59 Wang, S., Lin, Z., Niu, S. et al. (2013). Motion charged battery as sustainable flexible-power-unit. *American Chemical Society Nano* 7 (12): 11263–11271.

60 Yeo, H.G., Xue, T., Roundy, S. et al. (2018). Strongly (001) oriented bimorph PZT film on metal foils grown by rf-sputtering for wrist-worn piezoelectric energy harvesters. *Advanced Functional Materials* 28 (36): 1801327.

61 Roundy, S. (2005). On the effectiveness of vibration-based energy harvesting. *Journal of Intelligent Materials and structures* 16 (10): 809–823.

62 Zhu, G., Zhou, Y., Bai, P. et al. (2014). A shape-adaptive thin-film-based approach for 50% high-efficiency energy generation through micro-grating sliding electrification. *Advanced Materials* 26: 3788–3796.

63 Nozariasbmarz, A., Kishore, R.A., Poudel, B. et al. (2019). High power density body heat energy harvesting. *ACS Applied Materials & Interfaces* 11 (43): 40107–40113.

64 Magno, M. and Boyle, D. (2017). *Wearable energy harvesting: from body to battery.12th International Conference on Design & Technology of Integrated Systems in Nanoscale Era*, 1–6. Palma de Mallorca: *DTIS*.

65 Sun, C., Zhu, H., Baker, E.B. III, et al. (2013). Weavable high-capacity electrodes. *Nano Energy* 2 (5): 987–994.

7

Function Expansion of Smart Energy Fibers and Textiles

As previously discussed, smart energy fibers and textiles were used to power portable microelectronics, for which researchers mostly focused on the energy density. As the energy density of the smart energy fibers and textiles is getting higher and higher, it is expected to be applied to daily life in the future.

However, situations such as stretching, bending, or wetting exist in real life. Therefore, except for the energy density, it is still challenging to expand functions including stretchability, hydrophobic performance, stability, air permeability, and color-changeable ability, determining the stability, safety, and comfort of the smart energy fibers and textiles for daily life.

To meet different working environments and application requirements, people have done a lot around the function expansion. This chapter will introduce smart energy fibers and textiles' latest function expansion progress in many aspects, including stretchability, hydrophobicity, stability, air-permeability, and color-changeable ability.

7.1 Stretchability of Smart Energy Fibers and Textiles

With the rapid development of personal information platforms, smart electronic textiles (E-textiles) integrate into modern life. It will inevitably encounter problems such as stretching during the working environment, which affect the wearing application of the smart E-textiles [1]. The smart E-textiles are prepared by integrating with various kinds of functional electrodes and woven fabric fibers together. The electrode is mostly made by coating the functional conducting materials on the metal or elastomer substrate. Conventional conducting materials, including monocrystalline silicon, polycrystalline silicon, or metal oxide films, have inherent brittleness and rigidity that cannot be stretched, limiting their application for the smart E-textiles under a stretching working environment.

In recent years, many efforts have been made to develop new stretchable electrodes based on conductive materials and electrode structural designs, to make smart energy fibers and textiles more stable during the stretching working environment [2]. The concept of stretchability means the capability of maintaining conductive pathways under large strains (>>1%) and can restore their original

Textile-Based Energy Harvesting and Storage Devices for Wearable Electronics, First Edition.
Xing Fan, Nannan Zhang, and Yi Wang.

properties after releasing the induced strain, which has been touted as a key component of the smart E-textiles [3].

This section will first introduce the stretchable electrode details based on elastic conductive materials and special electrode structure designs. We also give a review of practical assembling approaches to realize the stretchable energy fiber and textiles.

7.1.1 Stretchable Electrode Based on Elastic Conductive Materials

The most direct way to fabricate stretchable electrodes is based on elastic conductive materials, maintaining their conductivity during stretching. Elastic conductive materials are usually synthesized by incorporating conductive materials as the additive into elastomer substrate, which can then be applied for preparing stretchable electrodes. Conductive materials, including organic (such as a conductive polymer) and inorganic materials (such as carbon nano-materials and metal nano-structure), have attracted extensive attention from academia and industry due to their tremendous potential stretchability. Compared with traditional conductive materials, elastic conductive materials exhibit superior potentials for a stretchable electrode. The elastic conductive materials will be introduced from two aspects: organic conductive additives and inorganic conductive additives.

In terms of organic conductive additives, conducting polymers have been mostly used in stretchable conductors due to the charge transport capability of conductive domains and polymer chains' mechanical flexibility. Poly(3,4-ethylenedioxythiophene):polystyrene sulfonate (PEDOT:PSS) [4], polyaniline (PANI) [5], polypyrrole [6] are regarded as promising nano-materials for the fabrication of stretchable electrode. In a recent study by Wang et al., various enhancers for stretchability and electrical conductivity were incorporated into PEDOT:PSS chains to soften the PSS domains and promote partial aggregation of the conductive poly-3,4-ethylenedioxythiophene (PEDOT) regions [7]. As a result, a high conductivity of 3390 S/cm under 100% strain, 4100 S/cm up to 600% strain could be achieved (Figure 7.1a). Another strategy to improve the conductivity and stretchability of conducting polymers involves the formation of block-copolymers. For example, PANI can be covalently grafted onto the backbone of a soft triblock-copolymer SEBS-g-MA (poly(styrene-co-ethylenebutylene-costyrene) grafted with maleic anhydride). The copolymer's overall stretchability could be further improved by blending it with additional SEBS [8]. The sample with 15% (v/v) PANI displayed an excellent elastic property of 4180% strain while maintaining a minimum conductivity of 100 S/cm (Figure 7.1b). Besides, Guo et al. used one conductor for both electrodes and leads of a stretchable polymeric multielectrode array using a polypyrrole/polyol-borate composite film [9]. The polypyrrole/PCTC composite film generates good electrical conductivity of the hydrophobic diols in polycaprolactone-block-polytetrahydrofuran block-polycaprolactone (PCTC). When assisted by the borate network, it could exhibit excellent electromechanical performance (electrical conductivity of 116.3 ± 7.8 S/cm) (Figure 7.1c). Another conducting polymer, poly(3-butylthiophene-2,5-diyl) (P3BT), has

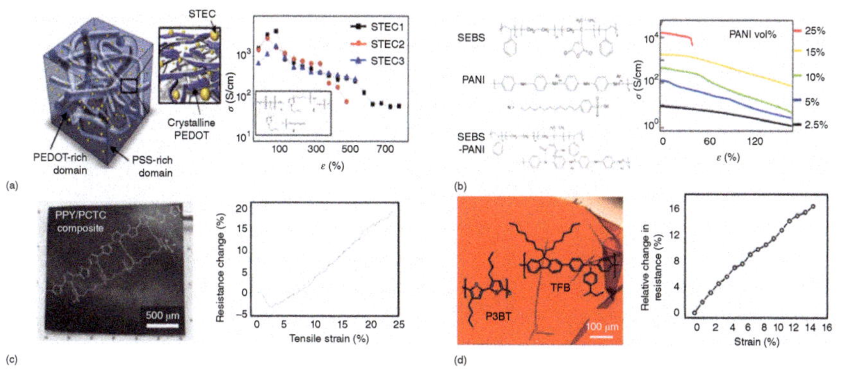

Figure 7.1 Conductivity for various organic conductive additives: (a) schematic illustration of the PEDOT/STEC (left). Conductivity change under strain for PEDOT incorporated with three different enhancers (right). Source: Wang et al. [7]. © 2017, American Association for the Advancement of Science. (b) Molecular formulae for PANI and SEBS-g-MA (left). Conductivity versus strain of 20/80 SEBS-g-MA/SEBS (right). Source: Stoyanov et al. [8]. © 2013, John Wiley & Sons. (c) Photograph of a PPy/PCTC composite film and its chemical structure (left). Resistance change versus tensile strain curve of the PPy/PCTC composite (right). Source: Adapted with permission from Guo et al. [9]. Copyright (2014) by WILEY-VCH Verlag GmbH & Co. KGaA, Weinheim. (d) Photograph of PthTFB and chemical structures of TFB and P3BT (left). Relative change in resistance under strain for PthTFB polymer nanosheets (right). Source: Guan et al. [10]. Copyright (2018) by WILEY-VCH Verlag GmbH & Co. KGaA, Weinheim.

also been utilized in stretchable conductors through nanoconfinement and strain engineering [10]. Through P3BT nanofibrils formation within the poly[(9,9dioctylfluorenyl-2,7-diyl)-co-(4,4′-(*N*-(4-sec-butylphenyl) diphenylamine)] (TFB) matrix, the resulting polymer (denoted PthTFB) exhibited high stretchability, mechanical robustness, and electrical conductivity with more than 1000-cycle durability tests (Figure 7.1d).

In terms of inorganic conductive additives, carbon-based materials and metal-based materials are widely investigated, as summarized in Figure 7.2a.

Carbon-based materials such as carbon blacks (CB) [19], carbon nanotubes (CNTs) [20], and graphene [21], were the most widely used. CB is a form of paracrystalline carbon. It can be the building block of conductive networks due to its high thermal/electrical conductivity and commercially available characteristics. For example, CB particles are mixed with polydimethylsiloxane (PDMS) to produce a conductive electrode [11]. However, such an electrode's conductivity is not very high due to the CB particles' low conductivity (Figure 7.2b). In another example, CB particles were filled into a poly(styrene–butadiene–styrene) (SBS) elastomer, as shown in Figure 7.2c, and the change in the electrical conductivity of the fabricated conductive electrode with cyclic deformation was investigated [12]. Compared to CB's electrical and mechanical properties, CNTs are more suitable materials to form the percolation network due to their high aspect ratios. Shin et al. prepared a conductive electrode using vertically aligned multi-walled carbon nanotubes (MWCNTs) in a polyurethane (PU) film [13]. The composite has a 3D

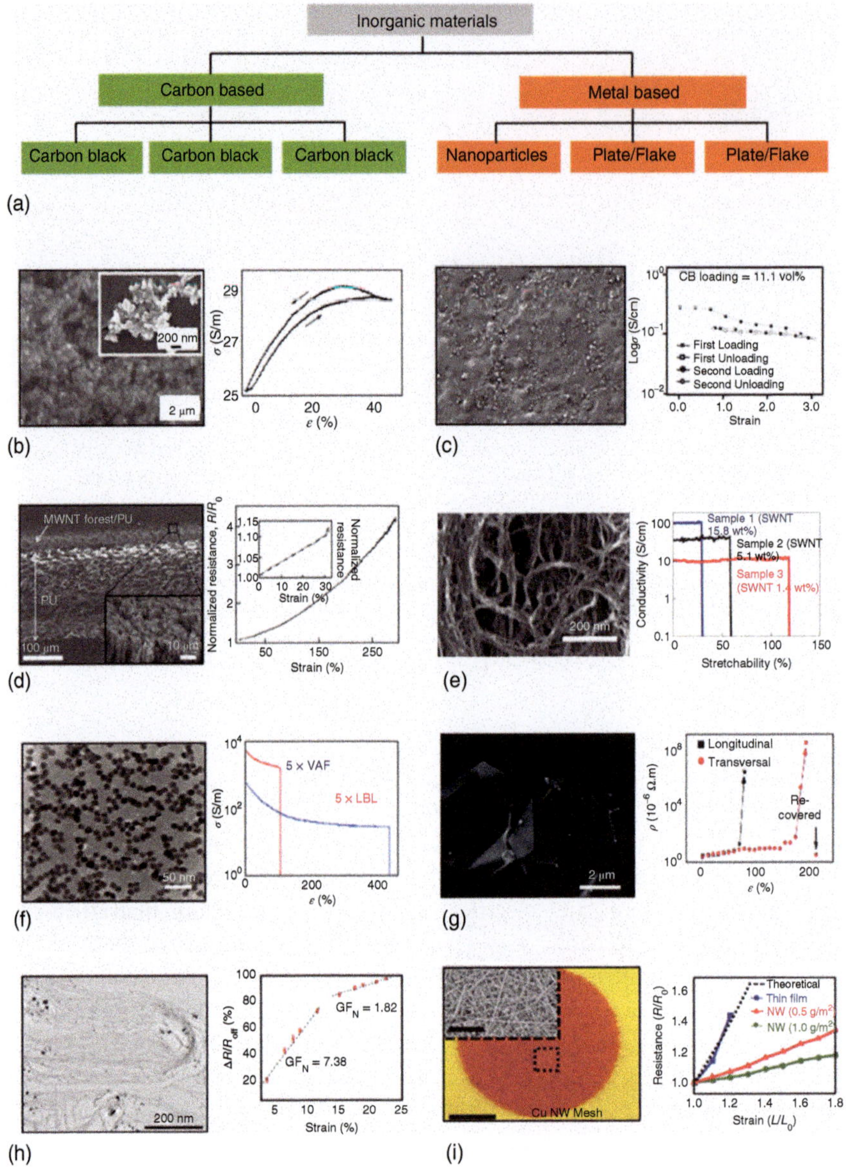
Inorganic materials
Carbon based
Metal based
Carbon black
Carbon black
Carbon black
Nanoparticles
Plate/Flake
Plate/Flake
(a)
200 nm
2 μm
σ (S/m)
ε (%)
(b)
CB loading = 11.1 vol%
First Loading
First Unloading
Second Loading
Second Unloading
Strain
(c)
MWNT forest/PU
PU
100 μm
10 μm
Normalized resistance, R/R0
Normalized resistance
Strain (%)
(d)
200 nm
Sample 1 (SWNT 15.8 wt%)
Sample 2 (SWNT 5.1 wt%)
Sample 3 (SWNT 1.4 wt%)
Conductivity (S/cm)
Stretchability (%)
(e)
50 nm
5 × VAF
5 × LBL
σ (S/m)
ε (%)
(f)
2 μm
Longitudinal
Transversal
Re-covered
ρ (10⁻⁶ Ω.m)
ε (%)
(g)
200 nm
GF_N = 1.82
GF_N = 7.38
Strain (%)
(h)
Cu NW Mesh
Theoretical
Thin film
NW (0.5 g/m²)
NW (1.0 g/m²)
Resistance (R/R0)
Strain (L/L0)
(i)

Figure 7.2 Conductivity for various types of conductive materials based on carbon- and metal-based materials: (a) flow chart for conductive materials based on carbon- and metal-based materials. (b) SEM image of carbon black (CB) powder (left) and its variation of the conductivity (right). Source: Adapted with permission from Niu et al. [11]. Copyright (2007) by WILEY-VCH Verlag GmbH & Co. KGaA, Weinheim. (c) SEM image showing the fractal morphology of CB aggregates in a CB-poly (styrene-butadiene-styrene) (SBS) elastomer composite (left) and its electrical conductivity (right). Source: Adapted with permission from Lv et al. [12]. Copyright (2008) by Taylor & Francis. (d) SEM image of graphene foam (left) and change in the composite's resistance (right). Source: Adapted with permission from Shin et al. [13]. Copyright (2010) by WILEY-VCH Verlag GmbH & Co. KGaA, Weinheim. (e) SEM image of SWNTs uniformly dispersed in rubber (left). Conductivity of printed elastic conductors as a function of stretchability corresponding to 15.8, 5.1, and 1.4 wt% SWNT loading (right). Source: Adapted with permission from Chen et al. [14]. Copyright (2012) by Springer Nature Limited.) (f) TEM image of Au nanoparticles (left). The conductivity as a function of the uniaxial strain of films deposited by the layer-by-layer assembly (LBL) and vacuum-assisted flocculation (VAF) (right). Adapted with permission from Kim et al. [15]. Copyright (2013) by Springer Nature Limited. (g) SEM image of Au nanosheets (left). Change in the electrical resistivity upon stretching the line-patterned Au nanosheets transferred onto Ecoflex along transverse and longitudinal directions (right). Source: Adapted with permission from Moon et al. [16]. Copyright (2013) by WILEY-VCH Verlag GmbH & Co. KGaA, Weinheim. (h) SEM image of an ultrathin Au NW (left). Electrical resistance changes under the strain of the Au NWs sandwiched between two polydimethylsiloxane sheets (right). Source: Adapted with permission from Gong et al. [17]. Copyright (2014) by Springer Nature Limited. (i) Optical image (left) and SEM image (left inset) of a highly networked mesh structure of the as-prepared Cu NWs. Normalized resistance vs strain for the nanocomposite elastomer, thin film-based elastomer, and bare elastomer (right). Source: Adapted with permission from Han et al. [18]. Copyright (2016) by WILEY-VCH Verlag GmbH & Co. KGaA, Weinheim.

accordion-like structure of the conductive CNT network held together by the PU binder. It exhibits a linear increase in the resistance up to 30% strain and a nonlinear increase in resistance at higher strain values (Figure 7.2d). Graphene-based materials, such as graphene flakes and 3D-structured graphene, have gained popularity due to their better electrical and mechanical properties than CB and CNTs. Chen et al. prepared graphene conductive electrodes by employing porous Ni foam with an interconnected structure as a 3D scaffold (Figure 7.2e). This kind of graphene foam has several advantages over chemically derived graphene sheets, such as a high charge carrier mobility and small resistance, leading to enhanced electrical and mechanical properties [14].

On the other hand, representative metal-based materials, including Au NPs [15], Au nanosheets [22], Ag flakes [23], and Ag nanowire (NWs) [24], have been used in a variety of stretchable electrode. For instance, Kim et al. used positively charged PU to complement the negative charge of citrate-capped Au NPs to obtain layer-by-layer (LBL) assembly of Au NPs as shown in Figure 7.2f, the maximum conductivity via the LBL is 11 000 S/cm [15]. Apart from that, Jeong and coworkers integrated 2D Au nanosheets with rubber to form a stretchable electrode in Figure 7.2g. An electrode consisting of multilayers of line-patterned Au nanosheets was transferred to a silicon rubber [16], Ecoflex, mechanical stability up to strain $e = 60$ and 140% in the longitudinal and transverse directions, respectively. Furthermore, metallic NWs, such as those of Ag, Au, and copper (Cu), has recently gained much attention because their high aspect ratio is an added advantage along with their high intrinsic conductivity and ductility. The rationale for the use of Au NWs is their stability against oxidation, despite the high production cost and the difficulties in their synthesis. An example of Au NW-based devices is a flexible pressure sensor electrode based on ultrathin Au NWs (Figure 7.2h). The NWs had a width of 2 nm and lengths of tens of micrometers and exhibited mechanical robustness while being flexible at the same time [17]. The gauge factor evaluation indicated that the sensor provides a higher linear slope of 7.38 at strain values below 14%, which drops to 1.82 in a high strain range (14–25%). Although Cu NWs have the highest susceptibility to oxidation among the representative metallic NWs, their production is relatively inexpensive because of Cu's natural abundance. Accordingly, many types of nanocomposites incorporating CuNWs as conductive materials have been prepared. For example, a nanocomposite with superior crack resistance, good contact adhesion, and enhanced normal/shear strength properties was obtained using Cu NWs with an average length of 50 mm (Figure 7.2i) [18]. Analysis of the resistance vs strain curve revealed that the stretchable electrode based on Cu NWs shows 25 times lower normalized electrical resistance than those estimated experimentally and theoretically for a Cu thin film used as a control.

Generally, although the aforementioned stretchable conductive materials have reasonable conductivity, a challenge for its device applications is the conductivity decrease at applied strain due to the reduced contact area among the conductive nanoparticles inside the elastic substrate under larger strain.

7.1.2 Stretchable Electrode Based Electrode Structural Designs

As we mentioned before, the conductive materials' conductivity will decrease under larger strain due to the reduced contact area among the nano-materials or the molecular chain's structure deformation. Another prevailing approach to achieve stretchability originates from specific geometrical configurations, such as buckled structure, open-mesh structure, coiled structure, and sponge structure.

1) The buckled structure is a promising design to enable the electrodes to be stretchable. CNT sheets (carbon nanotube sheets) were wrapped oriented in the fiber direction on stretched rubber fiber cores using the prestrain-release-buckling strategy [25]. After releasing, the fibers exhibited distinct short- and long-period sheath buckling that occurred reversibly out of phase in the axial and belt directions (Figure 7.3a). The fibers possessed a resistance change of less than 5% for a 1000% stretch, taking advantage of the hierarchically buckled structure.
2) An open-mesh structure could also enable the electrodes with high stretchability. For example, Wang et al. [26] used the open-mesh structure to engineer elasticity in intrinsically rigid nanocomposites, resulting in a significant increase of ultimate strain from 4% to 370% (Figure 7.3b).
3) The coiled structure was demonstrated by Sun and co-workers [27], which possessed an ultrahigh sensitivity to tensile strain (detection limitation of 0.2% strain), a wide maximal sensing range (up to 100% strain), a fast signal response

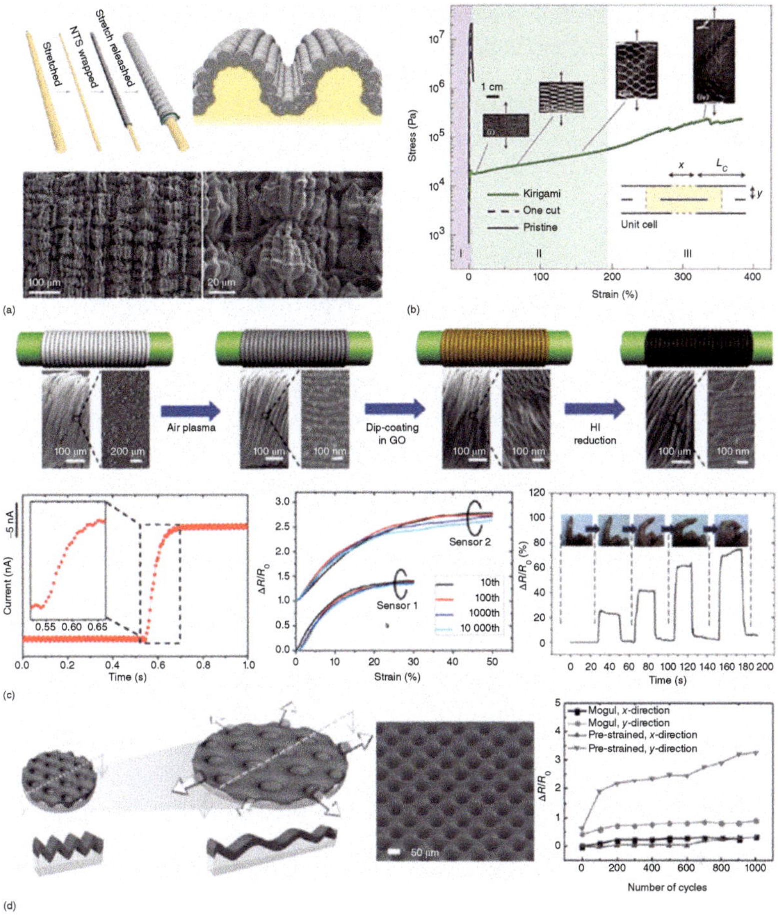

Figure 7.3 Stretchable electrode based on various types of stretchable structural designs: (a) hierarchically buckled sheath-core fibers. Source: Adapted with permission from Liu et al. [25]. Copyright (2015) by The National Academy of Sciences of the USA. (b) The effect of network topology on stretchability. Source: Wang et al. [26]. (c) Spring-like CNT ropes. Source: Adapted with permission from Cheng et al. [27]. Copyright (2015) by WILEY-VCH Verlag GmbH & Co. KGaA, Weinheim. (d) Mogul-patterned stretchable conductor. Source: Adapted with permission from Lee et al. [28]. Copyright (2016) by WILEY-VCH Verlag GmbH & Co. KGaA, Weinheim.

(<100 ms), and high reproducibility (up to 10 000 cycles) simultaneously (Figure 7.3c).

4) Sponge structure, such as the mogul-patterned stretchable structure, was designed by Lee et al. [28] for stretchable electronics. The mogul-patterned substrate can absorb eternal stress, resulting in multiple directional stretching of the structure layered on the substrate. More importantly, the bumps and valleys were regularly positioned in a hexagonal-closed-packed structure,

which enabled the layers and devices on the substrate to be more stable under multidirectional stretching (Figure 7.3d).

Overall, although the stress distribution in 3D coiled structures is more uniform than in other structures like 2D open-mesh structure, the fabrication processes of 3D coiled structures are more complex than those of other structures is probably the main obstacle toward large-scale applications and commercialization.

7.1.3 Assembling of Fiber-Type and Textile-Type Stretchable Devices

Stretchable electronic devices with stable performance under high tensile strain or large deformation have recently attracted increasing attention because of the multifunctional applications. The previous section discussed the elastic conductive materials and electrode structure design for fabricating a stretchable electrode. Next, we will discuss how to assemble fiber-type and textile-type stretchable devices from the above stretchable fiber electrode. We would mainly focus on several types of energy devices, such as stretchable supercapacitor (SC), stretchable solar cells textiles, stretchable sensors, and stretchable triboelectric nanogenerators (TENGs).

1) A highly stretchable SC has been developed with high performance by winding aligned CNT sheets on elastic fibers [29]. The high stretchability and specific capacitance have been simultaneously achieved by designing a coiled structure that favors high contact areas between the electrode and electrolyte besides the combined remarkable properties enabled by the aligned CNTs. Figure 7.4a shows that a fiber-shaped SC can be easily stretched by 100% without an obvious decrease in structural integrity.
2) Stretchable solar cells that can accommodate large strain and feature high cyclic mechanical endurance is challenging for wearable and skin-interfaced electronics application. Song and coworkers demonstrated such solar cells using an open-mesh structure like the kirigami design. Experiments and mechanical simulations showed that the kirigami structure effectively imparted stretchability to perovskite solar cells (PSCs) through out-of-plane deformation, significantly reducing the stress in devices [30]. The kirigami-based PSCs with optimal geometric parameters exhibited high mechanical deformability, including stretchability (strain up to 200%). More importantly, the kirigami PSCs revealed high mechanical endurance with almost unchanged performance even after 1000 repetitive stretching cycles (Figure 7.4b).
   Furthermore, stretchable photo-rechargeable energy fabric, based on the combination of stretchable solar cells and stretchable batteries, was developed via a couching embroidery process on an industrial sewing machine [31]. Applied strain could be measured by calculating the distances between the fixed edges before and after being stretched, respectively. As shown in Figure 7.4c, thanks to the textured structure of photo-rechargeable energy fabric, the fabric's output can stably maintain at I_{sc} of 123 μA and V_{oc} of 0.36 V at a strain of 100%.

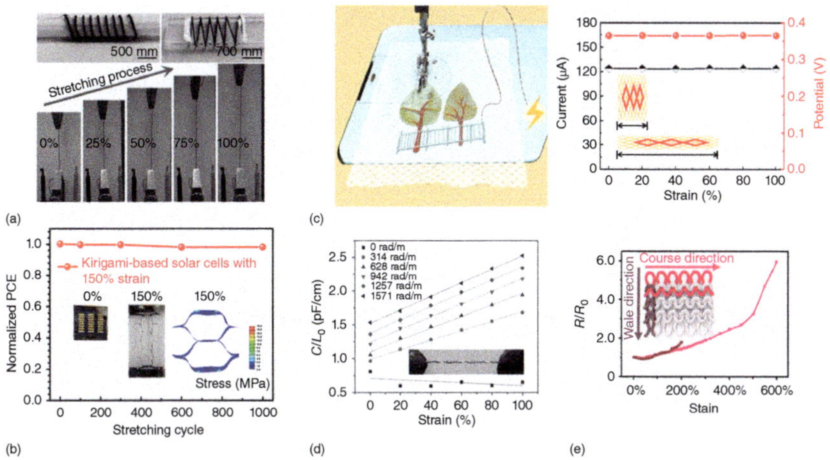

Figure 7.4 Stressing testing of various types of smart energy fibers and textiles: (a) photographs of two fiber-shaped supercapacitors being wound on different shapes of substrates and photograph of the fiber shaped supercapacitor with different strains of 0%, 25%, 50%, 75%, and 100%. Source: Adapted with permission from Yang et al. [29]. Copyright (2013) by WILEY-VCH Verlag GmbH & Co. KGaA, Weinheim. (b) Experimental tests of the kirigami-based PSCs under straining cycle. Source: Adapted with permission from Li et al. [30]. Copyright (2020) by American Chemical Society. (c) Performance of the photovoltaic tulle at different strains. Source: Adapted with permission from Huang et al. [31]. Copyright (2020) by American Chemical Society. (d) A plot of capacitance per initial length vs strain for fibers at various torsional levels with regression lines plotted for each level. Source: Adapted with permission from Christopher et al. [32]. Copyright (2017) by WILEY-VCH Verlag GmbH & Co. KGaA, Weinheim. (e) Resistance variation of the Ni-coated textile conductors along course and wale directions. Source: Adapted with permission from Cong et al. [33]. Copyright (2020) by American Chemical Society.

3) Stretchable sensors have been prepared by a helical structure, incorporated into smart E-textiles [32]. Liquid metals represent a promising class of materials for creating these sensors because they can undergo large deformations while retaining electrical continuity. The stretchable sensors are prepared by using hollow elastomeric fibers filled with liquid metal. The sensor electrode can also keep more stable under torsion and strain by helical structure design, which is also a type of coiled structure (Figure 7.4d).
4) The TENGs is also an effective approach to address wearable/portable/wireless smart devices' energy bottleneck. Wang and coworkers demonstrate a stretchable coplanar self-charging power textile (SCPT) with TENGs and micro-supercapacitors (MSCs), fabricated through a resist-dyeinganalogous method [33]. Due to the textured structure, the textile electrodes can maintain excellent conductivity at 600% and 200% tensile strain along the course and wale directions, respectively. The stretchable fabric-based TENG can output 49 V open-circuit voltage and 94.5 mW/m peak power density (Figure 7.4e).

To conclude, stretchable energy fibers and textile devices have largely depended on stretchable electrodes with high conductivity and stretchability. The stretchable electrodes' performance depends significantly on the conductive materials' properties and the electrode structure design. Various conductive materials, including conductive polymer, carbon nano-materials, and metal nano-structure, have been incorporated into elastomeric matrices to develop stretchable conductive electrodes with desired conductivity and stretchability. The electrode structure design proved that high stretchability could also be achieved through various structures, such as buckled structure, open-mesh structure, coiled structure, and sponge structure. These structures can further improve the overall stretchability of the textile-typed devices.

7.2 Hydrophobicity of Smart Energy Fibers and Textiles

Besides stretching, water intrusion or high humidity situations mostly happen in people's daily lives, which normally degrades smart energy fibers and textiles' performance. Hydrophobicity is also a crucial factor for wearable smart energy fibers and textiles, making wearable devices waterproof for daily-life applications.

On the hydrophobic property, a surface exhibits apparent contact angles (CAs) >150°, and low CA hysteresis with water are usually called superhydrophobic surfaces, which is very useful for waterproof [34]. Many strategies to create superhydrophobic surfaces have been put forward [35, 36]. These surfaces have attracted significant attention due to their unique water-repellent and their potential for smart energy fibers and textiles [37].

In this part, we will introduce the development history of conventional hydrophobic fabrics. And then, as a key to hydrophobic fibers and textiles, superhydrophobic coating technologies are reviewed from the aspects of both materials and preparation methods. After that, technologies for fabricating hydrophobic smart energy fibers and textiles are reviewed and compared with the conventional hydrophobic fabrics.

7.2.1 The History of Conventional Hydrophobic Fabrics

Since the 1930s, the development of hydrophobic textiles had become one of the hot topics in the textile industry. Coating hydrophobic agents fabricate conventional hydrophobic fabrics. The main hydrophobic agents are reactive aliphatic compounds, organosilicon, and fluorine-containing finishing agents [38]. For example, when aliphatic compounds are used as water repellent finishing agent to modify the surface of fabrics, the long carbon chain compounds with reactive groups at one end are used to modify the surface of textiles to obtain an endurable water-repellent effect [39].

Since the 1950s, polysiloxane entered the textile industry in the name of an organosilicon water-repellent agent, which was used to treat polyester cotton blended fabric and vinegar fiber fabric raincoat. For example, zirconium or titanium compounds are easy to hydrolyze. Thus, PDMS is used in combination with zirconium or titanium compounds and then coated on the fabric surface to make a hydrophobic fabric (Figure 7.5).

Since the 1960s, after organosilicon water-repellent agent entered the textile chemical processing field, 3M company of the United States has successively launched

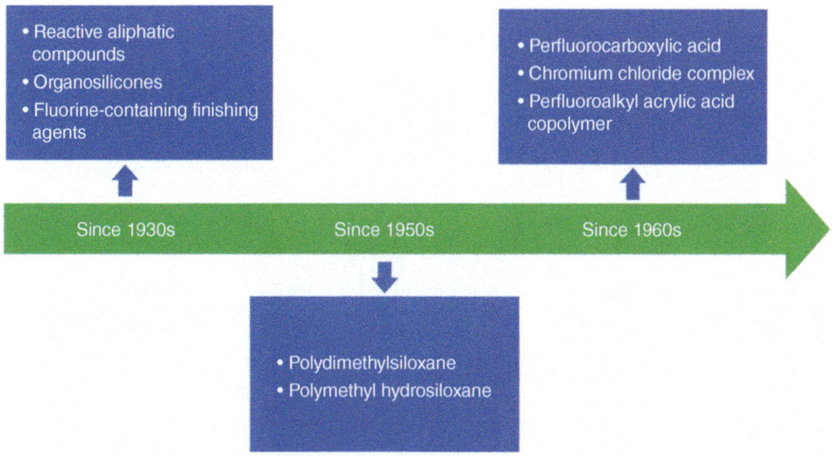

Figure 7.5 Development of traditional fabric hydrophobic treatment.

three commodities: perfluorocarboxylic acid, chromium chloride complex, and perfluoroalkyl acrylic acid copolymer. Since then, textiles have water repellency and a new function of oil repellency [40]. At the end of the 1980s, super water-repellent textiles were launched on the market. In the 1990s, researchers put forward a new concept of superhydrophobic, which has drawn people's attention from all walks of life.

7.2.2 The Development of Hydrophobic Coatings

The superhydrophobic coating technology is a key to achieving hydrophobic fibers and textiles, which can be obtained by coating different coating materials on fiber substrates. The coating materials can be divided into inorganic materials, organic materials, and inorganic–organic materials composite. The coating materials can be coated on the fiber substrates by various methods, including covalent LBL assembling, polymer roughening, chemical vapor depositing, sol–gel, and hydrothermal synthesis.

1) The inorganic superhydrophobic coating materials include silicon-based materials, carbon-based materials, and metal oxides materials. As shown in Figure 7.6a, a facile one-step ultrasound-assisted approach was developed to synthesize silica nanoparticles (SiNPs) functionalized with octyltriethoxysilane and their in situ incorporation into cotton fabrics [41]. The fabric showed stable superhydrophobicity with (water CA) and (water-shedding angle) SHA values of $152.8° \pm 2.6°$ and 8°, respectively. Carbon-based materials such as CNTs and graphene are also common inorganic materials used for superhydrophobic coatings. For example, Zhu et al. [42] have demonstrated a facile approach to fabricate a superhydrophobic CNTs coating with good reusability and easy repairability (Figure 7.6b). Endowing surfaces with good reusability and easy repairability opens a new avenue to extend the lifespan of superhydrophobicity.

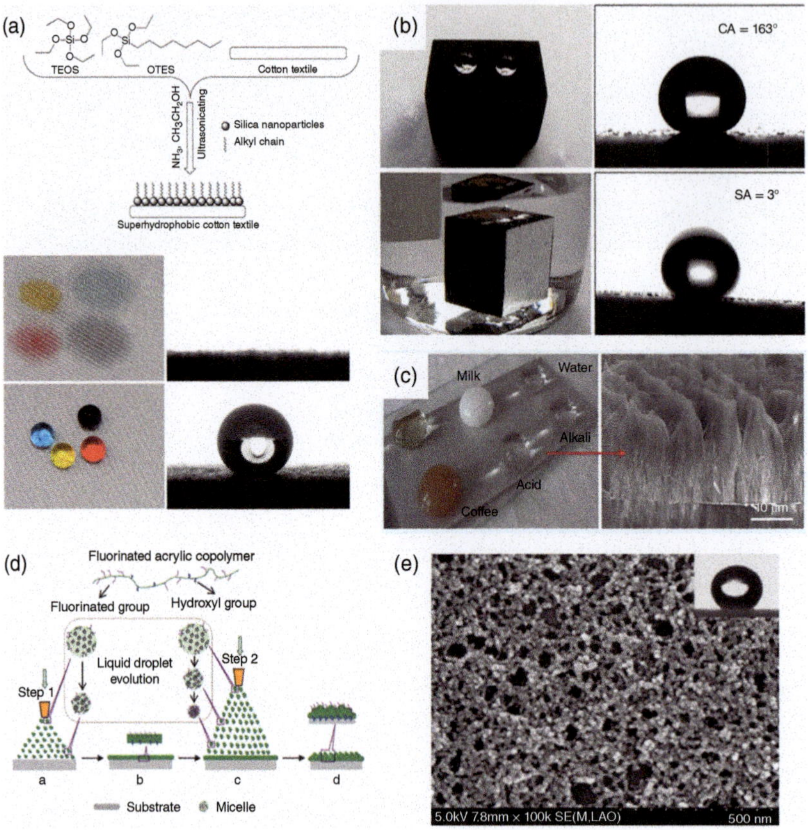

Figure 7.6 Materials used to prepare superhydrophobic coatings: (a) synthesis of silica nanoparticles (SiNPs) functionalized with octyltriethoxysilane and their in situ incorporation into cotton fabrics, and colored water droplets sitting on pristine and OTES-TEOS-treated cotton fabrics. Source: Adapted with permission from Mohammad and Mohammad [41]. Copyright (2013) by American Chemical Society. (b) Water droplets exhibit a spherical shape on the fluorinated CNTs coating. Source: Adapted with permission from Zhu et al. [42]. Copyright (2014) by Elsevier.) (c) Photographs of the diverse liquid droplets on the PDES-MS surface. Source: Adapted with permission from Peng et al. [43]. Copyright (2014) by American Chemical Society. (d) Schematic representation of the formation of SFAC coating. Source: Adapted with permission from Li et al. [44]. Copyright (2013) by Royal Society of Chemistry. (e) The glass treated with PEG/SiO_2. Source: Adapted with permission from Zhang et al. [45]. Copyright (2012) by Springer-Verlag.

Metal oxides were also applied for the superhydrophobic coatings. Peng et al. [43] have introduced a highly effective and easy-to-implement method to fabricate chemically stable and mechanically robust superhydrophobic alumina surface on Al substrate (Figure 7.6c). The rough hierarchical alumina pyramids-on-pores (HAPOP) surface can be obtained quickly at 0.3 M oxalic acid under room temperature with a current density of $0.16\,A/cm^2$ within 10 min with modifier stearic acid (STA) deposition, which could result in excellent nonwetting and extremely slippery behaviors.

2) Organic superhydrophobic coating materials are mainly polymer materials, such as polystyrene (PS), polyethylene (PE), polypropylene (PP), and fluorinated acrylic. These are the most commonly used for superhydrophobic coating. Through some simple optimization steps, a good superhydrophobic effect can be obtained [46]. For example, Li et al. [44] have synthesized a kind of fluorinated acrylic random copolymer which could aggregate into nanoscale micelles in suitable solvents and developed a facile process to fabricate superhydrophobic coatings through spraying the copolymer solution on an aluminum substrate (Figure 7.6d). The obtained superhydrophobic coating not only has resistance to low temperature and excellent durability underwater.
3) Inorganic–organic composite superhydrophobic coating materials are also significant for superhydrophobic coatings. It is prepared by coating inorganic nanoparticles with organic polymers, which has the advantages of both inorganic and organic materials. For example, Zhang et al. [45] first treated the glass plate surface with the organic–inorganic solution of polyethylene glycol (PEG) and SiO_2. It is further modified the glass plate with hexomethyldisilazane (HMDS) to prepare the organic–inorganic composite coating with superhydrophobic and transparent properties, with the water CA up to 168°.

These superhydrophobic coating materials can be coated on fiber or textile by various methods [47, 48], including covalent LBL assembly, polymer roughening, chemical vapor deposition, sol–gel processing, and hydrothermal synthesis.

Covalent LBL assembly is a versatile way to construct thin-film nanocomposites. This technology is popular because it is simple and forgiving [49]. After LBL assembly, the outer surface usually contains residual functional groups, marked by letters A and C in Figure 7.7a, which can be utilized to react with low-surface-energy agents for covalent hydrophilization.

Polymer roughening can be formed by the phase separation of a multicomponent mixture [50]. Suppose a stable mixture becomes unstable owing to changes such as cooling or solvent evaporation. In that case, it will begin to separate into two phases, one of which might be a solid or become solid at some point in the process. This creates a solid porous structure, or a lotus-like structure with micro-nanopapillae, as shown in Figure 7.7b. The size of the pores or papillae in the formed solid structures can be controlled if the rate of phase separation can be altered relative to solidification. Phase separation has received great interest in the fabrication of superhydrophobic surfaces owing to their low cost, ease of production, and the possibility of creating substrates with various shapes by casting and coating.

Chemical vapor deposition is a technique whereby gaseous reactants can be deposited onto a substrate to form a nonvolatile solid film. The film morphology is mostly dependent on the substrate's surface morphology, which is usually used as a template. As shown in Figure 7.7c, Zimmermann et al. [51] prepared superhydrophobic textile fabrics via a one-step gas-phase coating procedure by growing a layer of polymethylsilsesquioxane nanofilaments onto individual textile fibers. Eleven textile fabrics made from natural and synthetic fibers were successfully coated, and their superhydrophobic properties were evaluated.

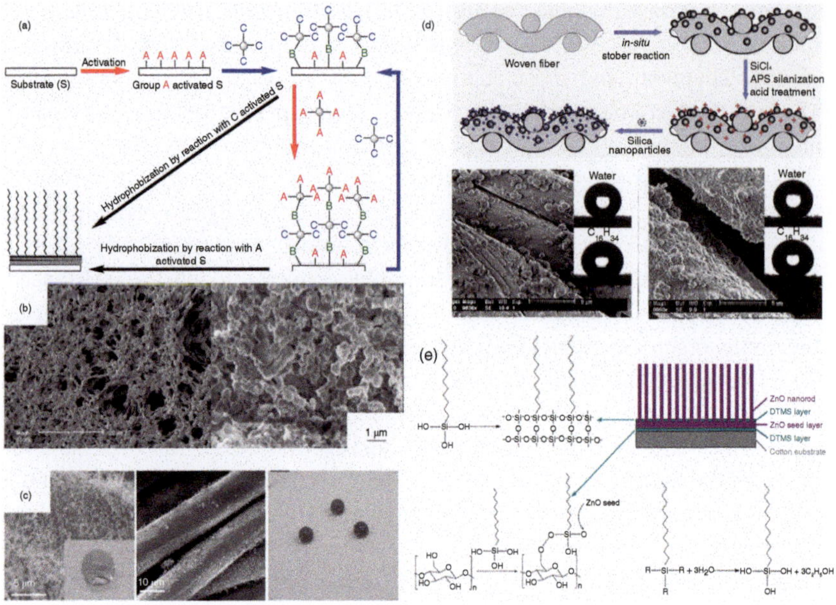

Figure 7.7 Preparation methods used to prepare superhydrophobic coatings: (a) steps used in the covalent layer-by-layer assembly of functional particles. Source: Adapted with permission from Xue et al. [49]. Copyright (2009) by Elsevier. (b) SEM images of porous membrane produced by the solvent casting of 17.9 mg/ml polypropylene solution using methyl ethyl ketone as the nonsolvent and superhydrophobic poly (vinyl chloride) film obtained by coating with a mixture of 2:1 (v/v) ethanol and H_2O as the nonsolvent. Source: Adapted with permission from Franco et al. [50]. Copyright (2008) by Elsevier. (c) SEM images of poly (ethylene terephthalate) textile fabric coated with silicon nanofilaments and coated samples after the abrasion test and drops of water showing the sample's wetting properties after the abrasion test. Source: Adapted with permission from Zimmermann et al. [51]. Copyright (2008) by WILEY-VCH Verlag GmbH & Co. KGaA, Weinheim. (d) Schematic illustration of the procedure for the preparation of dual-size structure on the surface of woven cotton fibers, combining an in situ Stöber reactions with silica nanoparticles' subsequent adsorption. Source: Adapted with permission from Leng et al. [52]. Copyright (2009) by the American Chemical Society. (e) The scheme shows a dodecyltrimethoxysilane-treated fiber structure for improved adhesion and stability of ZnO nanorods to a cotton substrate. Source: Adapted with permission from Wu et al. [53]. Copyright (2005) by the American Chemical Society.

Sol–gel processing is a well-recognized method of synthesizing gels and nanoparticles. Leng et al. [52] described superoleophobic cotton textiles' preparation based on a structure having multiple length scales. The produced surface had a high hexadecane CA and a low roll-off angle. The multiple length scales of the roughness originated from the woven structure having two additional layers of silica particles, covalently bonded to the fiber as schematically shown in Figure 7.7d.

Hydrothermal synthesis is a well-known method for the fabrication of nano/microscale materials [53]. There has been considerable research on the growth of ZnO materials with different substrates' structures to fabricate superhydrophobic surfaces. Figure 7.7e shows a superhydrophobic cotton textile by introducing a $ZnO@SiO_2$ nanorod array to the woven fiber network followed by hydrophobic surface modification. The $ZnO@SiO_2$ nanorod array coated cotton textile exhibited highly durable superhydrophobic properties.

7.2.3 Fabricating Technologies for Hydrophobic Smart Energy Fibers and Textiles

Various functional applications drive research on fabricating technologies for smart superhydrophobic energy fiber and textile surfaces. As for the fabricating technologies, it mainly includes two aspects: superhydrophobic coating materials and their corresponding coating methods. Most research is focused on developing new superhydrophobic coating materials, which is suitable for application in smart energy fibers and textiles energy device.

The following discussion will also cover how different types of superhydrophobic coatings materials improve smart energy fibers and textiles' hydrophobic performance.

Inorganic superhydrophobic coatings materials have gained wide attention. For example, Jeong et al. [54] developed an encapsulation barrier for photo-rechargeable energy fabric, which prevented moisture permeability even during the washing process by introducing a SiO_2-polymer composite capping layer. The chemical stability imparted by a reaction between SiO_2 and Al_2O_3 can prevent the phase transitions that cause degradation. It was found that the proposed encapsulation barrier exhibited little change in characteristics even after 20 washings with 10 minutes cycles (Figure 7.8a).

Organic superhydrophobic coatings materials were also used to prepare superhydrophobic coatings. This includes the heat-shrink tube (polyvinyl chloride [PVC], polyethylene terephthalate [PET]), polyacrylamide (PAM), polymethyl methacrylate (PMMA), and polyester fabrics. The solid-type SC was fabricated by weaving dendritic cable positive and negative electrodes together with cotton wires. The electrode is also encapsulated in a heat-shrink tube (Figure 7.8b). With good mechanical and charge–discharge stability, it could be charged up to 1.8 V, outputting an energy density of 0.1408 mWh/cm and power density of 3.01 mW/cm [55]. Li et al. [56] have successfully constructed a high-performance waterproof and stretchable yarn zinc ion battery (ZIB) by double-helix yarn electrodes and a cross-linked PAM-based polymer electrolyte (Figure 7.8c). The solid-state yarn ZIB also demonstrates superior waterproof capability (96.5% capacity retention after 12 h soaking). These properties enable the yarn ZIBs to serve as good and reliable energy storage devices for flexible and wearable applications by combining with high electrochemical performance.

What is more, as present in Figure 7.8d, Fan and coworkers present a solution to powering portable electronics in a wearable way by fabricating an all-solid photovoltaic textile. Each string in the textile is a new type of wire-shaped solar cell based on two twisted fiber electrodes. Both of their electrodes possess a excellent waterproof capability, which is encapsulated by PMMA [57]. In recently, polyester fabrics

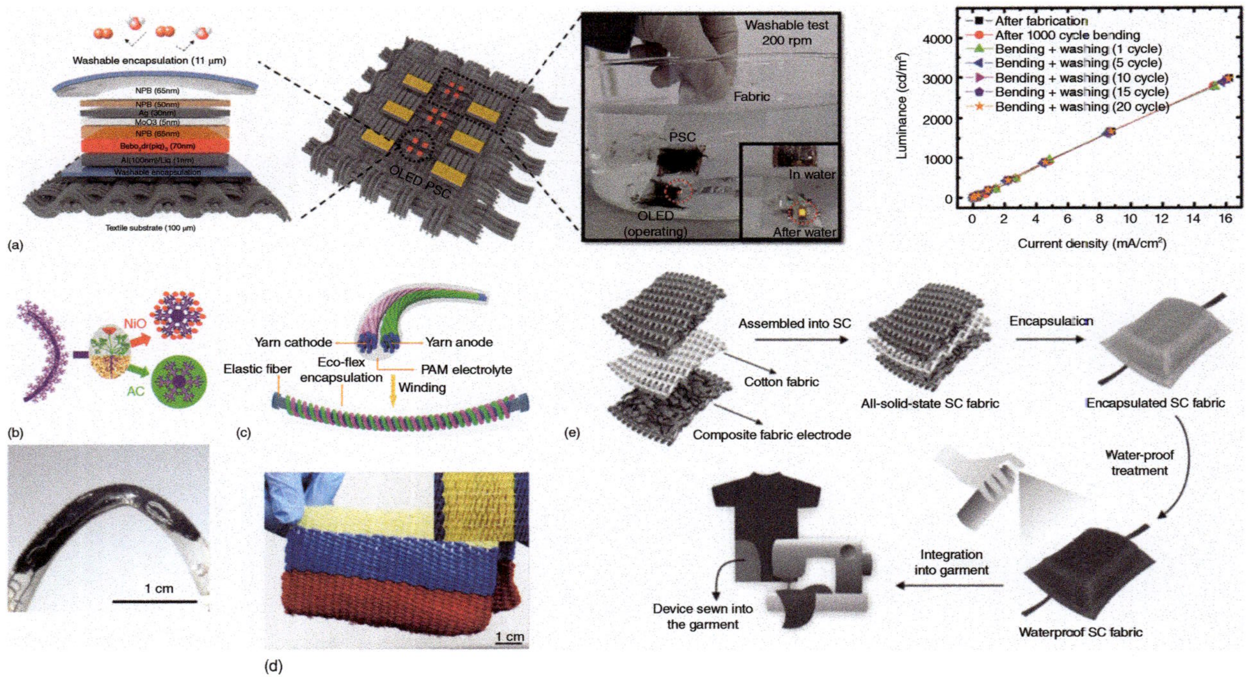

Figure 7.8 Different types of superhydrophobic coatings used for smart energy fibers and textiles: (a) construction of the photo-rechargeable energy fabric and the energy fabric's output stability under watering. Source: Adapted with permission from Jeong et al. [54]. Copyright (2019) by Royal Society of Chemistry. (b) Photo of a cable-like flexible supercapacitor packaged using a plastic tube. Source: Adapted with permission from Yang et al. [55]. Copyright (2017) by American Chemical Society. (c) Schematic diagram of fabrication and encapsulation of the yarn ZIB. Source: Adapted with permission from Li et al. [56]. Copyright (2018) by American Chemical Society. (d) A photograph of an as-fabricated photovoltaic textile mixed with colored wool wires. Source: Adapted with permission from Zhang et al. [57]. Copyright (2015) by WILEY-VCH Verlag GmbH & Co. KGaA, Weinheim. (e) Schematic illustration of the fabrication, encapsulation, and integration of all-solid-state fabric-type SCs. Source: Adapted with permission from Yang et al. [58]. Copyright (2017) by WILEY-VCH Verlag GmbH & Co. KGaA, Weinheim.

have received much attention for their good waterproof properties. For example, all-solid-state fabric-type SC devices encapsulated by polyester fabrics can reach a record-breaking performance of 2.7 F/cm at 20 mA/cm at the first charge–discharge cycle, and 10 h continuous underwater operation in Figure 7.8e [58]. The SC devices easily assemble into tandem structures and integrate into garments by simple sewing.

To conclude, hydrophobic performance is vital for the energy fibers and textiles devices for daily-life applications. Furthermore, the energy device's hydrophobic performance depends significantly on the hydrophobic coating materials and their corresponding coating methods. For the hydrophobic coating materials, various coating materials, including inorganic materials (silicon-based materials, carbon-based materials, and metal oxides materials), organic materials (PS, PE, PP, and fluorinated acrylic), and inorganic–organic materials composite and been prepared to enhance the hydrophobic performance of the fibers or fabrics. For the hydrophobic coating methods, it has been proved that high hydrophobic performance could also be achieved for various coating methods. These methods include covalent LBL assembly, polymer roughening, chemical vapor deposition, sol–gel processing, and hydrothermal synthesis, which can further improve the overall hydrophobic performance of the textile-typed devices. With multidisciplinary efforts, a waterproof wearable device's rapid development will be realized soon.

7.3 Endurability of Smart Energy Fibers and Textiles

Significant progress in smart energy fibers and textiles has been achieved for energy efficiency, stretchability, and hydrophobicity [59]. However, there is still a gap in practical applications on smart energy devices' stability even under a rough working environment.

Herein, we have discussed the potential difficulties and challenges for smart energy fibers and textiles' stability, including mechanical stability, chemical stability, and other working stability under a complex environment as below.

7.3.1 Mechanical Stability of Smart Energy Fibers and Textiles

In practical use, the smart energy fibers and textiles may encounter conditions of bending, folding, and other mechanical deformations, affecting their stabilities and service life. The effects of bending, twisting, and folding on the performance of smart energy fibers and textiles have been investigated by researchers [60].

1) The effects of bending on smart energy textile have been tested by many researchers [57]. As shown in Figure 7.9a, a colorful and flexible photovoltaic textile with various knitting patterns was demonstrated as a wearable power source for portable and flexible electronics. The photovoltaic textile still works very well, even at a bending angle of 120°. Figure 7.10b presents the CV curves of woven supercapacitor (W-SC) bend at 180° 1000 times at a scanning rate

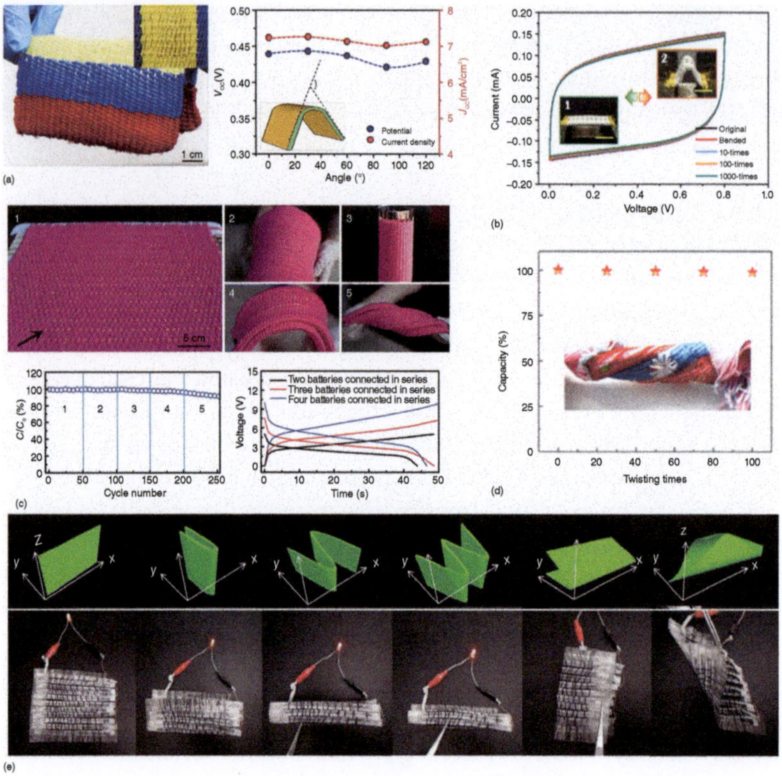

Figure 7.9 Mechanical stability of the smart energy textiles: (a) photovoltaic performance of the photovoltaic textile at different bending angles. Source: Adapted with permission from Zhang et al. [57]. Copyright (2016) by WILEY-VCH Verlag GmbH & Co. KGaA, Weinheim. (b) Effects of twisting and bending on the device resistances. Source: Adapted with permission from Chen et al. [61]. Copyright (2018) by Elsevier. (c) Energy textile woven with fiber-shaped LIBs under bending, folding and twisting. Source: Adapted with permission from Zhang et al. [62]. Copyright (2016) by Royal Society of Chemistry. (d) Performance of Zn/MnO_2 battery fabric with different twisting times. Source: Adapted with permission from Zhang et al. [63]. Copyright (2020) by Elsevier. (e) Photographs of the fabricated wearable Li-air battery powering a commercial LED under various bending and twisting conditions. Source: Adapted with permission from Liu et al. [64]. Copyright (2016) by WILEY-VCH Verlag GmbH & Co. KGaA, Weinheim.

of 50 mV/s [61]. According to the results, the proposed W-SC has excellent flexibility and mechanical durability.

2) The effects of twisting on smart energy textile have also been investigated by many researchers. As demonstrated in Figure 7.9c, a new fiber-shaped aqueous lithium-ion battery is developed using a polyimide/CNT hybrid fiber as the anode and $LiMn_2O_4$/CNT hybrid fiber the cathode [62]. This battery outputs a power density of 10 217.74 W/kg. It can be continuously fabricated through a solution process and woven into various flexible structures, such as textiles, that can be folded and twisted into various architectures. The electrochemical performance was also maintained under a variety of deformations. Besides, as summarized in

Figure 7.9d [63], Fan and coworkers have developed a photo-rechargeable fabric using an industrial shuttle-flying weaving process. After more than 100 repeats of continuous twisting, its capacity remains almost the same as its original value. With a collection of compelling features of lightweight, high flexibility, and softness, the energy textile paves a new way for wearable energy storage.

3) The effects of folding on smart energy textile have emerged as a focus in this area. Researchers have folded energy textile into various shapes and investigated their stability. As shown in Figure 7.9e, a wearable Li-air battery with ultrahigh energy density was fabricated based on an interlaced architecture [64]. It was found that the fabricated Li-air battery can not only be rolled up as easily as bamboo slips, but it can also be folded along other directions, meaning that the fabricated battery realizes flexible and wearable functions. Furthermore, it found that different shapes hardly influenced the discharge capacity.

As we discussed above, it can be found that many works have been done on the stability testing of intelligent energy fabrics, including the resistance to bending, folding, and twisting. In the next stage, intelligent energy fabrics can be applied to a harsher test to simulate the real-life, laying a solid foundation for daily application in the future.

7.3.2 Chemical Stability of Smart Energy Fibers and Textiles

Except for mechanical stability, the smart energy fibers and textiles' chemical stability is another issue that needs to be addressed. At present, the energy device's chemical stability is mostly characterized by the cycle number and performance retention duration.

In terms of cycle number, Zhong et al. tested the chemical stability of fiber-shaped SCs. In addition to the remarkable initial electrochemical performance, the SC fabric devices showed a positive gain in capacitance after cycling for 10 000 charge–discharge cycles at a large current density of 20 mA/cm, which was tested continuously for 26 days (Figure 7.10a). After 10 000 cycles, the fiber-shaped SC still reached a high areal capacitance of 3.2 F/cm [65]. Many researchers also tested the cycle number stability of energy textile-typed energy devices. For example, in the present study, Wang et al. [66] have designed a synchronous deposition strategy to continuously fabricate SCs energy textile based on aligned CNT composite fibers (Figure 7.10b). It exhibits the cyclic stability of the fiber-shaped SCs that retained high capacitance during 10 000 charge–discharge cycles.

In terms of the duration of performance retention, Peng and coworkers [67] tested wire-shaped PSCs' chemical stability. It is found that these wire-shaped PSCs were highly stable, which the energy conversion efficiencies had been maintained by over 70% without sealing in the air after 16 days (Figure 7.10c). Many researchers also tested the duration of performance retention of energy textile-typed energy devices. For example, Fan et al. tested the duration of performance retention of photo-charging energy textile. As presented in Figure 7.10d, the fabric could be charged up to 6.4 V under ambient solar irradiation and continue to provide electric power at a discharging current of 0.1 mA for two hours.

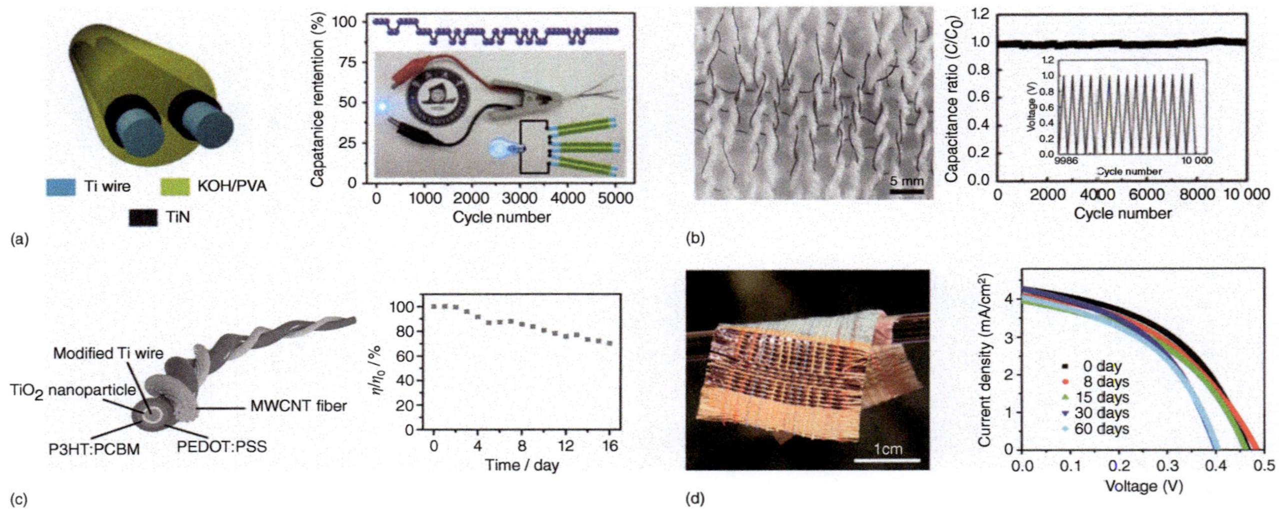

Figure 7.10 Chemical stability of the smart energy fiber and textiles: (a) schematic of the structure of TiN nanowire-based FSC and cycle performance at a current of 0.05 mA of the FSC. Source: Adapted with permission from Zhong et al. [65]. Copyright (2014) by American Chemical Society. (b) The as-fabricated supercapacitor fiber being woven into flexible textiles. Source: Adapted with permission from Wang et al. [66]. Copyright (2015) by WILEY-VCH Verlag GmbH & Co. KGaA, Weinheim. (c) Dependence of energy conversion efficiency on time for the PSC wire in the air. Source: Zhang et al. [67]. © 2014, John Wiley & Sons. (d) Structural design of the hybrid power textile and a photograph of the hybrid power textile mixed with colored wool wires. Source: Adapted with permission from Chen et al. [68]. Copyright (2016) by Macmillan Publishers Limited, parts of Springer Nature.

Furthermore, self-charging energy textile is very air-stable. The experimental results show that the energy textile can be stored for more than 60 days without significant voltage loss under the ambient environment. After charged by the solar light in the daytime, the energy fabric can directly power a cellphone [68].

The smart energy fibers and textiles' chemical stability issue should be thoroughly considered before entering our daily lives [69–74]. It is essential to develop smart energy fibers and textiles that possess highly chemical stability in a complex environment.

7.3.3 Other Working Stability Under Complicate Environment

Except for mechanical stability and chemical stability, smart energy fibers and textiles' stable performance also largely depends on various environmental conditions, such as temperature, humidity, and pressure [75, 76]. The stable performance of smart energy fibers and textiles will decrease quickly when the environmental condition is beyond the capacity of energy device can bear. Next, we will discuss the effect of various working environmental conditions such as high-temperature, low-temperature, and pressure conditions on smart energy fibers and textiles' stable performance.

1) Yin et al. reported the high-temperature condition test, which developed a highly safe flexible Li–O_2 (SFLO) battery with water/fireproof bifunctional properties (Figure 7.11a). A nail penetration experiment was executed in terms of safety, showing that the SFLO battery could release a great amount of heat to avoid an explosion. Even when exposed to flame, the SFLO battery could sustain the fire without combustion, highly confirming its extreme safety [77].
2) The low-temperature condition test was reported by Zhi and coworkers, which proposes an intrinsically freeze-resistant flexible zinc manganese-dioxide battery (Zn-MnO_2-B) [78]. This freeze-resistant battery exhibits excellent energy/power density, stable cycling performance, and remarkable flexibility at subzero temperatures, benefitting from the exceptional physicochemical properties of the anti-freezing hydrogel electrolyte (Figure 7.11b).
3) The compressive resistance condition test was reported by Zhi and coworkers [79], which successfully fabricated an extremely safe, wearable, and rechargeable solid-state ZIB using a novel designed gelatin and PAM-based hierarchical polymer electrolyte (Figure 7.11c). It can serve as a reliable power source that can work under various severe conditions, such as hammered, punctured. Equipped with surprising safety properties and outstanding electrochemical performance, the ZIB is believed to provide a new opportunity for flexible and wearable energy storage technologies.
4) Fan and coworkers reported the outdoor practical application condition test, which presented a flimsy, breathable, water-resistant, twistable, and stretchable photo-rechargeable energy fabric via a couching embroidery process on an industrial sewing machine [63]. The photo-rechargeable energy fabric was capable of providing power for electronic products sustainably. It is a green

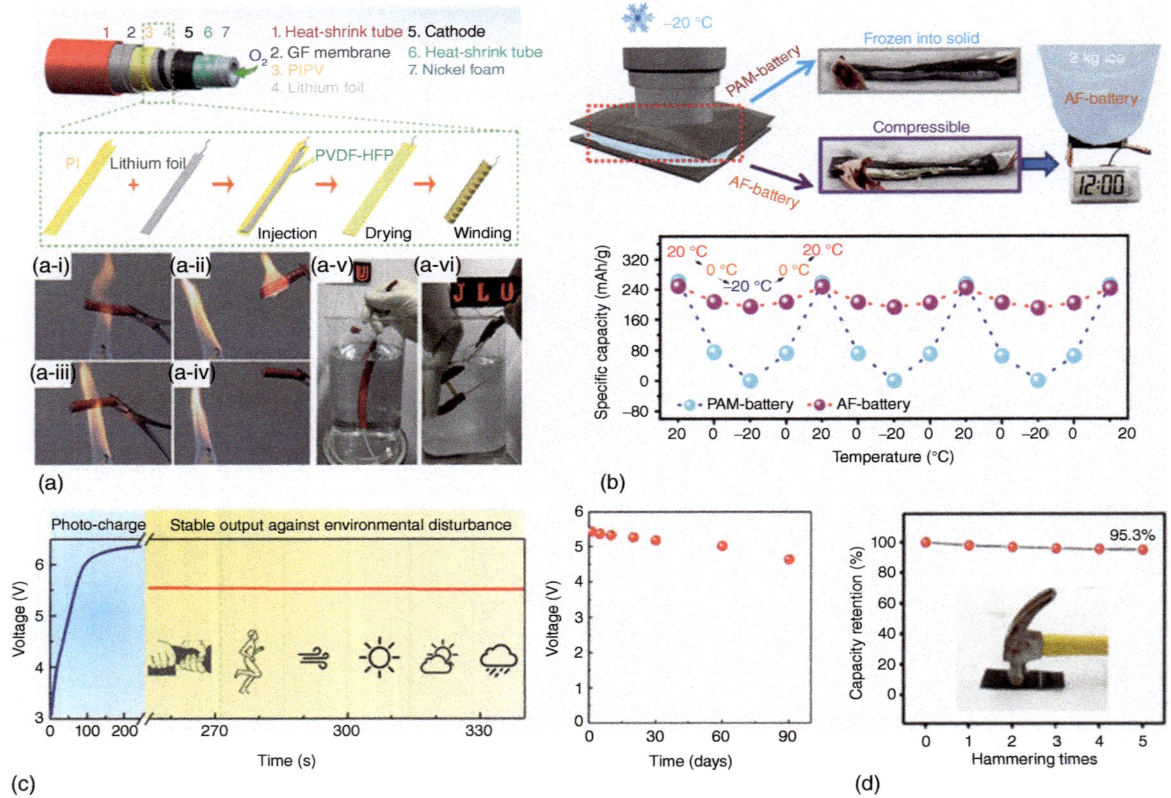

Figure 7.11 The working stability of smart energy device under a complex environment: (a) fireproof test and waterproof experiment of the SFLO battery. Source: Adapted with permission from Yin et al. [77]. Copyright (2018) by WILEY-VCH Verlag GmbH & Co. KGaA, Weinheim. (b) Photographs of the PAM-battery and cyclic testing of the PAM-battery under 20, 0, and −20 °C at 0.3 A/g. Source: Adapted with permission from Mo et al. [78]. Copyright (2019) by Royal Society of Chemistry. (c) Electrochemical performance of the solid-state rechargeable ZIB under hammering test. Source: Li et al. [79]. © 2018, Royal society of chemistry. (d) The Photo-rechargeable fabrics delivered stable electrical output under varying environmental disturbance. Source: Adapted with permission from Zhang et al. [63]. Copyright (2020) by Elsevier.

alternative to traditional portable energy harvesting and storage methods that expands the weather adaptability and wearing comfortability for applications, especially in summer (Figure 7.11d).

To conclude, smart energy fibers and textiles' stable performance should be thoroughly considered before they enter our daily life [80–82]. Moreover, the energy device's stability depends significantly on mechanical stability, chemical stability, and other working stability under complex environments. Various tested methods, including bending angle, bending times, twisting, twisting times, and folding into various shapes on smart energy fibers and textiles' stable performance, were investigated for mechanical stability. It is mostly characterized by the cycle number and the duration of the energy device's performance retention for chemical stability. The working stability under a complex environment, such as high-temperature, low-temperature, and pressure conditions, has been tested. By carefully checking the stability from all the above aspects, it is hopeful for the smart energy textile device to be applied in many wearables, skin-contacted, and implantation applications very soon.

7.4 Air Permeability of Smart Energy Fibers and Textiles

Traditional textiles can only meet people's basic daily wear needs, but it cannot meet additional functional needs such as air permeability. The air permeability of textile is the performance for air to pass through textile when there is a pressure difference between the two sides of the textile, which can be quantified by the volume of air per unit area flowing through the textile in unit time under the prescribed pressure difference, with a unit of mm/s. The excellent air permeability of the textile gives a particularly comfortable feeling for people. Because a wearable textile's impermeability will induce an inflammation reaction after long-time contact with the human skin [83], good air permeability is required, especially for summer wearable textiles. The air permeability of wearable textile is an issue that needs to be addressed.

In general, the air permeability of the textile can be affected by textile material and textile structure. Many studies have been done on the two aspects, so the following will introduce the material and structure that affect the textile's air permeability in detail.

7.4.1 The Influence of Textile Materials on Air Permeability

As we all know, the textile materials would largely affect the air permeability of the textile. Air-permeable textile materials can be divided into two categories, one is natural textiles, and another is synthetic textiles. Natural textiles, such as cotton textile, hemp textile, wool textile, usually have good air permeability. However, their material source is very limited in the natural environment. So, synthetic textiles were developed with higher air, and vapor permeability appears, mainly in nylon textile, polyester textile, and acrylic textiles.

Figure 7.12 The textiles made of different materials: (a) cotton textile [84]. (b) Hemp textile [85]. (c) Wool textile. (d) Nylon textile [86]. (e) Polyester textile [87]. (f) Acrylic textile [88].

We will discuss the air permeability of natural textiles. The first natural textile is cotton textiles (Figure 7.12a) (http://cn.dreamstime.com/%E9%BB%84%E8%89%B2%E6%A3%89%E7%BB%87%E7%89%A9%E6%9E%84%E9%80%A0%E4%BA%86%E8%83%8C%E6%99%AF%EF%BC%8C%E6%97%B6%E5%B0%9A%E6%A0%B7%E5%BC%8F%E8%AE%BE%E8%AE%A1-image110721606). It is made of cotton yarn, which is also known as cotton cloth. There are many kinds of cotton textiles, such as poplin textile, fine-spinning textile, velveteen textile. With an air permeability up to 271 mm/s, cotton textiles can be worn comfortably, widely used in clothing textiles, decorative textiles, and industrial textiles. The second natural textile is hemp textile (Figure 7.12b). Hemp textile (https://ss2.baidu.com/6ON1bjeh1BF3odCf/it/u=3090625978,2405432171&fm=27&gp=0.jpg) is made of hemp fiber. There are many kinds of hemp textiles, but only a few soft hemp textiles such as ramie, flax, and jute can be used as textile materials. Ramie is the most used hemp textile, which feels cool and would not stick to human skin, owing to its air permeability of >500 mm/s. Thus, it is generally used for summer clothing. The third natural textile is wool textile (Figure 7.12c). It is a textile made of wool. According to fineness and length, wool is divided into five kinds, which is fine wool, semi-fine wool, long wool, hybrid wool, and coarse wool. The wool textiles have been widely used in woolen cloths, wool blankets, and other wool textiles, taking advantage of the good air permeability (>356 mm/s), good warmth retention, and comfortable wearing,

We will also introduce the air permeability of synthetic textiles. The first air-permeable synthetic textile is a nylon textile (Figure 7.12d) (https://graph.baidu.com/thumb/v4/1319582352,2179427348.jpg). It is also called polyamide textile.

Table 7.1 The data of textile materials on the air permeability (mm/s).

Textile material	Air permeability (mm/s)
Cotton textile material	271
Hemp textile material	535
Wool textile material	356
Nylon textile material	255
Polyester textile material	249

Nylon is a linear structure polymer formed by amide bonds and several methylene groups. Nylon textile has certain air permeability, 255 mm/s, flexibility, and wear resistance advantageous in medical and knitwear, knitting, and silk industries. The second air-permeable synthetic textile is polyester textile (Figure 7.12e) (https://graph.baidu.com/thumb/v4/3766231601,2116922989.jpg). It is commonly known as "polyester." Polyester textile also has certain air permeability, which is 249 mm/s. Its shape preservation is excellent, with higher strength and elastic recovery ability, widely used civil and industrial textiles. Other synthetic textiles such as acrylic textile (Figure 7.12f) also have air permeability (https://graph.baidu.com/thumb/v4/2549795119,2543096601.jpg). However, there is little research focus on it, so there is no specific air permeability data.

Table 7.1 shows the air permeability of different textile materials [84]. As Table 7.1 shown, the air permeability of hemp textile is the largest, which can reach up to 535 mm/s.

In summary, textile materials are one of the main factors affecting textile air permeability. Moreover, natural textiles (cotton textile, hemp textile, wool textile) and synthetic textiles (nylon textile, polyester textile, and acrylic textile) are breathable. In general, natural textiles have better air permeability than synthetic textiles. Among them, the air permeability of natural textiles (hemp textile) reaches up to 535 mm/s, while the air permeability of synthetic textiles (nylon textile) is only 255 mm/s.

7.4.2 The Influence of Textile Structure Design on Air Permeability

In addition to the textile materials, textile structures also greatly influence the air permeability of textiles. The air permeability of textile can be affected by various textile structures, including woven structure made up of yarns with a cross relationship. A knitted structure is made up of yarns with a wound relationship, and an embroidered structure is made up of yarns embroidered on the textile following various decorative patterns.

The first textile structure is a woven structure. It is woven by warp and weft yarns on the loom by a specific law of the fabric, also known as shuttle weaving. According

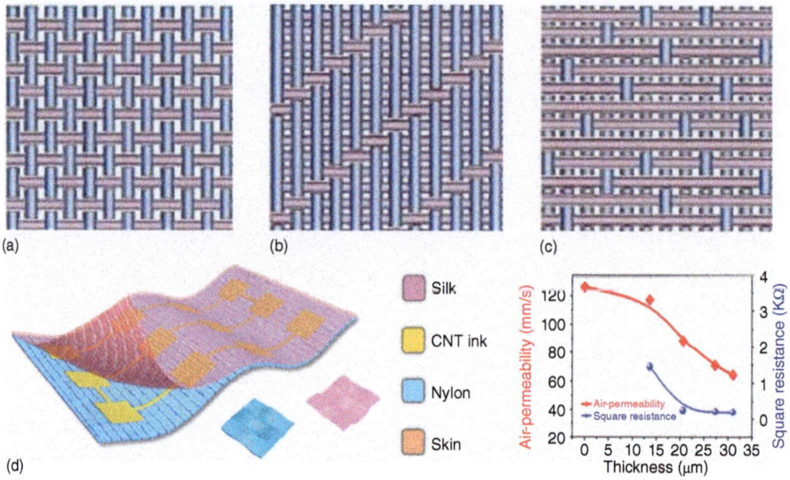

Figure 7.13 The texture structure of woven textiles: (a–c) plain textile, twill textile, stain textile.[1] (d) The structure design of the electronic textile and the air permeability of the textile. Source: Adapted with permission from Cao et al. [85]. © 2018, American Chemical Society.

to the different texture, woven textile structures can be divided into plain, twill, and satin.[1]

Plain (Figure 7.13a) is a basic tissue in which the longitude and latitude points alternate in a 1:1 ratio, so it is the simplest of all textile structures. Because it is made by interweaving the warp and weft one up and down, plain textiles have more interweave points, and their texture firm and surface level. Furthermore, the air permeability of plain structure textile among hemp textile is 506 mm/s.

Twill (Figure 7.13b) is another form of textile texture in which the longitude and latitude point alternate at certain angles. Twill textile is less of longitude and latitude than plain textile, so it is not as firm as plain textile, but the twill textile feels soft and smooth. Furthermore, its air permeability of twill textile among hemp textile is 535 mm/s.

Satin (Figure 7.13c) is also a form of textile texture in which the longitude and latitude points alternate are longer and have fewer interlacing points. Satin textile has a soft texture, smooth and even surface, and it is full of luster. More importantly, satin textile has excellent air permeability, 677 mm/s among hemp textile, so it is widely used in textiles. For example, Cao et al. [85] utilized conductive CNTs and screen-printing technology to fabricate an E-textile whose texture structure is woven, as illustrated in Figure 7.13d. Moreover, this kind of E-textile embraced high air permeability.

The second textile structure is the knitted structure. It refers to textile in which yarn is formed into a coil by knitting needles, and then the coils are interwoven in a series. According to the different modes of production, it is divided into weft knitting and warp knitting. Its basic structure is shown in Figure 7.14a. Because

1 https://baike.baidu.com/tashuo/browse/content?id=aa3243d2503c801aa42f9845&fr=vipping.

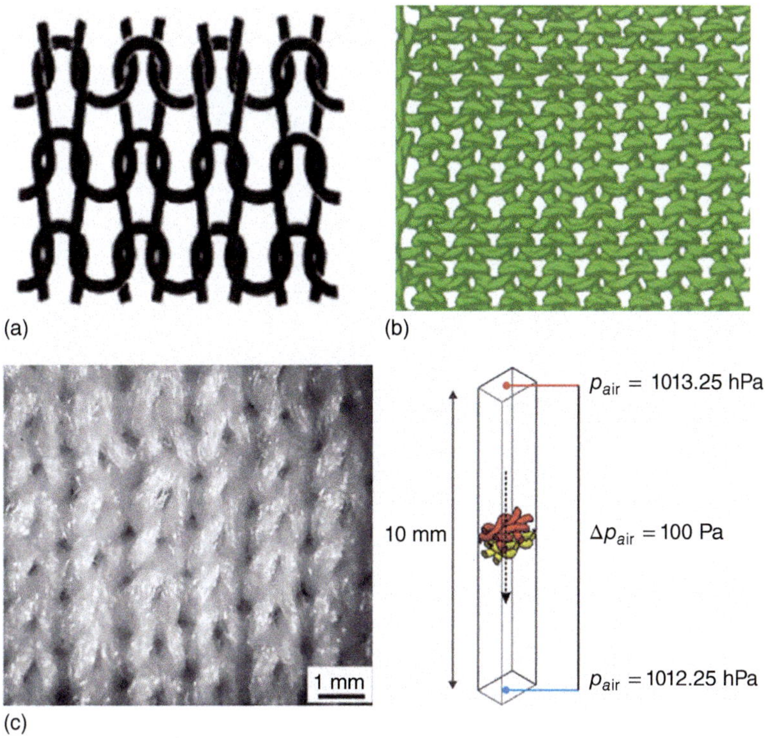

Figure 7.14 The structure of knitted textiles and its textile: (a, b) The basic structure of knitted textile and knitted textile.[2] (c) Optical microscopic images of double layer knitted textile and simulation conditions for calculating the air permeability of knitted textile (dashed line indicate the airflow). Source: Adapted with permission from Puszkarz and Krucińska [86]. Copyright (2018) by Autex Research Journal.

knitted textiles (Figure 7.14b)[2] are formed by porous coils, they have greater air permeability. For example, Puszkarz and coworker [86] researched that 10 double-layer knitted textiles with potential application in multilayer garments tested their air permeability among the textiles geometrical structure and different composition. The results showed that the knitted textile's air permeability strongly depends on the yarn's thickness and geometrical parameters (Figure 7.14c).

The third textile structure is the embroidered structure and the general term for various decorative patterns embroidered on the textile by needle and thread. The couching embroidery process can be divided into four steps:

1) A conductive thread pierced down through a bottom tulle following the lead of a needle;
2) The end of the conductive thread rotated in a circle with the assisting of a shuttle hook, during which the conductive thread would also circle the bottom thread;
3) The needle would lift above the bottom tulle again, and the conductive thread would be tightened with the bottom thread;

2 https://ss2.baidu.com/6ON1bjeh1BF3odCf/it/u=2204487042,2245886036&fm=15&gp=0.jpg.

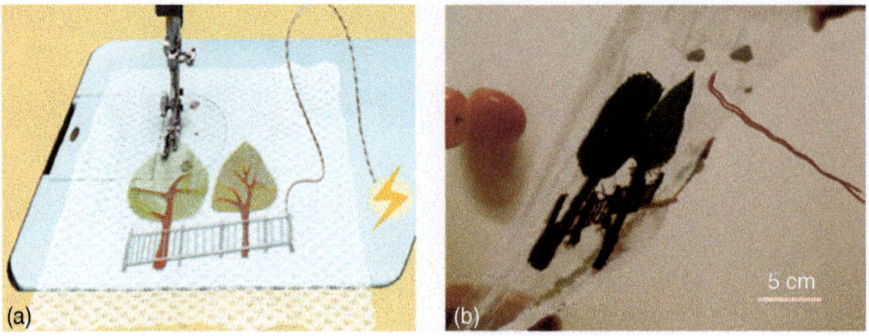

Figure 7.15 Embroidered textiles: structure design of a hybrid energy textile embroidered on a stitching machine and the energy textile's air permeability. Source: Adapted with permission from Huang et al. [87]. Copyright (2019) by American Chemical Society.

4) A cable photoanode fed in-between the conductive thread and the bottom tulle would move a small step together with the bottom tulle following a specified trace.

By repeating the embroidery circles, a photovoltaic fabric with arbitrary size and pattern can be constructed. The embroidered textile also has certain air permeability. For example, Huang et al. [87] proposed a reliable and industrial scalable electrode assembling strategy of couching embroidery to provide a light and breathable self-charging wearable power source. The energy textile is breathable to allow mild wind to blow through without any difficulties (Figure 7.15a).

In conclusion, textile materials and textile structures are important factors affecting textile air permeability. Textile materials include natural textiles (cotton textile, hemp textile, wool textile) and synthetic textiles (nylon textile, polyester textile, and acrylic textile). In general, natural textiles have better air permeability than synthetic textiles. Textile structures include woven structure, knitted structure, and embroidered structure. By varying the textile materials and textile structure, the air permeability can be largely adjusted and further optimized.

7.5 Color-Change Ability of Smart Energy Fibers and Textiles

Smart color-changeable textiles are increasingly investigated as a development direction of wearable devices, owing to their applications in no mitting displays, sensors, military camouflage, and ornaments [88]. The color-changeable textile can significantly change its color under an external electrical, optical, and thermal excitation. According to the principle of color change, color-changeable textiles can be divided into electrochromic textiles, photochromic textiles, and thermochromism textiles. Electrochromic textile has optical properties, such as reflectivity, transmittance, and absorptivity, that can reversibly change under an applied electric field. Photochromic textiles refer to the textile that can produce a color change after being excited by a light source with a different wavelength. Thermochromism textile refers to the textile that the color-change occurs when heated or cooled [89].

In general, color-changeable materials are the most important factor affecting the color-change ability of textiles. Therefore, the following will introduce the color-changeable materials and the research progress for applications in color-changeable textiles.

7.5.1 Color-Changeable Materials

There are three typical color-changeable materials: electrochromic materials, photochromic materials, and thermochromic materials.

Electrochromic materials can be divided into the inorganic type and organic type. For the inorganic electrochromic materials, inorganic electrochromic materials can be divided into two types. One is the anodic color-changing material such as the oxide of Ir, Rh, Ni, and Co, which is colorless at a low valence state and will change its color at a high valence state. The other is the cathodic color-changing materials such as the oxide of W, Mo, V, Nb, and Ti, which is colorless at a high valence state and will change its color at a low valence state. As a typical representative of electrode coloring material, WO_3 [90] thin film is the earliest and most thoroughly studied. For example, Ma et al. [91] had demonstrated that vertically aligned WO_3 nano-structure films could be fabricated on fluorine-doped tin oxide (FTO)-coated glass substrates using a template-free hydrothermal technique. These nano-structured films exhibited enhanced electrochromic performance (Figure 7.16a).

The organic electrochromic materials like polythiophene, poly-pyrrole, and PANI are mainly used due to their good optical quality, fast color conversion, and good reversibility of circulation.

1) Polythiophene is a common conductive polymer used in organic solar cells, chemical sensors, and electrochromic devices (ECDs). PEDOT, a derivative of polythiophene, is an important material in preparing ECDs due to its high conductivity and solution processability. Moreover, its color change can be applied to electrochromic textiles and devices. For example, Li et al. [92] described that two kinds of electrochromic materials, PEDOT and poly(3-methylthiophene), can be deposited on the surface of the stainless-steel wires (SSWs) by electrochemical polymerization. These electrochromic fibers showed different colors on varying the applied voltages (Figure 7.16b).
2) Poly-pyrrole is a kind of heterocyclic conjugated conductive polymer with high stability and electrical conductivity. It is widely used in electrochromic materials and SCs. For example, Huang et al. [93] created an electrochromic textile by selectively depositing poly-pyrrole on commercially available polyester-covered cotton textile. The color-changing properties of this new smart textile were examined by applying a particular current. When the textile was subjected to a current of 120 mA, the temperature increased gradually so that the letters “D” turned from green to colorless (Figure 7.16c).
3) PANI has special electrical and optical properties. One of the important characteristics of PANI is the electrochromic property. For example, Xin et al. [95] used

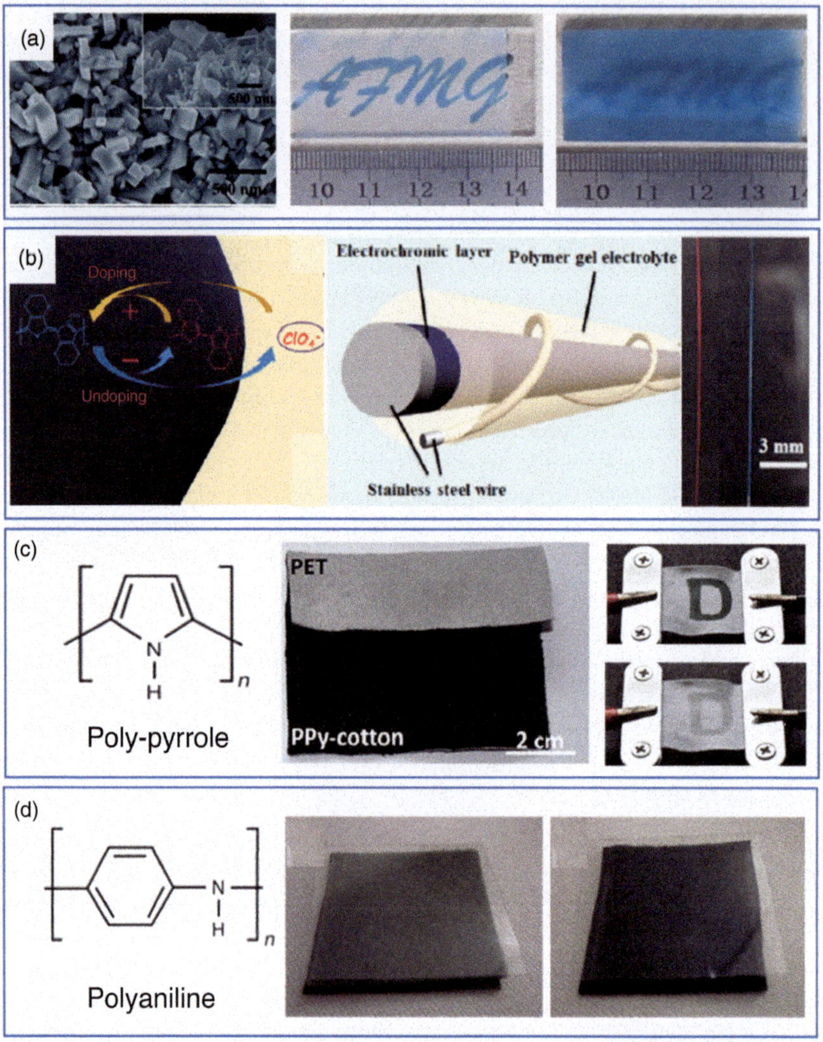

Figure 7.16 Chemical structure of various electrochromic materials: (a) FE-SEM image of WO_3 films and the digital photographs of the as-prepared cylindrical WO_3 nanorod array film at different stages: as-prepared and colored at −1.0 V. Source: Adapted with permission from Ma et al. [91]. Copyright (2012) by The Royal Society of Chemistry. (b) The fiber's electrochromic mechanism with PEDOT as an electrochromic layer and the structure diagram of the electrochromic fibers and the color change with the voltage. Source: Adapted with permission from Li et al. [92]. Copyright (2014) by American Chemical Society. (c) The poly-pyrrole structure and the photograph of a single-sided conductive fabric and photographs of the textile showing a smart color-changing effect by applying currents of 120/0 mA. Source: Adapted with permission from Huang et al. [93]. Copyright (2016) by The Royal Society of Chemistry. (d) The polyaniline structure and the flexible ECD containing viscose-polyaniline composite and the application of −3 V for 1 min. Source: Adapted with permission from Kelly et al. [94]. Copyright (2013) by Elsevier.

polyvinyl alcohol (PVA) as a spinning carrier and PANI to blend the synthesized PANI polyvinyl alcohol solution wet-spun the electrochromic fiber. Kelly et al. [94] had prepared a PANI composite with a flexible nonwoven textile spacer (viscose or PET). The prepared composite was employed within a flexible four-layer ECD. A reversible color change of the PANI-nonwoven composite is observed, from green to blue, when an electrical voltage (±3 V) is applied to the prepared device (Figure 7.16d).

Photochromic materials can be classified as screw hydrocarbons such as spiropyrans [96], dyes such as aromatic azo compounds [97], and metal halides such as AgX (X is a halogen) [98]. Thermochromic materials include CuI, PbO, Hg_4HgI_4, and Fe_2O_3. They have been widely applied in plate-type color-changeable devices. Although there is little research for photochromic and thermochromic materials in color-changeable fibers and textiles, they still exhibit potential for future applications.

In summary, color-changeable materials with different color-changing principles, including electrochromic, photochromic, and thermochromic, have been applied in the color-changeable textile. Electrochromic materials are mostly used for color-changeable textiles compared with photochromic and thermochromic materials. The electrochromic materials can be divided into inorganic electrochromic materials and organic electrochromic materials. Among inorganic electrochromic materials, WO_3 is the most widely used in color-changeable energy textiles. Among organic electrochromic materials, polymer electrochromic materials, such as polythiophene, poly-pyrrole, and PANI, are mainly used in color-changeable energy textiles. There are still many challenges for more color-changeable materials to be applied in textile.

7.5.2 Color-Changeable Textiles

Color-changeable textiles are very promising for use in daily life, such as none mitting displays, sensors, military camouflage, and ornaments. However, there are still some challenges in the application of color-changing textiles. For one thing, scalable encapsulation technologies are still in need of efficiently avoiding electrolyte leakage. At the same time, simultaneously maintain the advantages of the fiber-shaped or interwoven shaped substrate. The color transitions of some color-changeable materials are irreversible or very slow after the electricity stimuli. Many efforts have been devoted to solving these problems. Not only all-solid color-changeable textiles have been explored by developing new solid-type electrolyte to facilitate the encapsulation process, but also various color-changeable materials, such as polythiophene, poly-pyrrole, and PANI, have been explored and optimized to obtain color-changeable textile with good optical quality, fast color conversion, and good reversibility of circulation.

For example, Lu et al. [99] prepared a color-changeable textile based on polydiacetylene (PDA) composite fiber. The PDA composite fiber is produced by incorporating peptide-modified PDA into aligned CNTs on an elastic fiber substrate.

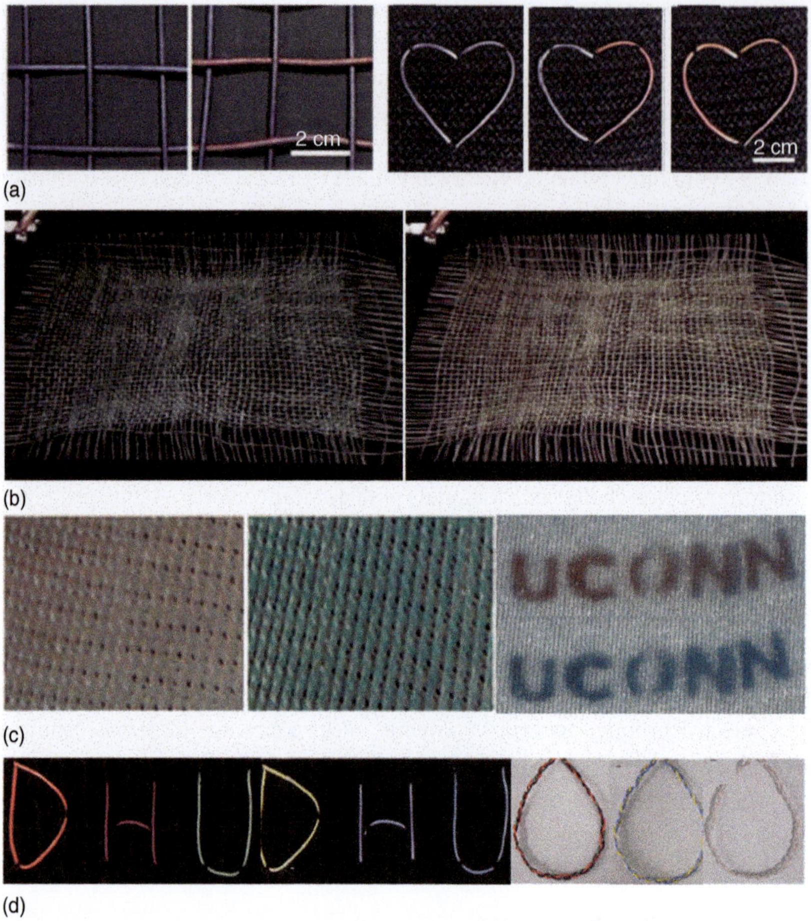

Figure 7.17 Different types of color-changing textiles: (a) photographs of an electrothermal chromatic textile with the transverse fibers changed from blue to red and electrothermal chromatic fibers woven into a textile in a heart shape displaying color changes selectively. Source: Adapted with permission from Lu et al. [99]. Copyright (2016) by The Royal Society of Chemistry. (b) Textiles based on the electrothermal effect. Source: Adapted with permission from Laforgue et al. [100]. Copyright (2014) by American Chemical Society. (c) Stretched electrochromic spandex textile electrode in different states. Source: Adapted with permission from Invernale et al. [101]. Copyright (2010) by the American Chemical Society. (d) A series of stretchable electrothermal chromatic fibers with patterns of letters "D," "H," and "U" are implanted into the textile, and the photographs of smart bracelets braided using various electrothermal chromatic fibers. Source: Adapted with permission from Li et al. [102]. Copyright (2017) by The Royal Society of Chemistry.

The color-changeable textile performs a rapid and reversible chromatic transition in response to electrical current. Due to their high flexibility and stretchability, these composite fibers can be integrated into different patterns and woven into smart textiles for displaying and sensing applications (Figure 7.17a)

Laforgue et al. [100] prepared a color-changeable textile based on electrochromic fibers. The electrochromic fibers can also be produced by coating a thermochromic layer onto the electrically conductive core. The electric current passage resulted in a Joule effect from the conductive core layer triggering a color change of the outer thermochromic layer. Electrochromic textiles showed a color change between green and beige on switching the electric current on and off (Figure 7.17b), and the reversible color change properties were also successfully demonstrated.

Invernale et al. [101] prepared an electrochromic textile made by spray-casting an electrochromic precursor onto one side of the conductive spandex and assembling two conductive spandex electrodes face-to-face that the modified layers were in contact. The electrochromic textile generated two colors, red and blue, due to the different redox states of the electrochromic polymer and the time taken for the electrochromic polymer to switch 0.3 s (Figure 7.17c).

Li et al. [102] prepared a color-changeable textile based on a stretchable and multicolor electrothermal chromatic fiber. The colors of the fibers can be switched within 15 s. At an applied current density of 284 mA/cm, the colorful smart fibers (patterned as "D," "H," and "U") quickly change surface color once the fiber temperature reaches 38 °C. As the temperature reaches 45 °C, the surface color will turn to white (Figure 7.17d). More importantly, they are easily woven into textiles and plaited into colorful hand chains.

7.6 Conclusions

In conclusion, textile materials are the most important factors affecting textile color-changeable. Among them, polythiophene, poly-pyrrole, and PANI are widely used in smart energy textiles. Nowadays, there are only a few research have been done on the aspect of electrochromic textiles. Besides, the possibility of combining different color-changeable materials with different color-changeable mechanisms is worth exploring, which may provide better environmental adaptability for complex application scenes in the future.

This chapter provides an overview of the function expansion of smart energy fibers and textiles. Before the smart energy fibers and textiles' approval for commercialization and biomedical use, many functions have been required, such as stretchability, hydrophobicity, stability, air permeability, and color-changing ability.

In terms of stretchability, various conducting materials, including carbon nano-materials, conductive polymer, metal nano-structure, have been incorporated into elastomeric matrices to develop the electrode with high stretchability. Various structures, such as buckled structure, open-mesh structure, coiled structure, and sponge structure, have been proposed to improve the overall stretchability of the textile-typed devices. Although many types of the stretchable electrode have been developed and investigated thus far, this is still much room for enhance the electrical conductivity of materials and simplify electrode structure design which is the main obstacle toward large-scale applications and commercialization.

Although superhydrophobic surfaces have been subjected to intensive investigations, further investigations are still necessary for both fundamental and applicational aspects of hydrophobic performance. For example, the fabrication process of superhydrophobic surfaces needs to be simplified. The durability and robustness are necessary to be improved for practical applications.

In terms of stability, it is essential for the wearable, skin-contacted, and implantation applications. All the materials and processes to fabricate the smart energy fibers and textiles should be cautiously examined to eliminate any risk.

In terms of air permeability, smart E-textiles' air permeability can be optimized by different patterns, which can satisfy various personal, comfortable requirements, especially for a cool and comfortable wearing experience in summer. Researchers have done a lot on breathable materials and the textile structure design. Nevertheless, developing some E-textiles with air permeability is a long and arduous task.

In terms of color-changing ability, textiles will go beyond their traditional protection and fashion roles and consequentially offer people more intelligent services. Among them, smart color-changing fibers or fabrics have become a hotspot for their various potential applications. However, the fabrication of flexible electrochromic fabrics is difficult since it requires real, flexible transparent electrodes to display the color changes and devices' encapsulation to avoid electrolyte leaks. Therefore, it is crucial to find a suitable way to change and control the fiber colors.

Smart E-textiles, equipped with multiple functions, like stretchability, hydrophobicity, stability, air permeability, and color-changing ability, will better meet people's needs and facilitate E-textile's wide application in the future.

References

1 Kang, S.K., Koo, J., Lee, Y.K., and John, A.R. (2018). Advanced materials and devices for bioresorbable electronics. *Accounts of Chemical Research* 51 (5): 988–998.

2 Cheng, T., Zhang, Y.Z., Lai, W.Y., and Huang, W. (2015). Stretchable thin-film electrodes for flexible electronics with high deformability and stretchability. *Advanced Materials* 27 (22): 3349–3376.

3 Hong, J.Y., Kim, W., Choi, D. et al. (2016). Omnidirectionally stretchable and transparent graphene electrodes. *ACS Nano* 10 (10): 9446–9455.

4 Lipomi, D.J., Lee, J.A., Vosgueritchian, M. et al. (2012). Electronic properties of transparent conductive films of PEDOT: PSS on stretchable substrates. *Chemistry of Materials* 24 (2): 373–382.

5 Ding, H., Zhong, M., Wu, H. et al. (2016). Elastomeric conducting polyaniline formed through topological control of molecular templates. *ACS Nano* 10 (6): 5991–5998.

6 Hur, J., Im, K., Kim, S.W. et al. (2014). Polypyrrole/agarose-based electronically conductive and reversibly restorable hydrogel. *ACS Nano* 8 (10): 10066–10076.

7 Wang, Y., Zhu, C., Pfattner, R. et al. (2017). A highly stretchable, transparent, and conductive polymer. *Science Advances* 3 (3): 1602076.

8 Stoyanov, H., Kollosche, M., Risse, S. et al. (2013). Soft conductive elastomer materials for stretchable electronics and voltage controlled artificial muscle. *Advanced Materials* 25 (4): 578–583.

9 Guo, L., Ma, M., Zhang, N. et al. (2014). Stretchable polymeric multielectrode array for conformal neural interfacing. *Advanced Materials* 26 (9): 1427–1433.

10 Guan, Y.S., Zhang, Z., Tang, Y. et al. (2018). Kirigami-inspired nanoconfined polymer conducting nanosheets with 2000% stretchability. *Advanced Materials* 30 (20): 1706390.

11 Niu, X.Z., Peng, S.L., Liu, L.Y. et al. (2007). Characterizing and patterning of PDMS-based conducting composites. *Advanced Materials* 19 (18): 2682–2686.

12 Lv, R.H., Xu, W.F., Na, B., and Chen, B.J. (2008). Insight into the role of filler network in the viscoelasticity of a carbon black filled thermoplastic elastomer: a strain dependent electrical conductivity study macromol. *Journal of Macromolecular Science Part B Physics* 47 (4): 774–782.

13 Shin, M.K., Oh, J., Marcio, L. et al. (2010). Elastomeric conductive composites based on carbon nanotube forests. *Advanced Materials* 22 (24): 2663–2667.

14 Chen, Z.P., Ren, W.C., Gao, L.B. et al. (2011). Three-dimensional flexible and conductive interconnected graphene networks grown by chemical vapour deposition. *Nature Materials* 10 (6): 424–428.

15 Kim, Y., Zhu, J., Yeom, B. et al. (2013). Stretchable nanoparticle conductors with self-organized conductive pathways. *Nature* 500 (7460): 59–63.

16 Moon, G.D., Lim, G.H., Song, J.H. et al. (2013). Highly stretchable patterned gold electrodes made of Au nanosheets. *Advanced Materials* 25 (19): 2707–2712.

17 Gong, S., Schwalb, W., Wang, Y. et al. (2014). A wearable and highly sensitive pressure sensor with ultrathin gold nanowires. *Nature Communication* 5: 3132.

18 Han, S., Kim, M.K., Wang, B. et al. (2016). Mechanically reinforced skin-electronics with networked nanocomposite elastomer. *Advanced Materials* 28 (46): 10257–10265.

19 Wu, G., Asai, S., Zhang, C. et al. (2000). A delay of percolation time in carbon-black-filled conductive polymer composites. *Journal of Applied Physics* 88 (3): 1480–1487.

20 Zhang, H., Liu, N.S., Shi, Y.L. et al. (2016). Piezoresistive sensor with high elasticity based on 3D hybrid network of sponge@CNTs@Ag NPs. *ACS Applied Materials & Interfaces* 8 (34): 22374–22381.

21 Luo, S.D. and Liu, T. (2013). SWCNT/graphite nanoplatelet hybrid thin films for self-temperature-compensated, highly sensitive, and extensible piezoresistive sensors. *Advanced Materials* 25 (39): 5650–5657.

22 Shin, M., Song, J.H., Lim, G.H. et al. (2014). Highly stretchable polymer transistors consisting entirely of stretchable device components. *Advanced Materials* 26 (22): 3706–3711.

23 Kim, S.H., Jung, S., Yoon, I.S. et al. (2018). Ultrastretchable conductor fabricated on skin-like hydrogel-elastomer hybrid substrates for skin electronics. *Advanced Materials* 30 (26): 1800109.

24 Kim, K.K., Hong, S., Cho, H.M. et al. (2015). Highly sensitive and stretchable multidimensional strain sensor with prestrained anisotropic metal nanowire percolation networks. *Nano Letter* 15 (8): 5240–5247.

25 Liu, Z.F., Fang, S., Moura, F.A. et al. (2015). Hierarchically buckled sheath-core fibers for superelastic electronics, sensors, and muscles. *Science* 349 (6246): 400–404.

26 Wang, C.F., Wang, C.H., Huang, Z.L., and Xu, S. (2018). Materials and structures toward soft electronics. *Advanced Materials* 30, 1801368. https://doi.org/10.1002/adma.201801368.

27 Cheng, Y., Wang, R., Sun, J., and Gao, L. (2015). A stretchable and highly sensitive graphene-based fiber for sensing tensile strain, bending, and torsion. *Advanced Materials* 27 (45): 7365–7371.

28 Lee, H.B., Bae, C.W., Duy, L.T. et al. (2016). Mogul-patterned elastomeric substrate for stretchable electronics. *Advanced Materials* 28 (16): 3069–3077.

29 Yang, Z., Deng, J., Chen, X. et al. (2013). A highly stretchable, fiber-shaped supercapacitor. *Angewandte Chemie International Edition* 52 (50): 13453–13457.

30 Li, H., Wang, W., Yang, Y. et al. (2020). Kirigami-based highly stretchable thin film solar cells that are mechanically stable for more than 1000 cycles. *ACS Nano* 14 (5): 1560–1568.

31 Huang, F., Yu, H., Xiang, S. et al. (2020). Embroidering a filmsy photorechargeable energy fabric with wide weather adaptability. *ACS Applied Materials & Interfaces* 12 (3): 3654–3660.

32 Christopher, B.C., Kuralamudhan, A., Liu, Y. et al. (2017). Stretchable capacitive sensors of torsion, strain, and touch using double helix liquid metal fibers. *Advanced Functional Materials* 27 (20): 1605630.

33 Cong, Z., Guo, W., Guo, Z. et al. (2020). Stretchable coplanar self-charging power textile with resist-dyeing triboelectric nanogenerators and microsupercapacitors. *ACS Nano* 14 (5): 5590–5599.

34 Guo, Z., Liu, W., and Su, B.L. (2011). Superhydrophobic surfaces: from natural to biomimetic to functional. *Journal of Colloid and Interface Science* 353 (2): 335–355.

35 Victor, J.J., Facchini, D., and Erb, U. (2012). A low-cost method to produce superhydrophobic polymer surfaces. *Journal of Materials Science* 47 (8): 3690–3697.

36 Liu, C., Zhu, L., Bu, W., and Liang, Y. (2018). Superhydrophobic surfaces: from nature to biomimetic through VOF simulation. *Micron* 107: 94–100.

37 Zhang, P. and Lv, F.Y. (2015). A review of the recent advances in superhydrophobic surfaces and the emerging energy-related applications. *Energy* 82: 1068–1087.

38 Higginbotham, R.S. (1964). Waterproofing and water-repellency. *Nature* 201 (4916): 228–229.

39 Barbara, R. and Mariachiara, Z. (2009). Adsorption study of several hydrophobic organic contaminants on an aquifer material. *American Journal of Environmental Sciences* 5 (4): 507–515.

40 Brewer, S.A. and Willis, C.R. (2008). Structure and oil repellency textiles with liquid repellency to hexane. *Applied Surface Science* 254 (20): 6450–6454.

41 Mohammad, E.Y. and Mohammad, S.K. (2013). One-step synthesis of superhydrophobic coating on cotton fabric by ultrasound irradiation. *Industrial and Engineering Chemistry Research* 52 (36): 12846–12854.

42 Zhu, X., Zhang, Z., Ge, B. et al. (2014). Fabrication of a superhydrophobic carbon nanotube coating with good reusability and easy repairability. *Colloids and Surfaces A: Physicochemical and Engineering Aspects* 444: 252–256.

43 Peng, S., Tian, D., Yang, X., and Deng, W. (2014). Highly efficient and large-scale fabrication of superhydrophobic alumina surface with strong stability based on self-congregated alumina nanowires. *ACS Applied Materials & Interfaces* 6 (7): 4831–4841.

44 Li, H., Zhao, Y., and Yuan, X. (2013). Facile preparation of superhydrophobic coating by spraying a fluorinated acrylic random copolymer micelle solution. *Soft Matter* 9 (4): 1005–1009.

45 Zhang, Y., Li, J., Huang, F. et al. (2013). Controlled fabrication of transparent and superhydrophobic coating on a glassmatrix via a green method. *Applied Physics A* 110 (2): 397–401.

46 Hong, D., Ryu, I., Kwon, H. et al. (2013). Preparation of superhydrophobic, long-neck vase-like polymer surfaces. *Physical Chemistry Chemical Physics* 15 (28): 11862–111867.

47 Anitha, C., Azim, S.S., and Mayavan, S. (2017). Fluorine free superhydrophobic surface textured silica particles and its dynamics-transition from impalement to impingement. *Journal of Alloys and Compounds* 711: 197–204.

48 Wang, S., Feng, L., and Jiang, L. (2006). One-step solution-immersion process for the fabrication of stable bionic superhydrophobic surfaces. *Advanced Materials* 18 (6): 767–770.

49 Xue, C., Jia, S., Zhang, J., and Tian, L. (2009). Superhydrophobic surfaces on cotton textiles by complex coating of silica nanoparticles and hydrophobization. *Thin Solid Films* 517 (16): 4593–4598.

50 Franco, J.A., Kentish, S.E., Perera, J.M., and Stevens, G.W. (2008). Fabrication of a superhydrophobic polypropylene membrane by deposition of a porous crystalline polypropylene coating. *Journal of Membrane Science* 318 (1–2): 107–113.

51 Zimmermann, J., Reifler, F.A., Fortunato, G. et al. (2008). A simple, one-step approach to durable and robust superhydrophobic textiles. *Advanced Functional Materials* 18 (22): 3662–3669.

52 Leng, B., Shao, Z., de With, G., and Ming, W. (2009). Superoleophobic cotton textiles. *Langmuir* 25 (4): 2456–2460.

53 Wu, X., Zheng, L., and Wu, D. (2005). Fabrication of superhydrophobic surfaces from microstructured ZnO-based surfaces via a wet-chemical route. *Langmuir* 21 (7): 2665–2667.

54 Jeong, E.G., Jeon, Y., Cho, S.H., and Choi, K.C. (2019). Textile-based washable polymer solar cells for optoelectronic modules: toward self-powered smart clothing. *Energy & Environmental Science* 12 (6): 1878–1889.

55 Yang, Y., Zhang, N., Zhang, B. et al. (2017). Highly-efficient dendritic cable electrodes for flexible supercapacitive fabric. *ACS Applied Materials & Interfaces* 9 (46): 40207–40214.

56 Li, H., Liu, Z., Liang, G. et al. (2018). Waterproof and tailorable elastic rechargeable yarn zinc ion batteries by a cross-linked polyacrylamide electrolyte. *ACS Nano* 12 (4): 3140–3148.

57 Zhang, N., Chen, J., Huang, Y. et al. (2016). A wearable all-solid photovoltaic textile. *Advanced Materials* 28 (2): 263–269.

58 Yang, Y., Huang, Q., Niu, L. et al. (2017). Waterproof, ultrahigh areal-capacitance, wearable supercapacitor fabrics. *Advanced Materials* 29 (19): 1606679.

59 Stoppa, M. and Chiolerio, A. (2014). Wearable electronics and smart textiles: a critical review. *Sensors* 14 (7): 11957–11992.

60 Wang, D., Zhang, Y., Lu, X. et al. (2018). Chemical formation of soft metal electrodes for flexible and wearable electronics. *Chemical Society Reviews* 47 (12): 4611–4641.

61 Chen, J., Guo, H., Pu, X. et al. (2018). Traditional weaving craft for one-piece self-charging power textile for wearable electronics. *Nano Energy* 50: 536–543.

62 Zhang, Y., Wang, Y., Wang, L. et al. (2016). A fiber-shaped aqueous lithium ion battery with high power density. *Journal of Materials Chemistry A* 4 (23): 9002–9008.

63 Zhang, N., Huang, F., Zhao, S. et al. (2020). Photo-rechargeable fabrics as sustainable and robust power sources for wearable bioelectronics. *Matter* 2 (5): 1260–1269.

64 Liu, Q.C., Liu, T., Liu, D.P. et al. (2016). A flexible and wearable lithium-oxygen battery with record energy density achieved by the interlaced architecture inspired by bamboo slips. *Advanced Materials* 28 (38): 8413–8418.

65 Zhong, J., Zhang, Y., Zhong, Q. et al. (2014). Fiber-based generator for wearable electronics and mobile medication. *ACS Nano* 8 (6): 6273–6280.

66 Wang, B., Fang, X., Sun, H. et al. (2015). Fabricating continuous supercapacitor fibers with high performances by integrating all building materials and steps into one process. *Advanced Materials* 27 (47): 7854–7860.

67 Zhang, Z., Yang, Z., Wu, Z. et al. (2014). Weaving efficient polymer solar cell wires into flexible power textiles. *Advanced Energy Material* 4 (11): 1301750.

68 Chen, J., Huang, Y., Zhang, N. et al. (2016). Micro-cable structured textile for simultaneously harvesting solar and mechanical energy. *Nature Energy* 1 (10): 16138.

69 Liu, R., Liu, Y., Chen, J. et al. (2017). Flexible wire-shaped lithium-sulfur batteries with fibrous cathodes assembled via capillary action. *Nano Energy* 33: 325–333.

70 Chai, Z., Zhang, N., Sun, P. et al. (2016). Tailorable and wearable textile devices for solar energy harvesting and simultaneous storage. *ACS Nano* 10 (10): 9201–9207.

71 Gao, Z., Lou, Z., Chen, S. et al. (2018). Fiber gas sensor-integrated smart face mask for room temperature distinguishing of target gases. *Nano Research* 11 (1): 511–519.

72 Zhang, Y., Jiao, Y., Lu, L. et al. (2017). An ultraflexible silicon-oxygen battery fiber with high energy density. *Angewandte Chemie International Edition* 129 (44): 13929–13934.

73 Cui, N., Liu, J., Gu, L. et al. (2015). Wearable triboelectric generator for powering the portable electronic devices. *ACS Applied Materials & Interfaces* 7 (33): 18225–18230.

74 He, W., Wang, C., Wang, H. et al. (2019). Integrated textile sensor patch for real-time and multiplex sweat analysis. *Science Advances* 5 (11): eaax0649.

75 Gao, H., Zhao, Z., Cai, Y. et al. (2017). Adaptive and freeze-tolerant heteronetwork organohydrogels with enhanced mechanical stability over a wide temperature range. *Nature Communication* 8: 15911.

76 Deng, T., Zhang, W., Zhang, H., and Zheng, W. (2018). Anti-freezing aqueous electrolyte for high-performance $Co(OH)_2$ supercapacitors at −30 °C. *Energy Technology* 6 (4): 605–612.

77 Yin, Y.B., Yang, X.Y., Chang, Z.W. et al. (2018). A water/fireproof flexible lithium-oxygen battery achieved by synergy of novel architecture and multifunctional separator. *Advanced Materials* 30 (1): 1703791.

78 Mo, F., Liang, G., Meng, Q. et al. (2019). A flexible rechargeable aqueous zinc manganese-dioxide battery working at −20 °C. *Energy and Environmental Science* 12 (2): 706–715.

79 Li, H., Han, C., Huang, Y. et al. (2018). An extremely safe and wearable solid-state zinc ion battery based on a hierarchical structured polymer electrolyte. *Energy and Environmental Science* 11 (4): 941–951.

80 Jung, Y., Jeong, Y.C., Kim, J.H. et al. (2016). One step preparation and excellent performance of CNT yarn based flexible micro lithium ion batteries. *Energy Storage Materials* 5: 1–7.

81 Zhang, Y., Wang, L., Guo, Z. et al. (2016). High-performance lithium-air battery with a coaxial-fiber architecture. *Angewandte Chemie International Edition* 128 (14): 4563–4567.

82 Dong, K., Wang, Y., Deng, J. et al. (2017). A highly stretchable and washable all-yarn based self-charging knitting power textile composed of fiber triboelectric nanogenerators and supercapacitors. *ACS Nano* 11 (9): 9490–9499.

83 Miyamoto, A., Lee, S., Cooray, N.F. et al. (2017). Inflammation-free, gas-permeable, lightweight, stretchable on-skin electronics with nanomeshes. *Nature Nanotechnology* 12 (9): 907–913.

84 Zhang, J., Wang, G., Cui, J. et al. (2009). Breathability test for textiles. *Printing and Dyeing* 35 (23): 38–40.

85 Cao, R., Pu, X., Du, X. et al. (2018). Screen-printed washable electronic textiles as self-powered touch/gesture tribo-sensor for intelligent human-machine interaction. *ACS Nano* 12 (6): 5190–5196.

86 Puszkarz, A. and Krucińska, I. (2018). Modeling of air permeability of knitted fabric using the computational fluid dynamics. *Autex Research Journal* 18 (4): 364–376.

87 Huang, F., Yu, H., Xiang, S. et al. (2020). Embroidering a filmsy photo-rechargeable energy fabric with wide weather adaptability. *ACS Applied Materials & Interfaces* 12 (3): 3654–3660.

88 Gauvreau, B., Guo, N., Schicker, K. et al. (2008). Color-changing and color-tunable photonic bandgap fiber textiles. *Optics Express* 16 (20): 15677–15693.

89 Gong, Y. and Chen, Q. (1995). Thermochromic materials and their applications. *Journal of Northwestern University (Natural Science Edition)* 25 (5): 537–541.

90 Buckley, D.N., Burke, L.D., and Mukahy, J.K. (1976). The oxygen electrode. Part seven-influence of some electrical and electrolyte variables on the charge capacity of iridium in the anodic region. *Journal of the Chemical Society, Faraday Transactions1: Physical Chemistry in Condensed Phases* 72 (1): 1896–1902.

91 Ma, D., Shi, G., Wang, H. et al. (2012). Morphology-tailored synthesis of vertically aligned 1D WO_3 nano-structure films for highly enhanced electrochromic performance. *Journal of Materials Chemistry A* 1 (3): 684–691.

92 Li, K., Zhang, Q., Wang, H., and Li, Y. (2014). Red, green, blue (RGB) electrochromic fibers for the new smart color change fabrics. *ACS Applied Materials & Interfaces* 6 (15): 13043–13050.

93 Huang, G., Liu, L., Wang, R. et al. (2016). Smart color-changing textile with high contrast based on a single-sided conductive fabric. *Journal of Materials Chemistry C* 4 (32): 7589–7594.

94 Kelly, F.M., Meunier, L., Cochrane, C., and Koncar, V. (2013). Polyaniline: application as solid state electrochromic in a flexible textile display. *Displays* 34 (1): 1–7.

95 Xin, B., Tian, Y., Yang, Z. et al. (2013). Development of polyaniline electrochromic fabrics. *Progress in Textile Science&Technology* 3: 1–4.

96 Zhang, X., Deng, X., and Bian, D. (2007). Application and development of photochromic materials in textiles. *Proceedings of the 7th Symposium on Functional Textiles and Nanotechnology*, vol. 4, pp. 409–412.

97 Jin, C. and Wang, D. (1997). Photochromic materials and their applications. *Modern Trade Industry* 9: 36–37.

98 Shen, Q., Lu, C., and Xu, Z. (2005). Research and application of photochromic materials. *Material Review* 19 (10): 31–35.

99 Lu, X., Zhang, Z., Sun, X. et al. (2016). Flexible and stretchable chromatic fibers with high sensing reversibility. *Chemical Science* 7 (8): 5113–5117.

100 Laforgue, A., Rouget, G., Dubost, S. et al. (2014). Multifunctional resistive-heating and color-changing monofilaments produced by a single-step coaxial melt-spinning process. *ACS Applied Materials & Interfaces* 4 (6): 3163–3168.

101 Invernale, M., Ding, Y., and Sotzing, G. (2010). All-organic electrochromic spandex. *ACS Applied Materials & Interfaces* 2 (1): 296–300.

102 Li, Q., Li, K., Fan, H. et al. (2017). Reduced graphene oxide functionalized stretchable and multicolor electrothermal chromatic fibers. *Journal of Materials Chemistry C* 5 (44): 11448–11453.

8

Emerging Electronic Fibers and Textiles

The development of textiles reflects human civilization to a certain extent: our ancestors connected various leaves to form an early textile model, then natural materials such as silk and cotton were woven into a warmer and more comfortable reality textiles, and since the last century, various man-made fiber materials (such as Nylon and Kevlar) have gradually appeared, greatly improving our lives. Recently, with the development of electronics and the Internet, textiles are facing new challenges. In addition to making people feel warm and comfortable, textiles are now expected to show more functions. There are also hopes or requirements to make electronic devices into textiles in order to use electricity more efficiently in many fields such as displays and sensing. Electricity dominates our lives and everywhere, which means that the development of smart electronic textiles is very important and may be practically applied. Indeed, interest in this direction is increasing, and significant progress has been made in recent years. Researchers assembled conductive materials and composites, such as carbon nanotube (CNT), reduced graphene oxide (rGO), Au, Ag nanoparticles (AgNPs), Ag nanowires (AgNWs), and conductive polymers (CPs) on fibers or fabrics to give them sensing functions and develop an emerging electronic fibers and textiles.

Compared with traditional bulky or flat structures, textile structures have unique and promising advantages. Bulky electronic devices are usually rigid, and planar electronic devices can be made into flexible films, but they usually deform when bent in one dimension and often fail under twisting or other severe deformation. In contrast, similar to clothes, electronic textiles are usually composed of fibers that can effectively adapt to severe and complex deformations with high stability. Electronic textiles are also soft and can closely contact curved surfaces, which is challenging for their bulky and flat counterparts. This attribute is essential for wearable applications. In addition, electronic textiles have a low weight, which is important in microelectronics. Electronic textiles are also breathable due to the voids formed when the fibers are assembled [1].

This chapter mainly focus on the emerging electronic fibers and textiles. We examine systems for stress sensing textile, strain sensing textiles, chemical sensing textiles, and other function coupled textiles.

Textile-Based Energy Harvesting and Storage Devices for Wearable Electronics, First Edition.
Xing Fan, Nannan Zhang, and Yi Wang.

8.1 Stress Sensing Textiles

Stress sensor is a device that senses a pressure signal and converts the pressure signal into a usable output electrical signal according to a certain law. In recent years, traditional stress sensors have been limited by the nature of their own materials, flexibility, and stretchability and are increasingly unsuitable in many emerging areas where ductility is required. Therefore, the research and development of the ductility of stress sensor parts is very rapid. It is under these conditions that textile-based stress sensors, namely, stress sensing textile, which introduce the smart clothing concept, have been an interesting topic in recent years. Stress sensing textiles are prepared by assembling the active material (rGO, CNT, AgNPs, AgNW, etc.) on the fiber or fabric according to the change of its resistance, capacitance, and other properties, which show great application prospects in the fields of artificial intelligence, bionics, flexible robots, health monitoring, and human–machine interaction due to its accuracy and high agility in detecting external stress.

According to their working mechanisms, the stress sensing textiles are mainly divided into three categories: piezoresistive stress sensing textiles, capacitive stress sensing textiles, and other stress sensing textiles.

8.1.1 Piezoresistive Stress Sensing Textiles

The piezoresistive stress sensing textile is a sensor that uses the piezoresistive effect. After the conductive materials are subjected to a force, the resistivity changes, and an electrical signal output proportional to the force change can be obtained through the measurement circuit (Figure 8.1). Moreover, the structure of the device is also very important for signal output. The development of piezoresistive stress sensing textiles is inseparable from high-performance fiber electrode materials. Its development is always accompanied by the development of new fiber electrodes and the innovation of device structures.

The development of the piezoresistive stress sensing textiles is shown in Figure 8.2. It mainly can be divided into two stages.

In the first stage, piezoresistive stress sensing textiles are mainly based on interwoven strips. Someya et al. [3] demonstrated a graphite-containing rubber pressure sensor layer integrated with organic transistors to form a very wide area structure, providing an ideal solution to realize artificial skin with low-cost processing technology. Later, Fan et al. [4] reported a large-scale integration of nanowires

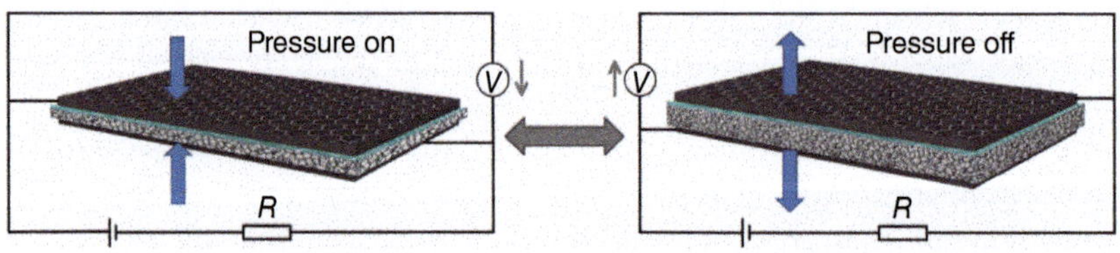

Figure 8.1 The working mechanism of piezoresistive stress sensor. Source: Lou et al. [2]. Licenced under CC BY 4.0.

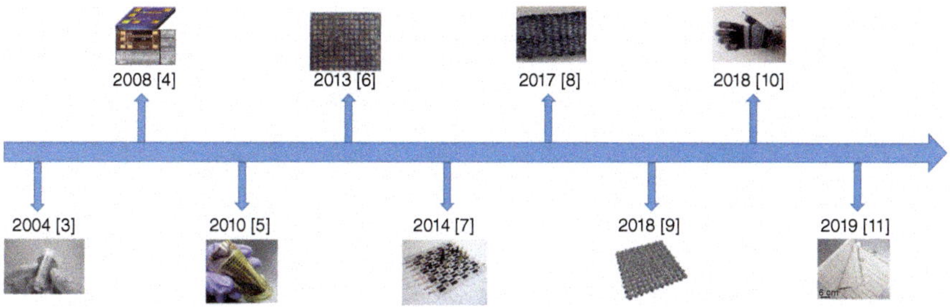

Figure 8.2 The development of piezoresistive stress sensing textile. Source: Adapted with permission from Someya et al. [3]. Copyright (2004) by The National Academy of Sciences of the USA; Adapted with permission from Fan et al. [4]. Copyright (2008) by The National Academy of Sciences of the USA; Adapted with permission from Kuniharu et al. [5]. Copyright (2010) by Macmillan Publishers Limited; Adapted with permission from Yao et al. [6]. Copyright (2013) by WILEY-VCH Verlag GmbH & Co. KGaA, Weinheim; Adapted with permission from Pan et al. [7]. Copyright (2014) by Macmillan Publishers Limited; Adapted with permission from Zhou et al. [8]. Copyright (2017) by Springer Nature Limited; Adapted with permission from Cao et al. [9]. Copyright (2018) by Elsevier; Adapted with permission from Francisco et al. [10]. Copyright (2018) by MDPI; Adapted with permission from Honda et al. [11]. Copyright (2019) by WILEY-VCH Verlag GmbH & Co. KGaA, Weinheim.

(NWs) for heterogeneous, multifunctional circuitry that utilizes both the sensory and electronic functionalities of single-crystalline nanomaterials in 2008. Kunikaru et al. [5] demonstrated macroscale ($7 \times 7\,\text{cm}^2$) integration of parallel NW arrays as the active-matrix backplane of a flexible pressure sensor array (Figure 8.3a). The integrated sensor array effectively functions as an artificial electronic skin, capable of monitoring applied pressure profiles with high spatial resolution. Yu and coworkers [6] reported a flexible and highly pressure-sensitive graphene/polyurethane (PU) sponge based on fractured microstructure design. Figure 8.3b shows the effects of fiber fracture structures generated by pre-compression treatment on the variation of the contact area of the modified fiber network with compressive stress. Bao and coworkers [7] presented an ultrasensitive resistive pressure sensor based on an elastic, microstructured conducting polymer thin film, which enables the detection of pressures of less than 1 Pa and exhibits a short response time, good reproducibility, excellent cycling stability, and temperature-stable sensing (Figure 8.3c). Chiseon et al. [12] demonstrated a pressure sensor array capable of mapping out the applied pressure over an area using a printed active-matrix thin film transistor (TFT) backplane. Figure 8.3d shows the pressure response of a representative pixel with an applied drain voltage of 5 V in the array. A stretchable piezoresistive stress sensor using a microstructure array was developed for the first time in 2014 by Choong et al. [13]. The sensor has a higher pressure sensitivity with elongation than typical stretchable electronics, achieving a sensitivity of $10.3\,\text{kPa}^{-1}$ when stretched by 40% (Figure 8.3e). However, the fabrication of the above surface microstructures relies on the fabrication of Si molds that are complex, requiring multiple steps, high cost. Wang et al. [14] presented a simple and low-cost method for fabrication of large-area patterned poly(dimethylsiloxane) (PDMS) conducting thin films with uniformly microstructured patterns and constructed a

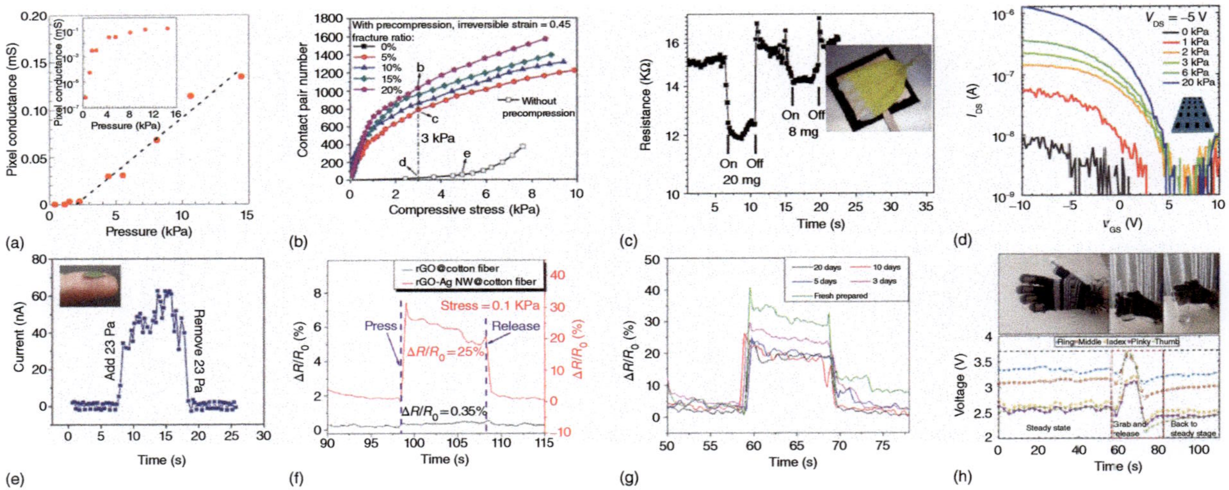

Figure 8.3 The performance of piezoresistive stress sensing textile. (a) Nanowire-based macroscale flexible pressure sensor devices. Source: Adapted with permission from Kuniharu et al. [5]. Copyright (2010) by Macmillan Publishers Limited. (b) The plots of contact pair number–compressive stress response curves for the fiber network. Source: Adapted with permission from Yao et al. [6]. Copyright (2013) by WILEY-VCH Verlag GmbH & Co. KGaA, Weinheim. (c) Transient response to the application and removal of a 20 mg weight and a flower petal (8 mg) on the micro-patterned device. Source: Adapted with permission from Pan et al. [7]. Copyright (2014) by Macmillan Publishers Limited. (d) Transfer curves of a representative pixel with varying amounts of applied pressure. Source: Chiseon et al. [12]. © 2015, John Wiley & Sons. (e) Characteristics of pressure sensor at 40% stretching. Source: Adapted with permission from Choong et al. [13]. Copyright (2014) by WILEY-VCH Verlag GmbH & Co. KGaA, Weinheim. (f) Resistance changes of rGO@cotton fiber and rGO-Ag NW@cotton fiber-based piezoresistive sensors under the pressure of 0.1 kPa. (g) The stability of rGO-Ag NW@cotton fiber-based piezoresistive sensor. Source: Adapted with permission from Cao et al. [9]. Copyright (2018) by Elsevier. (h) Measured voltage data of five constructed sensors placed in the fingertips of a glove holding a glass. Source: Adapted with permission from Francisco et al. [10]. Copyright (2018) by Multidisciplinary Digital Publishing Institute.

flexible pressure sensor. Su et al. [15] demonstrated a simple, efficient, low-cost, and scalable direct-molding method to flexible pressure sensors that are sensitive enough to mimic mimosa leaves.

In the second stage, piezoresistive stress sensing textiles are mainly based on interwoven fibers. It is important to make the sensors and communication systems as lightweight, flexible, and thin as possible. Lou et al. [2] developed a flexible pressure sensor based on the piezoresistive effect of multilayer graphene films on polyester textile. Cui and coworkers [8] demonstrated a design of highly sensitive, self-powered, and wearable electronic skin based on a pressure-sensitive nanofiber woven fabric sensor fabricated by weaving poly(vinylidene fluoride) (PVDF) electrospun yarns of nanofibers coated with poly(3,4-ethylenedioxythiophene) (PEDOT). Cao et al. [9] presented a flexible and wearable rGO-Ag NW@cotton fiber-based piezoresistive sensor, which can be well used in detecting human activities such as finger pressing, bending, twisting, walking, and pulse waves, showing high sensitivity, fast response, fast recovery, and high stability (Figure 8.3f,g). At the same year, Francisco et al. [10] also presented an easy-to-build textile pressure-resistive sensor created from low-cost conventional antistatic sheets and conductive woven fabrics. They evaluated the proposed sensor design on a classroom application consisting of a smart glove that measured the pressure applied by each finger (Figure 8.3h).

For the second stage, the development of active materials and device structures has been extremely rapid in recent years.

In terms of active materials, for piezoresistive stress sensors, these are based on elastomeric composites containing conductive fillers, such as rGO, CNT, metal particles, and CPs, which are incorporated into elastomers (e.g. PDMS and PU) to create piezoresistive behavior. In addition, metal films, NW, and conductive fibers have been used as active materials for piezoresistive stress sensors. Cui and coworkers [8] demonstrated a pressure-sensitive nanofiber woven fabric sensor fabricated by weaving PVDF electrospun yarns of nanofibers coated with PEDOT (Figure 8.4a). Yang and coworkers [9] synthesized the cotton fiber that is coated with reduced graphene oxide (GO) doped with silver NW (rGO-Ag NW@cotton fiber) via a simple and environmentally friendly hydrothermal method (Figure 8.4b). Zhang and coworkers [18] reported an omnidirectional bending and pressure sensor based on stretchable CNT-PU sponge. Cheng and coworkers [16] reported a wearable and highly sensitive pressure sensor with ultrathin gold NWs (Figure 8.4c). Chung and coworkers [13] used a conductive elastomeric composite of PEDOT:PSS and a polyurethane dispersion (PUD); a stretchable piezoresistive electrode coated on a micropyramidal PDMS array and a counter electrode that makes contact with a piezoresistive electrode were introduced. Yuan and coworkers [17] presented a simple, energy-saving, and large-scale capable process to fabricate flexible skin-like pressure sensors based on graphene/polyamide interlocking fabric (Figure 8.4d).

In terms of device structure, the structure of piezoresistive stress sensing textiles mainly has two types: network structure and sandwich structure. Cui and coworkers [8] presented a nanofiber woven fabric sensor with multileveled network structure, which significantly induced the change in contact area under ultralow load (Figure 8.5a). Yang and coworkers [9] developed a wearable rGO-Ag

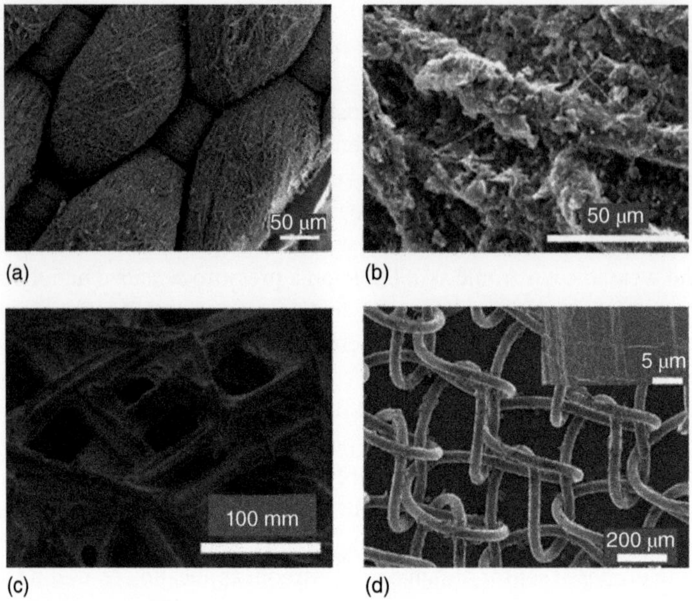

Figure 8.4 Active materials of piezoresistive stress sensing textile. (a) SEM images of rGO-Ag NW @cotton fiber film. Source: Adapted with permission from Zhou et al. [8]. Copyright (2017) by Springer Nature Limited. (b) Schematic of fabrication of rGO@cotton fiber and rGO-Ag NW@cotton fiber. Source: Adapted with permission from Cao et al. [9]. Copyright (2018) by Elsevier. (c) SEM images of the morphology of gold nanowire-coated tissue fibers. Source: Adapted with permission from Gong et al. [16]. Copyright (2014) by Macmillan Publishers Limited. (d) SEM images of the polyamide interlocking fabric after assembling the chemically reduced graphene (CRG) sheets. Source: Adapted with permission from Yin et al. [17]. Copyright (2018) by The Royal Society of Chemistry.

NW@cotton fiber piezoresistive sensor with network structure (Figure 8.5b). Cheng and coworkers [16] reported an efficient, low-cost fabrication strategy to construct a highly sensitive, flexible pressure sensor by sandwiching ultrathin gold NW-impregnated tissue paper between two thin polydimethylsiloxane sheets. Yuan and coworkers [17] fabricated a flexible skin-like pressure sensor based on graphene/polyamide interlocking staggered fabric (Figure 8.5c).

8.1.2 Capacitive Stress Sensing Textiles

Capacitive stress sensing textile is a sensing device that is promising for human activity monitoring and personal healthcare. The working mechanism of capacitive stress sensor is shown in Figure 8.6. A capacitor is a passive electronic component that stores energy in the form of an electrostatic field. Capacitors consist of two conducting plates separated by an insulating material (dielectric). The capacitance C is given by the equation $C = \varepsilon_0\varepsilon_r A/d$, where ε_0 is the electric constant, ε_r is the relative static permittivity of the dielectric, A is the area of the overlap of the two plates, and d is the thickness of the dielectric. For capacitive stress sensors, ε_r, A, and d can be sensitive to changes in pressure. Changes in d are commonly used to for the

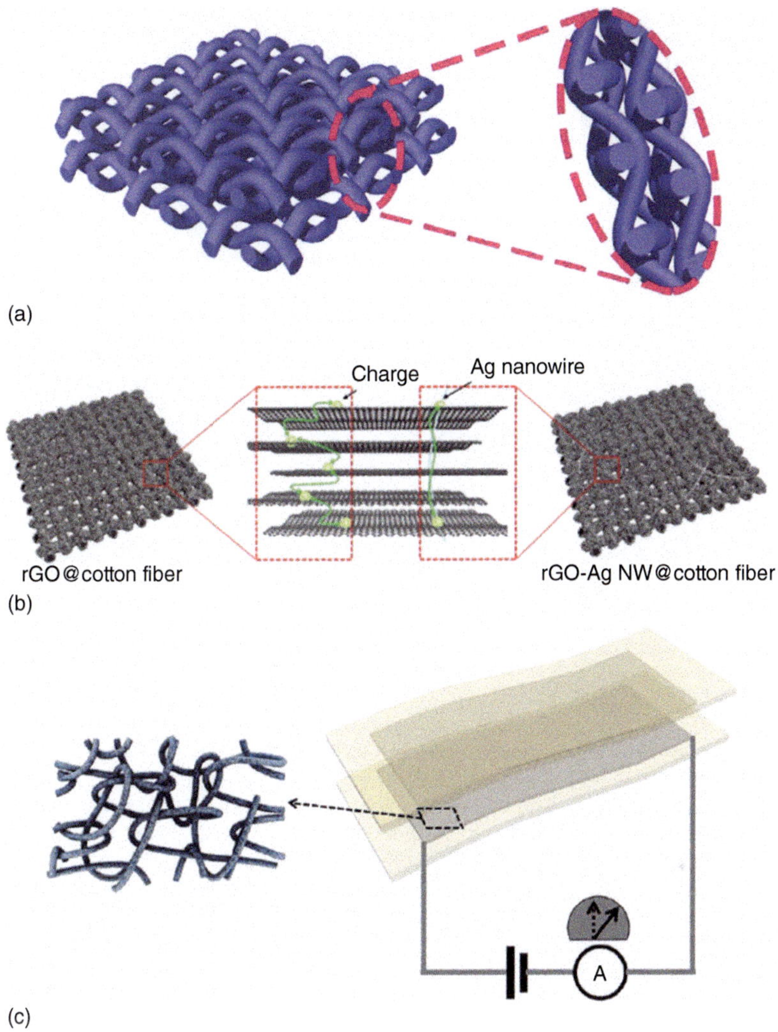

Figure 8.5 The structure of piezoresistive stress sensing textile. (a, b) Network structure. Source: Zhou et al. [8]. Licenced under CC BY 4.0.; Cao et al. [9]. © 2018, Elsevier. (c) Staggered structure. Source: Yin et al. [17]. © 2018, Royal Society of Chemistry.

detection of pressure [20]. Similarly, different active materials and device structures will affect the performance of the capacitive stress sensing textiles.

The development of the capacitive stress sensing textiles is shown in Figure 8.7. It mainly can be divided into two stages.

In the first stage, the early development of capacitive stress sensing textiles is mainly based on thin films. In 2010, Bao and coworkers [21] demonstrated a flexible capacitive stress sensor with unprecedented sensitivity and very short response times that can be inexpensively fabricated over large areas by microstructuring of thin films of the biocompatible elastomer polydimethylsiloxane. Bao

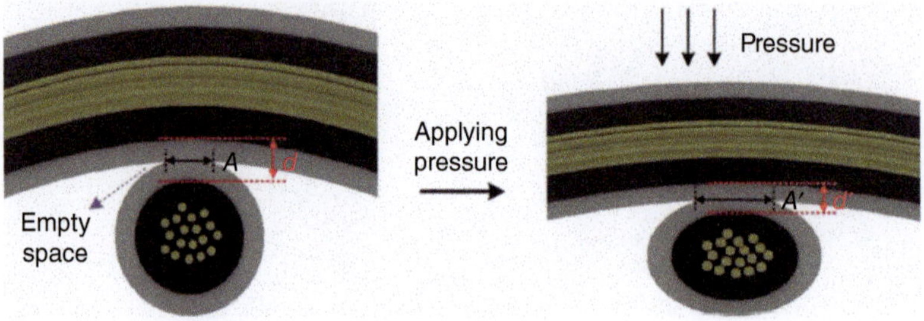

Figure 8.6 The working mechanism of capacitive stress sensor. Source: Lee et al. [19]. © 2015, John Wiley & Sons.

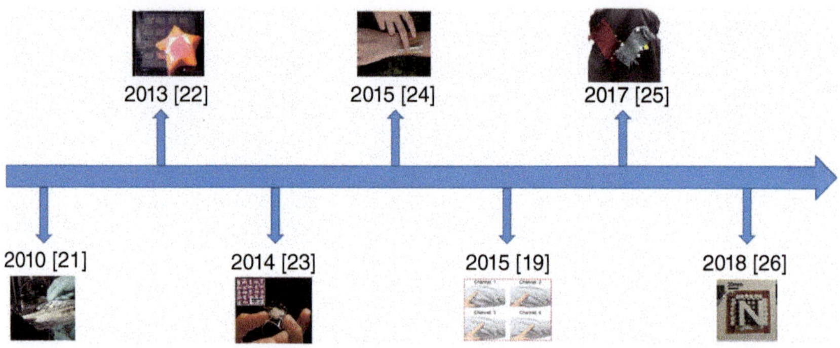

Figure 8.7 The development of capacitive stress sensing textile. Source: Adapted with permission from Lee et al. [19]. Copyright (2015) by WILEY-VCH Verlag GmbH & Co. KGaA, Weinheim; Adapted with permission from Stefan et al. [21]. Copyright (2010) by Macmillan Publishers Limited; Adapted with permission from Gregor et al. [22]. Copyright (2013) by Macmillan Publishers Limited; Adapted with permission from Lucie et al. [23]. Copyright (2014) by WILEY-VCH Verlag GmbH & Co. KGaA, Weinheim; Adapted with permission from Zang et al. [24]. Copyright (2015) by Macmillan Publishers Limited; Adapted with permission from Minjeong et al. [25]. Copyright (2017) by Elsevier; Adapted with permission from Kapoor et al. [26]. Copyright (2018) by WILEY-VCH Verlag GmbH & Co. KGaA, Weinheim.

and coworkers [22] reported the fabrication of flexible pressure-sensitive organic TFTs with a maximum sensitivity of $8.4\,kPa^{-1}$, a fast response time of $<10\,ms$, high stability over $>15\,000$ cycles, and a low power consumption of $<1\,mW$ (Figure 8.8a). Lucie et al. [23] developed a flexible three-axial force sensor for soft and highly sensitive artificial touch. At the same year, Bao and coworkers [27] proposed a retractable capacitive pressure sensor that combines capacitive and lateral strain sensing, as well as a triboelectric generator for distinguishing multiple mechanical stimuli. By constructing an air gap between the porous PDMS film and the single-walled carbon nanotube (SWCNT) film, the device achieves highly sensitive pressure detection. Zhu and coworkers [24] presented a flexible suspended gate organic thin film transistors (SGOTFTs) as a model platform that enables ultrasensitive pressure detection. An unprecedented sensitivity of $192\,kPa^{-1}$, a low limit-of-detection pressure of $<0.5\,Pa$, and a short response time of 10 ms were successfully realized, allowing the real-time detection of acoustic waves (Figure 8.8b). Tang and coworkers [28] reported a high sensitivity, ultrathin,

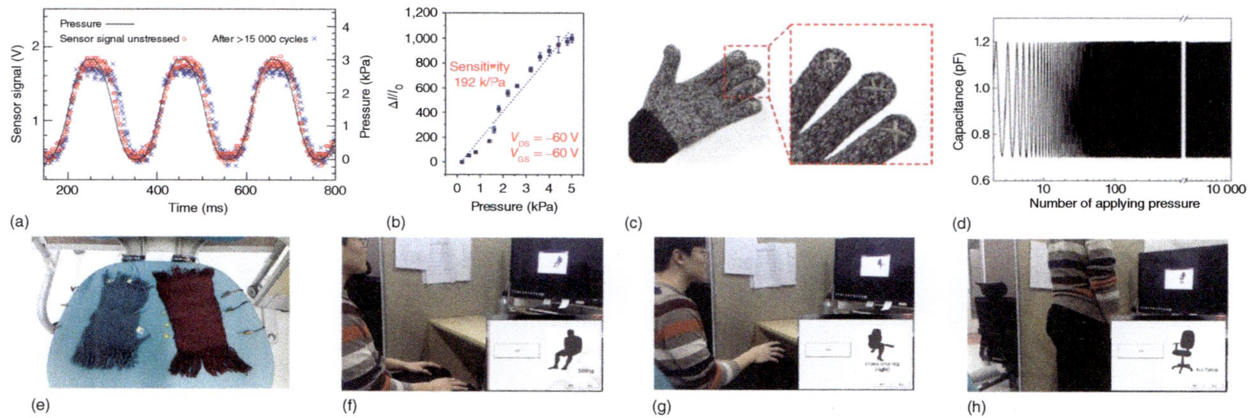

Figure 8.8 The performance of capacitive stress sensing textile. (a) Periodic fast pressure changes are used to determine the response time taken for the unstressed sensor and after > 15 000 pressure cycles. Source: Gregor et al. [22]. © 2013, Springer Nature. (b) Pressure response of the source–drain current. Source: Zang et al. [24]. Licenced under CC BY 4.0. (c) Photograph showing the smart glove with the textile-based pressure sensors on the ends of four fingers. (d) Stability of capacitive response of the pressure sensor to a load of 1.7 N over 10 000 cycles. Source: Adapted with permission from Lee et al. [19]. Copyright (2015) by WILEY-VCH Verlag GmbH & Co. KGaA, Weinheim. (e) Textile pressure sensors and connection with Arduino unit attached to chair. Different postures were displayed on computer monitor in real time according to changes in postures from (f) sitting, (g) crossing the leg, and (h) standing. Source: Adapted with permission from Minjeong et al. 2017 [25]. Copyright (2017) by Elsevier.

and transparent pressure sensor based on wrinkled graphene prepared by a facile liquid-phase shrink method.

In the second stage, compared with conventional pressure sensors, the applicability of textile pressure sensors will vary according to the surface coming in contact with the human body and the ease of integration and washing. Textile pressure sensors can offer much more comfortable touch feeling and economical prices. Xu et al. [29] successfully demonstrated a textile-based sensing system, called Smart Cushion, which analyzes the sitting posture of human being accurately and noninvasively in 2013. In order to develop a high-performance textile-based pressure sensor, conductive fibers with excellent electrical conductivity and stability against external deformation are fundamentally essential. In 2015, Lee et al. [19] described a textile-based pressure sensor with unprecedented sensitivity, excellent durability, a fast response, and a relaxation time based on highly conductive fibers coated with dielectric rubber materials (Figure 8.8c). The obtained capacitive type of textile pressure sensor exhibited high sensitivity (0.21 kPa^{-1}), very fast response times in the millisecond range, and high stability over more than 10 000 cycles (Figure 8.8d). For the purpose of the posture monitoring system using textile pressure sensors commercially available, the development of textile pressure sensors with high durability against mechanical deformations and the washing process are still challenged. In 2017, Minjeong et al. [25] presented a real-time sitting posture correction system based on highly durable and washable textile pressure sensors (Figure 8.8e). The sensing performance was 100% maintained even upon repeated sitting action over 1000 times and harsh washing with detergent. Also, they demonstrated the real-time display on a monitor of changes in sitting posture (Figure 8.8f–h).

For the second stage, the innovation of materials and the diversification of device structures have greatly promoted the development of capacitive stress sensing textiles.

In terms of active materials, over the past 10 years, a number of conductive materials and composites, such as indium tin oxide (ITO), CNT, Au, AgNPs, AgNW, and CPs, have been explored for their potential use as the electrode material of capacitive stress sensors. A variety of low modulus dielectric materials, such as PDMS, iontronic film, Ecoflex, sebacate, AgNW-PU composites, poly-(methyl methacrylate), GO, and CNT/PDMS, have been studied as the dielectric medium. Lee et al. [19] described a textile-based pressure sensor based on highly conductive fibers coated with dielectric rubber materials. The conductive fibers were fabricated by coating poly(styrene-block-butadiene styrene) (SBS) polymer on the surface of poly(*p*-phenylene terephthalamide) (Kevlar) fiber, followed by converting a huge amount of silver (Ag) ions into Ag nanoparticles directly in the SBS polymer (Figure 8.9a). Kim et al. [25] demonstrated a textile pressure sensor consisted of conducting Ni–Ti alloy fiber with an excellent fatigue resistance and a pressure-sensitive PU elastomer (Figure 8.9b).

In terms of device structure, the commonly used structures of capacitive stress sensing textiles are mainly staggered structure and twisted structure. Lee et al. [19] presented a textile-based pressure sensor by coating PDMS as dielectric layers on the surface of the conductive fibers and stacking the two PDMS-coated fibers

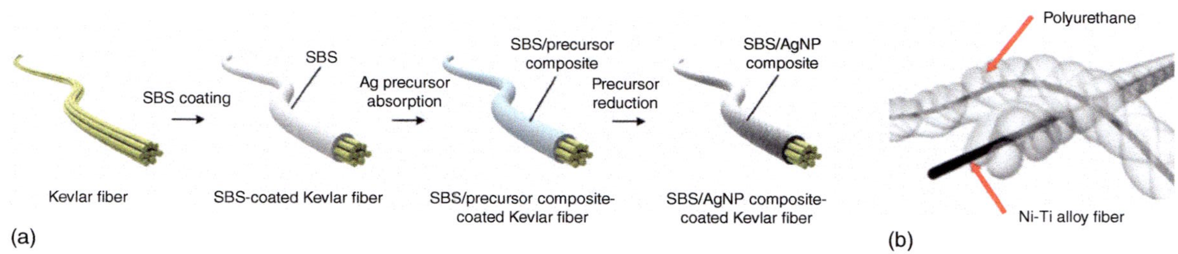

Figure 8.9 Active materials of capacitive stress sensing textile. (a) Schematic illustration of the fabrication of a conductive fiber. Source: Lee et al. [19]. © 2015, John Wiley & Sons. (b) Schematic illustration of textile pressure sensor based on Ni–Ti alloy fiber coated with polyurethane rubber. Source: Minjeong et al. [25]. © 2017, Elsevier.

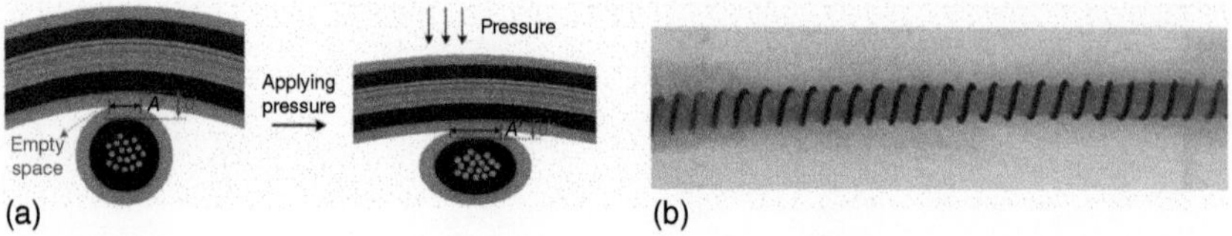

Figure 8.10 The structure of capacitive stress sensing textile. (a) Staggered structure. Source: Lee et al. [19]. © 2015, John Wiley & Sons. (b) Twisted structure. Source: Adapted with permission from Minjeong et al. [25]. Copyright (2017) by Elsevier.

perpendicularly to each other; a capacitive type of textile pressure sensor was successfully fabricated (Figure 8.10a). Kim et al. [25] developed a textile pressure sensor in which three PU fibers were twisted and covered around the Ni–Ti fiber, and then this assembly of the threads was instantly heated to 150 °C to fix their structure as shown in Figure 8.10b.

8.1.3 Other Stress Sensing Textiles

In addition to the above sensors, stress sensing textiles have also developed the following sensors, mainly including piezoelectric stress sensing textiles, triboelectric stress sensing textiles, and color-based stress sensing textiles.

In terms of piezoelectric stress sensing textiles, a piezoelectric material can generate electricity in response to mechanical forces. Piezoelectric stress sensing textiles are stress sensing fabric based on piezoelectric effect. Drean and coworkers [30] developed P(VDF-TrFE)/copper coaxial filaments for piezoelectric coaxial filament textile structure sensors. An example of electrical signal delivered by integrated filament in textile when subjected to stress actions is shown in Figure 8.11a. Luana et al. [31] introduced a large-area, flexible piezoelectric material that consists of sheets of electrospun fibers of the polymer poly(vinylidenefluoride-co-trifluoroethylene). The resulting material offers exceptional piezoelectric characteristics to enable ultrahigh sensitivity for measuring pressure, even at exceptionally small values (0.1 Pa) (Figure 8.11b). Quantitative analysis provides detailed insights into the pressure sensing mechanisms and establishes engineering design rules. Potential applications range from self-powered micro-mechanical elements to self-balancing robots and sensitive impact detectors.

In terms of triboelectric stress sensing textiles, triboelectric effect can be also to build stress sensors. A triboelectric sensor is usually made of two distinct materials having opposite tendencies to acquire charges. The contact and separation motion between the two materials triggers charge transfer, leading to a measurable alternative current or voltage in the external circuit. The magnitude of the electrical signals can be correlated with the force applied on the sensor. Triboelectric stress sensing textile is a stress-sensing fabric based on triboelectric effect. Wang and coworkers [32] presented a screen-printed washable electronic textile as self-powered stress tribo-sensors utilizing conductive CNTs and screen-printing technology (Figure 8.11c). The e-textile exhibits highly sensitive touch/gesture sensing

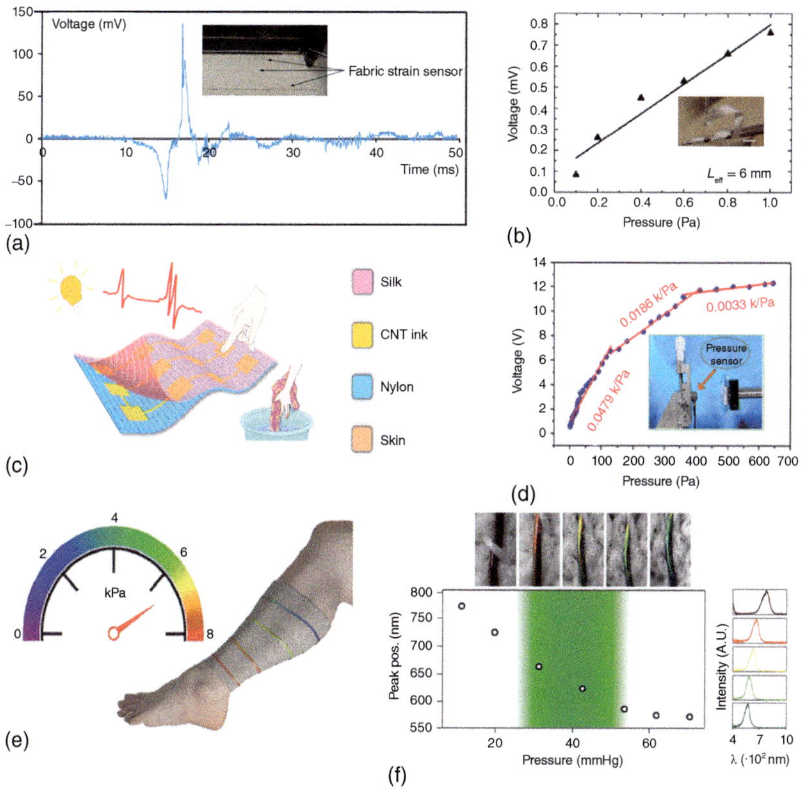

Figure 8.11 Other stress sensing textiles. (a) Voltage variation when pressure performed at the surface of the fabric. Source: Adapted with permission from Kechichea et al. [30]. Copyright (2013) by Elsevier. (b) Experimental pressure response curve in the low-pressure regime. Source: Adapted with permission from Luana et al. [31]. Copyright (2013) by Macmillan Publishers Limited. (c) Schematic illustration of washable electronic textile (WET) with an array of CNT electrodes by screen printing. (d) Pressure sensitivity of the WET. Source: Adapted with permission from Cao et al. [32]. Copyright (2018) by American Chemical Society. (e) Depiction of the proposed use of fibers as colorimetric sub-bandage pressure indicators in compression therapy. Source: Adapted with permission from Joseph et al. [33]. Copyright (2018) by WILEY-VCH Verlag GmbH & Co. KGaA, Weinheim. (f) Peak wavelengths of a fiber for varying applied sub-bandage pressure. Source: Adapted with permission from Joseph et al. [33]. Copyright (2018) by WILEY-VCH Verlag GmbH & Co. KGaA, Weinheim.

performance and has potential applications for human–machine interfacing (Figure 8.11d).

In terms of color-based stress sensing textiles, as discussed before, all the sensors producing electrical signals require electrometers to function. In this regard, colorimetric sensors are more favorable as the color of the sensor is directly readable by the user. Joseph et al. [33] indicated a colorimetric sensor that can change color upon stretch in the compressive bandage. The sensor is in fact a photonic fiber that is made by cladding a thin bilayer of PDMS and polystyrene–polyisoprene triblock copolymer a few times around an extruded PDMS fiber (Figure 8.11e). This layered structure enables the fiber to reflect visible light according to the applied axial stress. After obtaining the color-stress relation, the fiber sensor is woven onto compressive bandages to monitor the sub-bandage pressure. The stretch of the bandage gave rise to the pressure on the leg. The accuracy of the sensor was validated through the comparison with bandages having geometric stress indicators (Figure 8.11f).

8.2 Strain Sensing Textiles

Strain sensor is a sensor that measures the strain generated by an object due to force deformation. The traditional strain sensor is generally composed of a large metal strain gauge. This strain sensor tends to occupy a large space and is not easy to carry. The strain sensor should have the advantages of lightweight, good flexibility, easy carrying, high measurement accuracy, and high sensitivity. Textile-based strain sensor, namely, strain sensing textiles, is prepared by assembling the active material (rGO, CNT, PEDOT:PSS, etc.) on the fiber or fabric according to the change of its resistance, capacitance, and other properties. At the same time through reasonable device structure design, the flexible strain sensors capable of sensing the deformation of an object to be measured and converting it into an easily transmitted, processed electrical signal are mounted on the skin or integrated into clothing, thereby demonstrating the potential of the fiber strain sensors as candidates for electronic textiles, wearable electronics, and biomedical engineering.

According to their measurement mechanisms, the strain sensing textiles are divided into piezoresistive strain sensing textiles, capacitive strain sensing textiles, and triboelectricity strain sensing textiles.

8.2.1 Piezoresistive Strain Sensing Textiles

Piezoresistive strain sensing textile is a wearable piezoresistive strain sensor that must meet with certain criteria including high stretchability and flexibility, low hysteresis, and high sensitivity. The working mechanism of piezoresistive strain sensor is shown in Figure 8.12. When conductive materials are subjected to mechanical deformation, their electrical properties change. This electromechanical response is known as the piezoresistive effect. Due to the Poisson ratio (ν), materials that are elongated also contract in the transverse direction of elongation. Consequently, the resistance R of a conductive material will change, as shown by the following equation: $R = \rho L/A$, where ρ is the resistivity, L is the length, and A is the

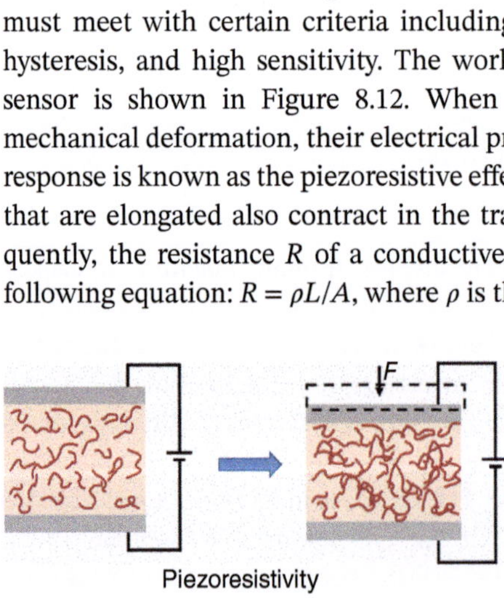

Figure 8.12 The working mechanism of piezoresistive strain sensing textile. Source: Heikenfeld et al. [34]. © 2018, Royal Society of Chemistry.

cross-sectional area of the conductor [34]. With the innovation of materials and the diversification of device structures, piezoresistive strain sensing textiles have developed rapidly in a short period of time.

The development of the piezoresistive strain sensing textiles is shown in Figure 8.13. It can be divided into three stages.

In the first stage, traditionally, piezoresistive strain sensors based on metals or Si have high sensitivities and can be cost-effective. Recently, interest has increased in the development of flexible and stretchable strain sensors that can be attached on clothing or on the human body for applications in e-skin and human motion monitoring in medicine. For example, in 2010, Ahn and coworkers [35] presented a flexible and stretchable piezoresistive strain sensor in which gauge factor was ~6.1. Takeo et al. [36] reported a class of wearable and stretchable strain sensors fabricated from thin films capable of measuring strains up to 280% (Figure 8.14a). Yan et al. [45] reported an innovative method for the fabrication of high-strain sensors that exhibited a gauge factor of 7.1 at 100% strain, which is >10 times higher than stretchable CNT and AgNW sensors (Figure 8.14b).

In the second stage, various commercial polymer fibers such as polyamide 6, Lycra, polyolefin, and PU are used as the elastic substrate. For example, in 2014, Mohammad et al. [49] demonstrated the production of conducting elastomeric composite fibers with wide strain sensing capabilities that may be suitable for applications in wearable bionics and stretchable electronics. In 2015, Lee et al. [46] reported styrene–butadiene–styrene fibers embedded in silver NWs that has a conductivity of up to 2450 S/cm and a strain at break of 900% (Figure 8.14c). However, polymer-based fiber sensors have low stability during sensing. Due to the high mechanical and electrical properties of its combination, CNT have been extensively studied for decades. For example, spring-like CNT fibers were prepared by Cao and coworkers [47], which exhibited high fracture strains of up to 285% and reversible and linear changes in resistance to strain. Resolving an entangled yarn and releasing its twists could simultaneously rotate a heavy object (30 000 times the yarn weight) for more than 1000 cycles at high speed (Figure 8.14d). However, the scale-up manufacturing of high-quality CNT fibers remains challenging. Currently, fiber sensors prepared by synthetic methods are competing candidates for large-scale smart textiles.

In the third stage, the bases of piezoresistive strain sensing textiles are mainly based on interlaced fibers. A scaled-up fiber wet-spinning production of electrically conductive and highly stretchable PU/PEDOT : PSS fiber is demonstrated in 2015 for the first time by Shayan et al. [39]. The knitted textile sensor can distinguish different magnitudes of applied strain with cyclically repeatable sensor responses at applied strains of up to 160% (Figure 8.14e). Zhang and coworkers [40] demonstrated for the first time that carbonized silk fabrics (CSFs) could be used for the fabrication of ultra-stretchable and highly sensitive strain sensors, which showed superior performance in detecting both large and subtle human motions. Cai et al. [41] used a simple method to fabricate flexible strain sensing fabrics (FSSFs) on elastic nylon/PU fabric, exhibiting good washability. Zhang and coworkers [48] fabricated a strain sensor with wide workable strain range through

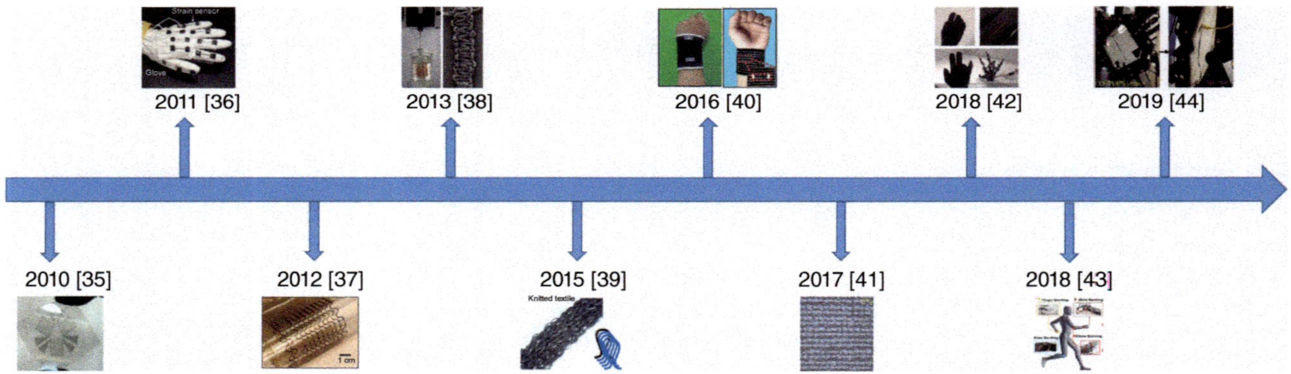

Figure 8.13 The development of piezoresistive strain sensing textile. Source: Adapted with permission from Lee et al. [35]. Copyright (2010) by American Chemical Society; Adapted with permission from Takeo et al. [36]. Copyright (2011) by Springer Nature Limited; Adapted with permission from Lu et al. [37]. Copyright (2012) by WILEY-VCH Verlag GmbH & Co. KGaA, Weinheim; Adapted with permission from Sun et al. [38]. Copyright (2013) by American Chemical Society; Adapted with permission from Shayan et al. [39]. Copyright (2015) by American Chemical Society; Adapted with permission from Wang et al. [40]. Copyright (2016) by WILEY-VCH Verlag GmbH & Co. KGaA, Weinheim; Adapted with permission from Cai et al. [41]. Copyright (2017) by Elsevier; Adapted with permission from Jaehong et al. [42]. Copyright (2018) by American Chemical Society; Adapted with permission from Yang et al. [43]. Copyright (2018) by American Chemical Society; Adapted with permission from Montazerian et al. [44]. Copyright (2019) by Elsevier.

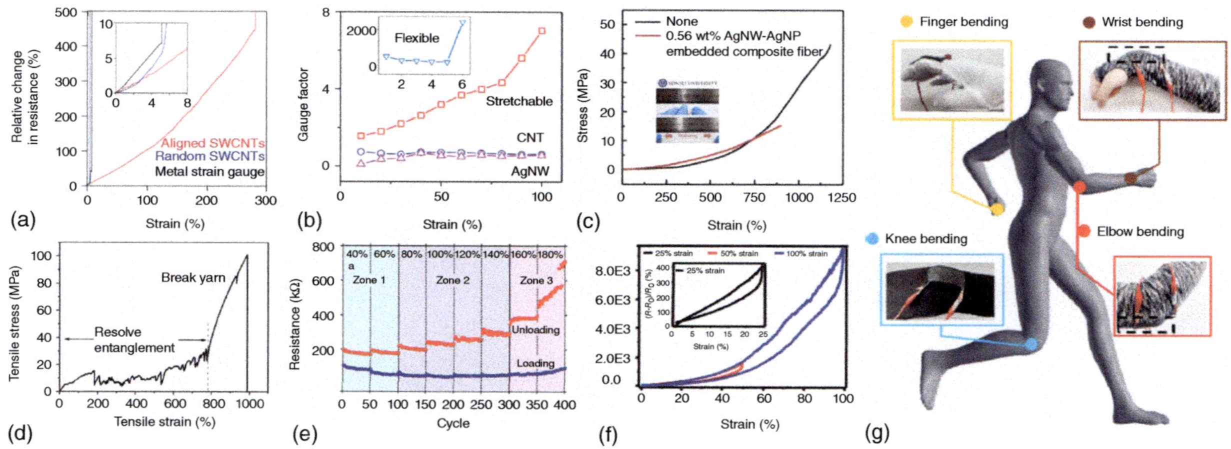

Figure 8.14 The performance of piezoresistive strain sensing textiles. (a) Relative change in resistance versus strain for the strain sensor. Source: Takeo et al. [36]. © 2011, Springer Nature. (b) Plots of gauge factors versus strain for flexible nanopaper, stretchable nanopaper, and stretchable CNT and AgNW electrodes. Source: Yan et al. [45]. © 2014, John Wiley & Sons. (c) Stress–strain curves for tensile tests on SBS fiber and 0.56 wt% AgNW–AgNP-embedded composite fiber. Source: Adapted with permission from Lee et al. [46]. Copyright (2015) by WILEY-VCH Verlag GmbH & Co. KGaA, Weinheim. (d) Tensile stress–strain curve of a CNT entanglement showing a failure strain of 985%. Source: Li et al. [47]. © 2013, American chemical society. (e) Strain sensing properties of knitted four-ply PU/PEDOT:PSS fibers and commercial Spandex yarn between 20% and 180% strain for 50 cycles. Source: Shayan et al. [39]. © 2015, American Chemical Society. (f) Resistance for different strains during the loading–unloading cycle. Source: Wang et al. [48]. © 2017, Royal Society of Chemistry. (g) Detection of various human motions using the wearable graphene textile strain sensor. Source: Adapted with permission from Yang et al. [43]. Copyright (2018) by American Chemical Society.

a facile and large-scale process. The obtained strain sensors exhibit remarkable combined features of ultrahigh sensitivity in a wide sensing range (average gauge factor of 29.7 within 40% strain and of 173.0 for a strain of 60–100%), ultralow detection limit (0.01% strain), high durability and stability (10 000 stretching cycles at 100% strain), fast response (<70 ms), and translucence, which could be ascribed to the unique woven structure of silk georgette (Figure 8.14f). Jaehong et al. [42] presented a facile approach for fabricating highly stretchable and sensitive fiber strain sensors that was integrated into a glove to control a hand robot and effectively applied to monitor the large volume expansion of a balloon and a pig bladder for an artificial bladder system. At the same year, Yang et al. [43] fabricated a close-fitting and wearable graphene textile strain sensor based on a graphene textile without polymer encapsulation. Compared with the previously reported strain sensors, the strain sensor exhibits a distinctive negative resistance variation with increasing strain. In addition, the graphene textile strain sensor can be knitted on clothing for detecting both subtle and large human motions, showing the tremendous potential for applications in wearable electronics (Figure 8.14g).

For the third stage, the development of active materials and device structures has been extremely rapid in recent years.

In terms of active materials, for piezoresistive strain sensing textile, these contain conductive fillers, such as rGO, CNT, metal particles, and PEDOT:PSS, which are incorporated into elastomers (e.g. PDMS and PU) to create piezoresistive behavior. For example, Cao and coworkers [47] produced an elastic strain sensor-based twisted CNT yarns (Figure 8.15a). Wang and coworkers [41] used a simple method to fabricate FSSFs through the coating of GO nanosheets on elastic nylon/PU fabric, followed by reduction of GO with sodium borohydride (Figure 8.15b). Ren and coworkers [43] also fabricated a close-fitting and wearable graphene textile strain sensor based on a graphene textile without polymer encapsulation (Figure 8.15c). Compared with CNT and graphene, carbon-based materials from natural biomaterials have received extensive research interests, owing to their good electrical conductivity, large-scale and low-cost production capability, and environmental benignity. Zhang and coworkers [40, 48] demonstrated ultra-stretchable and highly sensitive strain sensors based on CSFs and carbonized silk georgette respectively (Figure 8.15d,e). Wallace et al. [39] demonstrated a scaled-up fiber wet-spinning production of electrically conductive and highly stretchable PU/PEDOT:PSS fibers for the first time. The knitted textiles based on PU/PEDOT:PSS fibers exhibit strain sensing properties that were dependent upon the number of PU/PEDOT:PSS fibers used in knitting (Figure 8.15f). Lee et al. [42] presented a facile approach for fabricating highly stretchable and sensitive fiber strain sensors by embedding Ag nanoparticles into a stretchable fiber (Figure 8.15g).

In terms of device structure, the commonly used structures of piezoresistive strain sensing textiles are mainly network structure and twisted structure. Wang and coworkers [41] fabricated an FSSFs with network structure (Figure 8.16a). Zhang and coworkers [48] also demonstrated a strain sensor with another woven network structure (Figure 8.16b). Introducing twists into CNT yarns could produce hierarchical architectures and extend their application areas. Cao and coworkers [47]

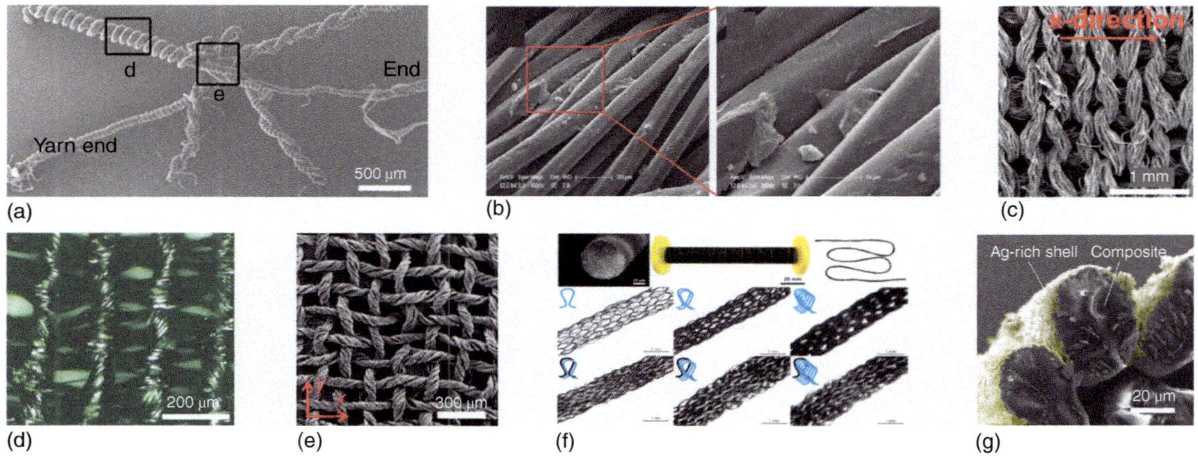

Figure 8.15 Active materials of piezoresistive strain sensing textile. (a) SEM images of two CNT entanglements. Source: Adapted with permission from Li et al. [47]. Copyright (2013) by American Chemical Society. (b) RGO treated fabrics. Source: Adapted with permission from Cai et al. [41]. Copyright (2017) by Elsevier. (c) SEM image of the graphene textile in the *x*-direction. Source: Adapted with permission from Yang et al. [43]. Copyright (2018) by American Chemical Society. (d) The carbonized silk fabric. (e) SEM images of the CSG. Source: Adapted with permission from Wang et al. [40]. Copyright (2016) by WILEY-VCH Verlag GmbH & Co. KGaA, Weinheim; Adapted with permission from Wang et al. [48]. Copyright (2017) by The Royal Society of Chemistry. (f) SEM images of PU/PEDOT:PSS fiber. Source: Adapted with permission from Shayan et al. [39]. Copyright (2015) by American Chemical Society. (g) Cross-sectional SEM images of the fiber strain sensor comprise a Ag-rich shell. Source: Adapted with permission from Jaehong et al. [42]. Copyright (2018) by American Chemical Society.

Figure 8.16 The structure of piezoresistive strain sensing textile. (a, b) Network structure. Source: (a) Cai et al. [41]. © 2017, Elsevier; (b) Adapted with permission from Wang et al. [48]. Copyright (2017) by The Royal Society of Chemistry. (c) Twisted structure. Source: Adapted with permission from Li et al. [47]. Copyright (2013) by American Chemical Society.

utilized such twists to produce elastic strain sensors over large strain (up to 500%) and rotation actuators with high energy density (Figure 8.16c).

8.2.2 Capacitive Strain Sensing Textiles

Capacitive strain sensing textiles are a strain sensing fabric based on capacitance effect. Its working mechanism is shown in Figure 8.17. The capacitive change is governed by the classic equation $C = \varepsilon A/d$, where ε is the permittivity of the cavity between two plates and A and d represent the overlap area and the distance between two plates, respectively. As the distance, area, or permittivity is altered by the external loads, it leads to the change of capacitive readouts, which can be measured either using a passive capacitor or through modifying the response curve of an active component, such as using field-effect transistors (FET) [34]. Among them, active materials and device structures have a certain influence on the performance and application development of capacitive strain sensing textile.

The development of the capacitive strain sensing textiles is shown in Figure 8.18. It mainly can be divided into two stages.

In the first stage, capacitive strain sensing textiles are mainly based on interwoven strips. In 2011, Bao and coworkers [50] produced a strain sensor on a PDMS substrate that accommodates strains of up to 150% and demonstrates conductivities as high as 2200 S/cm in the stretched state. Daniel et al. [51] presented a highly elastic strain gauge based on capacitive sensing of parallel, CNT-based percolation

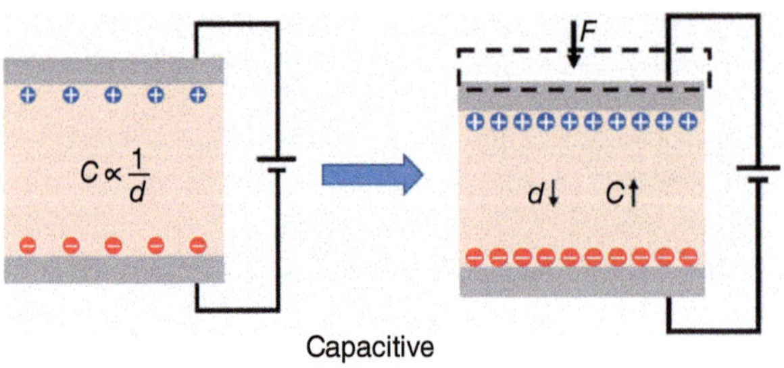

Figure 8.17 The working mechanism of capacitive strain sensing textile. Source: Heikenfeld et al. [34]. © 2018, Royal Society of Chemistry.

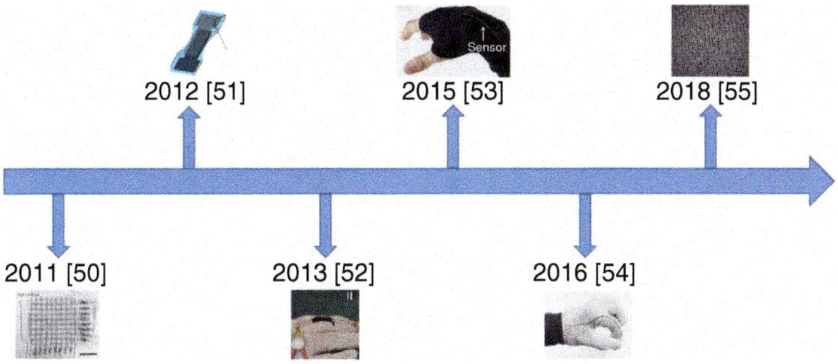

Figure 8.18 The development of capacitive strain sensing textile. Source: Adapted with permission from Darren et al. [50]. Copyright (2011) by Springer Nature Limited; Adapted with permission from Daniel et al. [51]. Copyright (2012) by American Chemical Society; Adapted with permission from Cai et al. [52]. Copyright (2013) by Springer Nature Limited; Adapted with permission from Andreas et al. 2015 [53]. Copyright (2015) by WILEY-VCH Verlag GmbH & Co. KGaA, Weinheim; Adapted with permission from Changsoon et al. [54]. Copyright (2016) by American Chemical Society; Adapted with permission from Ozgur [55]. Copyright (2018) by MDPI.

electrodes separated by a dielectric elastomer. In addition, in 2013, Cai et al. [52] designed a new multifunctional CNT-based capacitive strain sensors that can detect strains up to 300% with excellent durability even after thousands of cycles (Figure 8.19a). Furthermore, Yao and co-worker [56] reported a capacitive strain sensor in which AgNWs were used as electrodes and Ecoflex as a dielectric layer in 2014. The devices were demonstrated for several wearable applications including monitoring thumb movement, sensing the strain of the knee joint in patellar reflex, and other human motions such as walking, running, and jumping from squatting (Figure 8.19b).

In the second stage, capacitive strain sensing textiles are mainly based on interwoven fibers. Many methods have been explored to fabricate capacitive textile sensors. Among them, it is the easiest to weave a capacitive fiber sensor into commercial textiles. Andreas et al. [53] created textile-mounted capacitive soft strain sensor (CS3) fibers for detecting elongational strains (Figure 8.19c). These customizable sensors can be readily integrated with textiles and may find specific application in wearable electronics. Changsoon et al. [54] developed and characterized novel sandwich fibers as highly deformable strain sensors, which generated approximately 115.7% capacitance changes during stretching (200%) (Figure 8.19d,e). Ozgur et al. [55] developed a new sensor design is presented to create stretchable, capacitance-based strain sensors for human motion tracking with high response time (Figure 8.19f). This involves the use of stretchable, conductive-knit fabric within the silicone elastomer matrix, as interdigitated electrodes.

For the second stage, the development of active materials and device structures has been extremely rapid in recent years.

In terms of active materials, for capacitive strain sensors, these are based on CNT networks and metal NW networks such as CNT and AgNWs. Changsoon et al. [54] introduced a strain sensor with functionalized buckled CNT electrodes

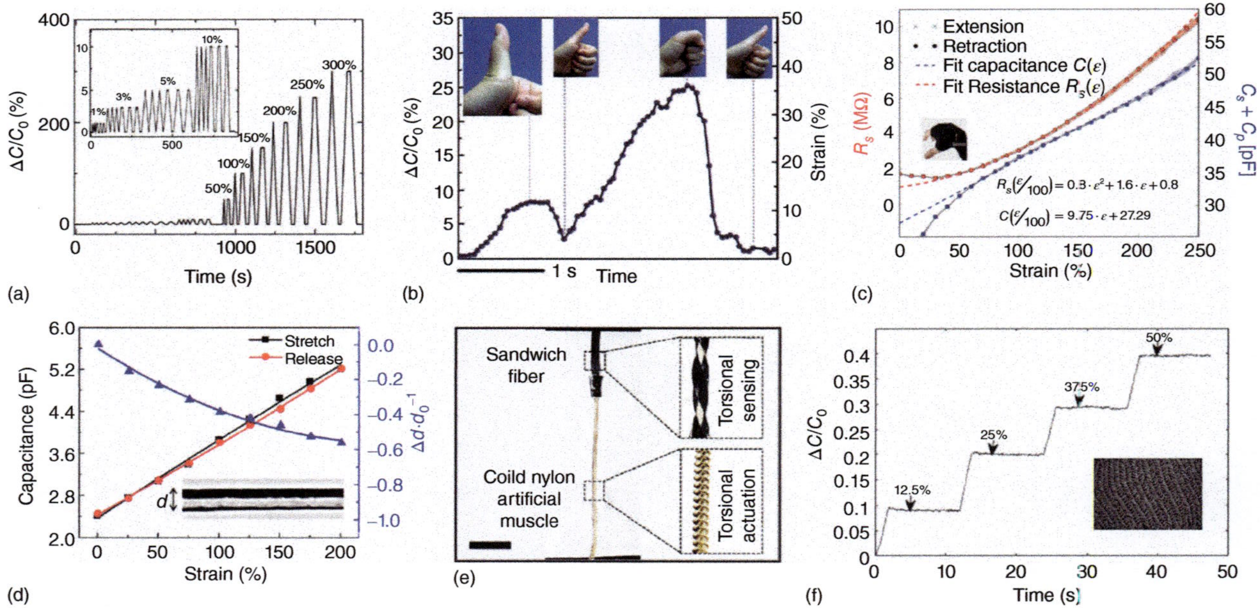

Figure 8.19 The performance of capacitive strain sensing textile. (a) Relative changes in capacitance of a strain sensor. Source: Cai et al. [52]. © 2013, Springer Nature. (b) Relative capacitance change and strain associated with thumb flexure. Source: Adapted with permission from Yao and Zhu [56]. Copyright (2014) by The Royal Society of Chemistry. (c) Model predictions, sensor resistance, and total capacitance of the sensor up to 250% strain. Source: Adapted with permission from Andreas et al. [53]. Copyright (2015) by WILEY-VCH Verlag GmbH & Co. KGaA, Weinheim. (d) Capacitance and fractional change in fiber thickness versus tensile strain. (e) Photo image showing an 8 cm-long coiled nylon torsional artificial muscle and attached sandwich sensor. Source: Adapted with permission from Changsoon et al. [54]. Copyright (2016) by American Chemical Society. (f) Drift of the sensor under constant strain levels. Source: Adapted with permission from Ozgur [55]. Copyright (2018) by Multidisciplinary Digital Publishing Institute.

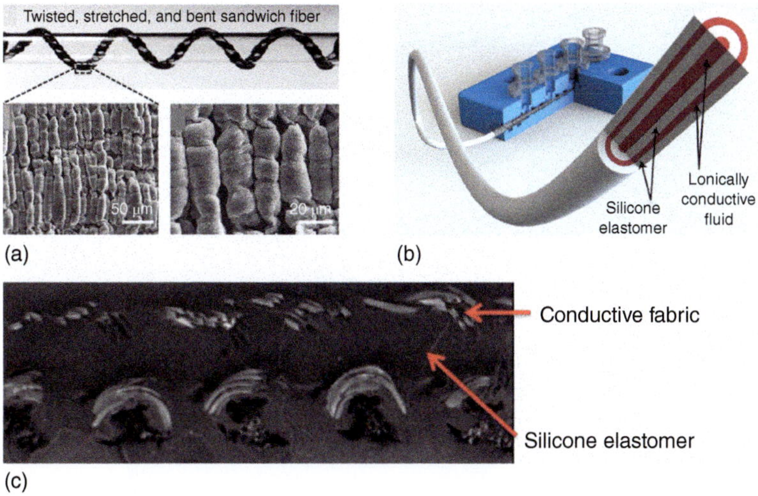

Figure 8.20 Active materials of capacitive strain sensing textile. (a) SEM images showing microscopic CNT buckles at two levels of magnification. Source: Adapted with permission from Changsoon et al. [54]. Copyright (2016) by American Chemical Society. (b) Schematic illustration of multicore–shell printing process. Source: Andreas et al. [53]. © 2015, John Wiley & Sons. (c) SEM images show a cross section of the conductive fabric. Source: Adapted with permission from Ozgur [55]. Copyright (2018) by Multidisciplinary Digital Publishing Institute.

(Figure 8.20a). On the microscopic scale, the buckled CNT electrodes effectively absorb tensile or shear stresses, providing an essentially constant electrical conductance. Zhu and co-worker [56] developed highly stretchable multifunctional sensors. The reported sensors utilize the capacitive sensing mechanism, where silver NWs are used as electrodes (conductors) and Ecoflex is used as a dielectric. The silver NW electrodes are screen-printed. In addition, Lewis et al. [53] presented a strain sensor based on ionically conductive fluid and a silicone elastomer that serve as the conductor and dielectric/encapsulant, respectively (Figure 8.20b). Caddesi et al. [55] showed a strain sensor that combines conductive-knit textile and silicone elastomer. The conductive properties of the electrodes affect the overall performance of the capacitive sensor, and it is crucial to use electrodes with good conductive performance under repetitive loading (Figure 8.20c).

In terms of device structure, the structures of the capacitive strain sensors are mainly multicore–shell structure, twistable sandwich structure, and network structure. Lewis et al. [53] created textile-mounted CS3 fibers with multicore–shell structure for detecting elongational strains (Figure 8.21a), where the sensor architecture can be customized by varying the respective inner (nested) and outer nozzle sizes, ink flow rates, print speed, and print path length. A weavable fiber strain sensor is introduced by Kim et al. [54], which comprise a dielectric layer sandwiched between functionalized buckled CNT electrodes (Figure 8.21b). Caddesi et al. [55] also presented a textile-based, interdigital, capacitive, soft strain sensor with network structure (Figure 8.21c). We believe that this type of sensor is

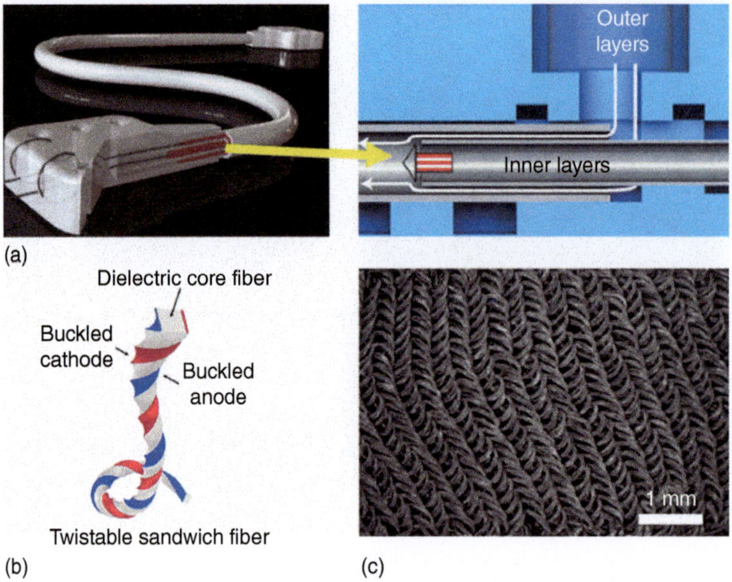

Figure 8.21 The structure of capacitive strain sensing textile. (a) Multicore–shell structure. Source: Adapted with permission from Andreas et al. [53]. Copyright (2015) by WILEY-VCH Verlag GmbH & Co. KGaA, Weinheim. (b) Twistable sandwich structure. Source: Changsoon et al. [54]. © 2016, American chemical society. (c) Network structure. Source: Adapted with permission from Ozgur [55]. Copyright (2018) by Multidisciplinary Digital Publishing Institute.

a suitable candidate for monitoring human body movements for medical, sports, or entertainment applications.

8.2.3 Triboelectricity Strain Sensing Textiles

Triboelectricity strain sensing textiles are that triboelectric fibers are mechanically strong to be woven into a commercial textile for textile-based strain sensors, which can detect magnitude and direction of the motion. Its working mechanism is shown in Figure 8.22. When the device is bent or compressed, the inner surfaces of the two different material layers will be in close contact, and charge transfer occurs, thereby respectively carrying positive and negative charges. After the pressure is released, the opposite frictional charges and electric field will be generated between the two surfaces, thereby forming a potential difference between the two electrodes. In order

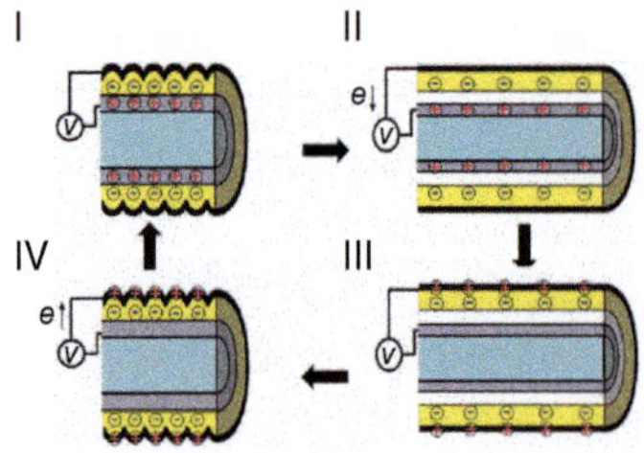

Figure 8.22 The working mechanism of triboelectricity strain sensing textiles. Source: Hyeon et al. [57]. Licenced under CC BY 4.0.

to shield this potential difference, electrons will be driven to flow from one electrode to the other through an external circuit until the potentials of two electrodes are equal again [57].

As an emerging technology, triboelectricity strain sensing textiles have experienced a fast development in wearable electronics. For example, Hyeon et al. [57] introduced a new type of stretchable and weavable triboelectric fiber with micro-diameter dimensions. The stretchable triboelectric fiber can be reversibly stretched up to 50% in tensile direction while generating voltage output proportional to the applied tensile strain (Figure 8.23a). Gong et al. [58] presented a stretchable sheath–core structural triboelectric fiber (SSCTEF) developed to serve as a self-powered multifunctional kinematic sensor. Owing to the advanced built-in wavy structure design, the fiber-like sensor exhibits an ultrahigh working strain (100%) and demonstrates high sensitivity in response to not only stretching but also bending and compressing (Figure 8.23b). He et al. [59] developed a smart textile fabricated with a simple dip-coating method with physical strain sensing. A height-varying multi-arch strain sensor with a large strain sensing ranging from

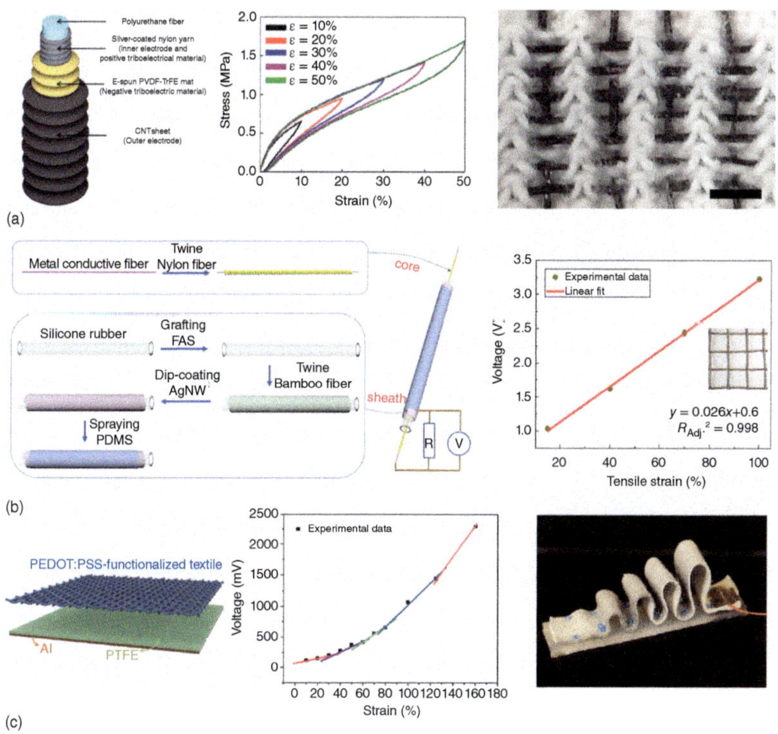

Figure 8.23 Triboelectricity strain sensing textiles. (a) Stretchable triboelectric fiber structure and performance. Source: Adapted with permission from Hyeon et al. [57]. Copyright (2016) by Springer Nature Limited. (b) The stretchable sheath–core structural triboelectric fiber (SSCTEF). Source: Gong et al. [58]. (c) Triboelectric property characterization of the smart strain sensor textile-based TENGs. Source: Adapted with permission from He et al. [59]. Copyright (2019) by Elsevier.

10% to 160% is developed (Figure 8.23c). Furthermore, the arch-shaped strain sensors mounted on the human fingers are demonstrated to monitor hand gestures for American Sign Language interpretation and robotic hand control. Additionally, four textile-based sensors located at different body parts are employed to track human activities.

8.3 Chemical Sensing Textiles

Chemical sensors are instruments that are sensitive to various chemicals (for example, Na^+, Ca^{2+}, NH_3, NO_2, etc.) and convert their concentrations into electrical signals for detection. As chemical sensors are increasingly used in human daily life, wearable chemical sensors have been gradually developed, namely, chemical sensing textiles. Different from the majority of flexible sensors that aim to gain the information of physical activities or vital signs, textile-based chemical sensors provide a more straightforward manner for noninvasive personal health monitoring at the molecular level. It has a wide range of applications in medical remote diagnosis and real-time monitoring, anti-theft, security alarm, and energy saving.

According to different chemical detection objects, chemical sensing textiles can be divided into ion sensing textiles, humidity sensing textiles, and gas sensing textiles.

8.3.1 Ion Sensing Textiles

Human sweat is an important and easily accessible body fluid containing a wealth of chemicals that can reflect an individual's physiological state. Eccrine sweat, which is easily accessible noninvasively, is excreted directly onto the surface of the skin and is composed of water and various ions (e.g. Na^+, K^+, Ca^{2+}). It has been shown that abnormal health conditions can significantly affect sweat composition by varying the concentrations of sweat analytes. The transition from blood analysis to in situ sweat analysis via a wearable chemical sensing textile could provide a noninvasive and attractive means of dynamic health assessment.

Peng and coworkers [60] reported a new and general strategy of making electrochemical fabric from sensing fiber units (Figure 8.24a,b). These units efficiently detect a variety of physiological signals such as glucose, Na^+, K^+, Ca^{2+}, and pH. The electrochemical fabric is highly flexible and maintains structural integrity and detection ability under repeated deformations, including bending and twisting. Figure 8.24c,d shows the open-circuit potentials (OCP) of Na^+-sensing fibers in the electrolyte solutions with physiologically relevant concentrations of $(10–160)\times10^{-3}$ mM Na^+. Excellent sensing performances were found for Na^+-sensing fibers with sensitivities of 45.8 mV/dec of concentration. Zhang and coworkers [61] reported a flexible sweat analysis patch based on a silk fabric-derived carbon textile for Na^+ detection (Figure 8.24e,f). The SilkNCT-based Na^+ sensors showed linearly increased OCP with concentrations of 5–100 mM for Na^+ corresponding to sensitivity of 51.8 mV per decade of concentration for Na^+

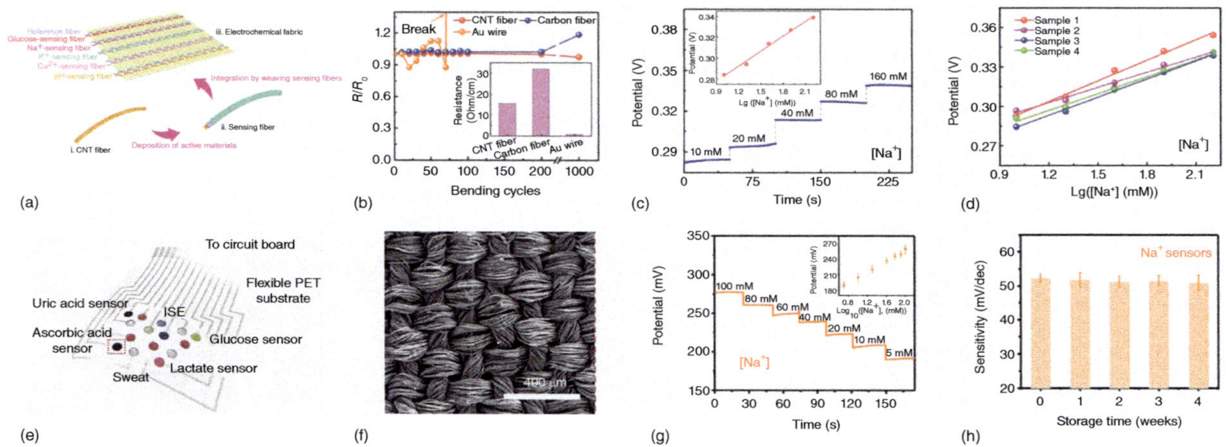

Figure 8.24 Ion chemical sensing textiles. (a) Schematic illustration for the fabrication of the electrochemical fabric. (b) Dependence of electrical resistance on the fiber electrode from bending cycle number at a bending angle of 90°. (c) The open-circuit potential responses of the Na^+. (d) The reproducibility of Na^+. Source: Wang et al. [60]. © 2018, John Wiley & Sons. (e) Schematic illustration of wearable sweat analysis patch. (f) SEM images of the carbonized silk. (g) OCP responses of the SilkNCT-based Na^+ sensor. (h) Electrochemical responses of the Na^+ sensors. Source: (f) Adapted with permission from He et al. [61]. Copyright (2019) by the American Association for the Advancement of Science, (h) He et al. [61]. © 2019, American Association for the Advancement of Science.

(Figure 8.24g). In addition, the electrochemical responses of the SilkNCT-based sensors showed negligible changes over four weeks (Figure 8.24h), with the relative standard deviations (RSDs) of 2.9% for Na^+. The results revealed the long-term stability of the SilkNCT-based sensors, which could be attributed to the good stability of the graphitic nanocarbon structure of SilkNCT.

Later, Peng and coworkers [60] reported an electrochemical fabric from sensing fiber units. Figure 8.25a,b shows the OCP of K^+-sensing fibers in the electrolyte solutions with physiologically relevant concentrations of $(2–32) \times 10^{-3}$ mM K^+. Excellent sensing performances were found for K^+-sensing fibers with sensitivities of 35.9 mV/dec of concentration. In 2018, an electrochemical fabric from sensing fiber units is reported by Peng and coworkers [60]. These units efficiently detect a variety of physiological signals (for example, Ca^{2+}). Figure 8.25c,d shows the OCP of Ca^{2+}-sensing fibers in the electrolyte solutions with physiologically relevant concentrations of $(0.5–2.53) \times 10^{-3}$ mM Ca^{2+}. Excellent sensing performances were found for Ca^{2+}-sensing fibers with sensitivities of 52.3 mV/dec of concentration (Figure 8.25c,d). The electrochemical fabric is highly flexible and maintains structural integrity and detection ability under repeated deformations, including bending and twisting. They demonstrate the capacity to monitor health conditions of human body in real time with high efficiency.

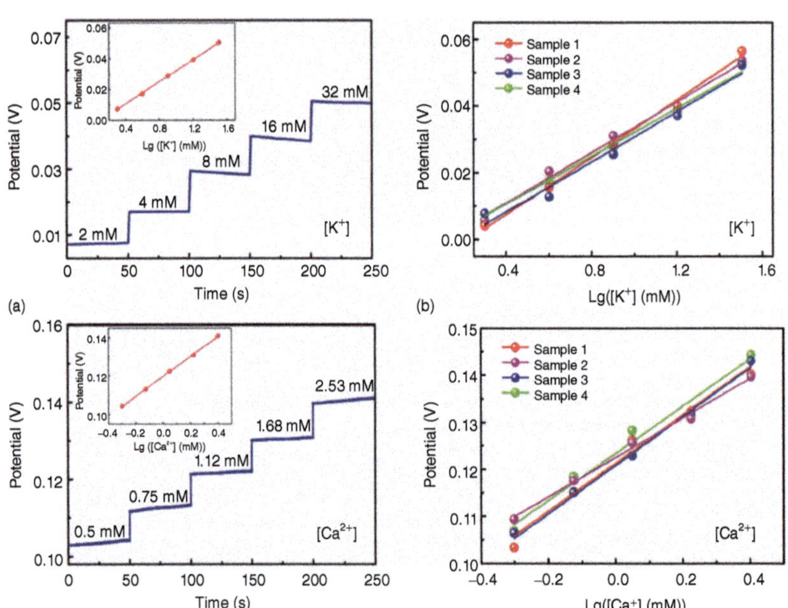

Figure 8.25 The open-circuit potential responses of the (a) K^+-sensing and (c) Ca^{2+}-sensing fibers to the respective analyte solutions. The reproducibility of (b) K^+ and (d) Ca^{2+}. Source: Wang et al. [60]. © 2018, John Wiley & Sons.

8.3.2 Humidity Sensing Textiles

Humidity plays a vital role to the survival of animals and plants, human comfort, industrial process control, and high-tech instruments. Thus, the measurement of humidity is significantly important. The principle of humidity sensing is intuitive, as a higher water content will result in a better conductance.

Lu and coworkers [62] fabricated a silk textile-based humidity sensor by successive electroless plating of conductive interdigital electrodes and spray coating of a GO sensing layer (Figure 8.26a). Surface morphologies of the devices were characterized to show the construction of the sensor. The uniform surface distribution of GO nanosheets was exhibited using the Raman mapping method (Figure 8.26b). After optimization, the as-prepared device could accurately detect the human respiration rate and effectively differentiate normal, deep, and fast breathing. The flexibility test indicates that the sensor can tolerate 2500 times of bending and twisting without influencing its performance (Figure 8.26c). Moreover, by designing the spray-coating mask, GO layers with various patterns could be deposited to achieve both functionality and aesthetics (Figure 8.26d). This work not only develops a silk fabric-based sensor for potential applications in assessing of human basic health status but also provides a facile approach to preparing textile-based wearable electronic devices.

8.3.3 Gas Sensing Textiles

Rapid and accurate determination of chemical gas is of practical importance in the workplace and as a safety measure in an unknown environment. To track the gas concentration for personal wellness and security surveillance, wearable gas sensors, namely, gas chemical sensing textiles, were developed to detect the concentration of gases such as nitrogen dioxide (NO_2), ammonia (NH_3), acetic acid (CH_3COOH), hydrogen sulfide (H_2S), hydrogen (H_2), ethanol (C_2H_5OH), dimethyl methylphosphonate ($C_3H_9O_3P$, DMMP), acetone (CH_3COCH_3), methanol (CH_3OH), and tetradecane ($C_{14}H_{30}$). Increase or decrease in resistance is typically detected, depending on whether the gaseous stimulus is an electron donator or an electron acceptor.

NO_2 has an effect on the lungs by lining inflammations, which can lead to wheezing, coughing, colds, flu, and bronchitis due to lung infection. Elevated NO_2 exposure is dangerous to people with asthma because of more frequent and intense attacks. In addition, it can increase the risk of otitis media, eczema, ear/nose/throat infections, and sensitization to food allergens in children, as well as increased blood coagulability in adults. The measurement of NO_2 was achieved by Lee and coworkers using a wearable fiber-based gas sensor (Figure 8.27a) [63]. Reduced GO was selected as the biomarker targeting component due to strong affinities to NO_2 gas. When exposed to NO_2, the gas sensor, which was built as a p-type semiconductor, will undergo a resistance reduction toward elevated hole concentration. The electrical resistance was measured using a semiconducting parameter analyzer. The developed NO_2 gas sensor has been revealed to obtain selectivity to other

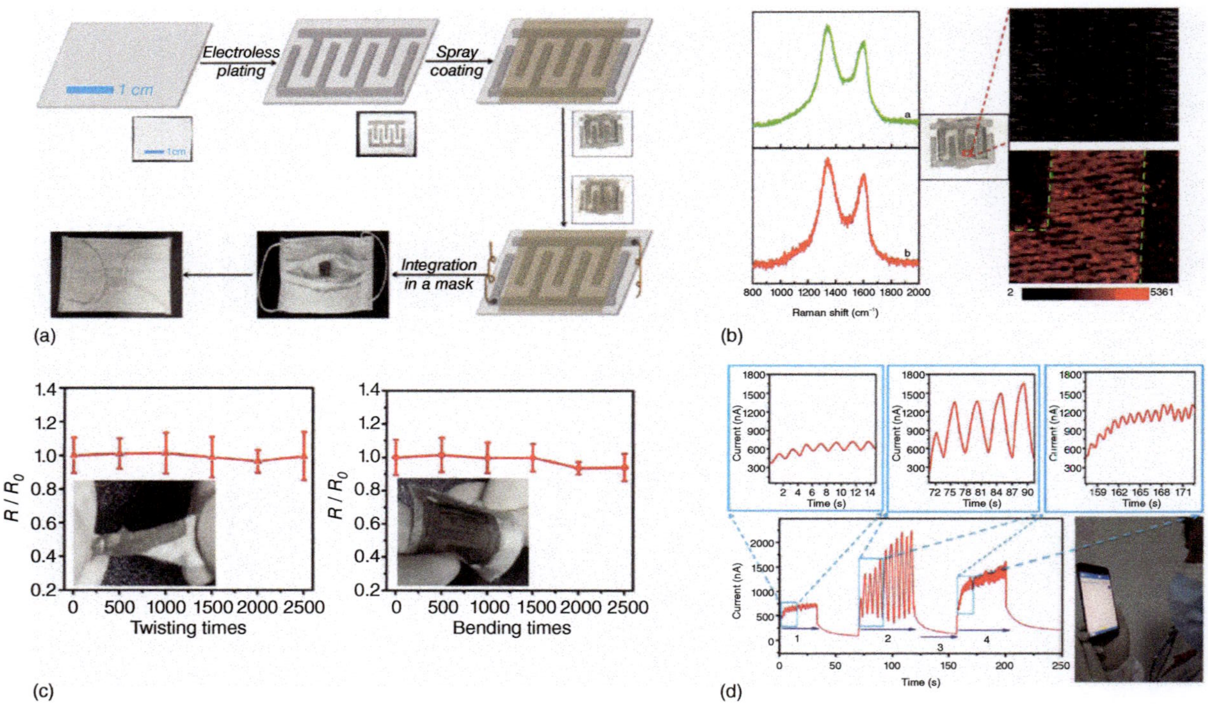

Figure 8.26 Humidity chemical sensing textiles. (a) Fabrication process of a silk fabric-based humidity sensor. (b) Raman spectra and optical image of silk fabric. (c) Effects of twisting and bending on the device resistances. (d) Response curves corresponding to various breathings. Source: Adapted with permission from Li et al. [62]. Copyright (2018) by Royal Society of Chemistry.

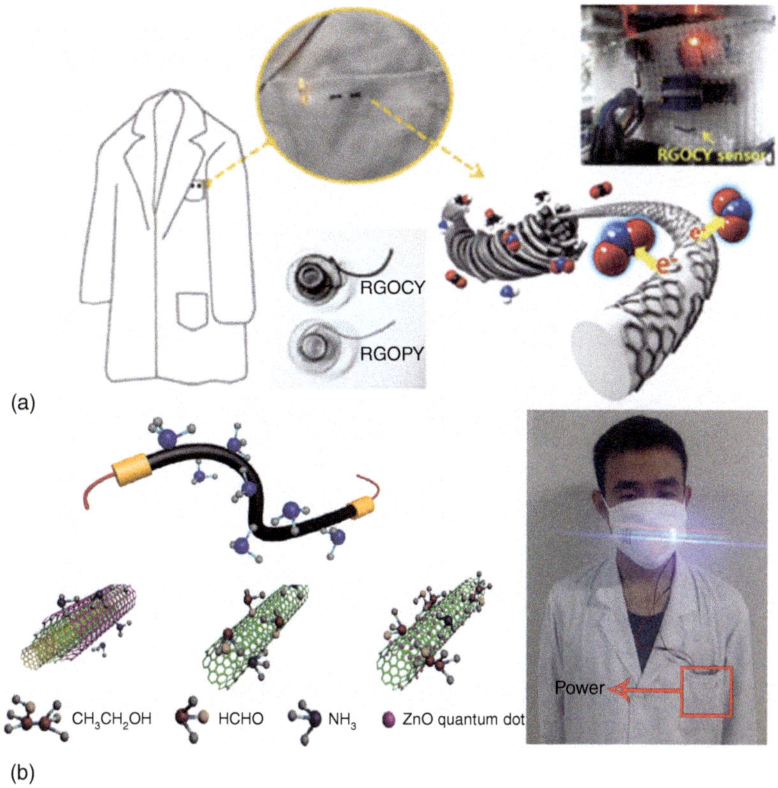

Figure 8.27 Scheme of the gas chemical sensing textiles. (a) A wearable fiber-based gas sensor enabled by NO_2 detection functions. Source: Adapted with permission from Yong et al. [63]. Copyright (2015) by Springer Nature Limited. (b) A face mask with embedded fiber gas sensors for poisonous gas (HCHO, NH_3, and C_2H_5OH) sensing. Source: Adapted with permission from Gao et al. [64]. Copyright (2017) by Springer.

gases such as CH_3COCH_3, C_2H_5OH, ethylene, and CO_2. Another example of a face mask embedded with three fiber gas sensors was developed by Shen and coworkers for toxic gas detection (Figure 8.27b) [64]. Nylon ropes were functionalized with SWCNTs, MWCNTs, and ZnO quantum dot-decorated SWCNTs.

Accordingly, the face mask was enabled to sense NO_2 and NH_3 by different LED colors. For example, Byungjin et al. [65] reported that two-dimensional (2D) metal ($NbSe_2$)–semiconductor (WSe_2)-based flexible, wearable, and launderable gas sensors can be prepared through simple one-step chemical vapor deposition of prepatterned WO_3 and Nb_2O_5. Compared with a control device with an Au/WSe_2 junction, gas-sensing performance of the 2D $NbSe_2$/WSe_2 device was significantly enhanced, which might have resulted from the formation of a $Nb_xW_{1-x}Se_2$ transition alloy junction lowering the Schottky barrier height (Figure 8.28a). This would make it easier to collect charges of channels induced by molecule adsorption, improving gas response characteristics toward chemical species including NO_2 and NH_3 (Figure 8.28b,c). 2D $NbSe_2$/WSe_2 devices on a flexible substrate provide

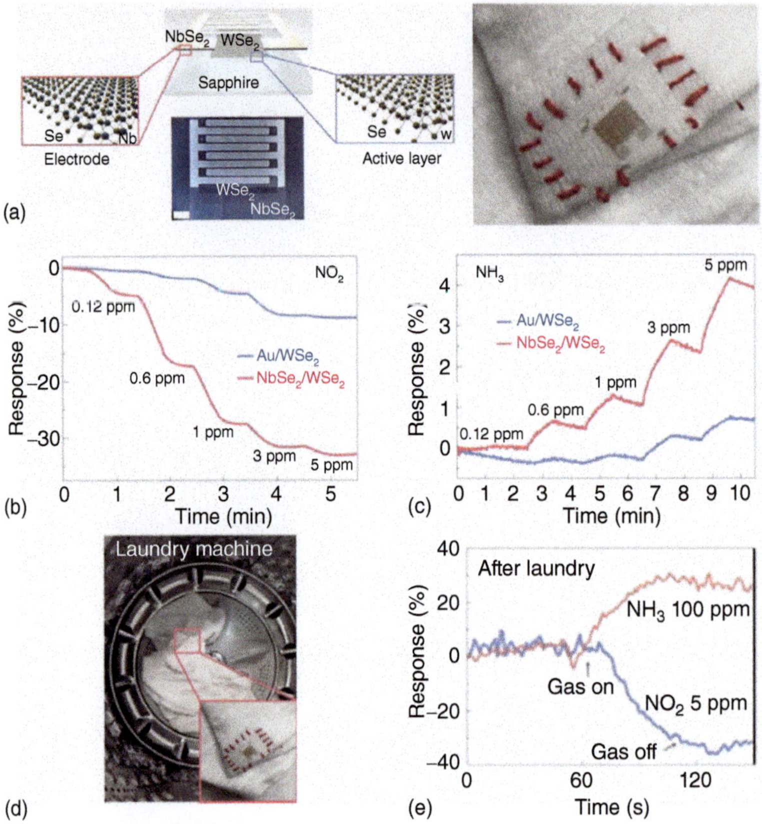

Figure 8.28 The performance of gas chemical sensing textiles. (a) Schematic image of the $NbSe_2/WSe_2$ gas-sensing device and fabric sensor. (b) NO_2 and (c) NH_3 analyte gases for both $NbSe_2/WSe_2$ and Au/WSe_2. (d) Gas-sensing device stitched on a T-shirt with red threads and placed in a laundry machine. (e) Gas responses of the launderable device showing chemical and mechanical durabilities. Source: Adapted with permission from Byungjin et al. [65]. Copyright (2016) by American Chemical Society.

gas-sensing properties with excellent durability under harsh bending. Furthermore, the device stitched on a T-shirt still performed well even after conventional cleaning with a laundry machine, enabling wearable and launderable chemical sensors (Figure 8.28d,e). These results could pave a road toward futuristic gas-sensing platforms based on only 2D materials.

8.4 Other Function Coupled Textiles

In addition to the single-function sensors described above, as people's production and life needs increase, sometimes sensors with multiple functions are needed. In order to realize the multimodality of sensor-coupled wearable platforms for applications in human activity monitoring, personal healthcare, and human–machine interfaces, the integration of flexible and stretchable multiple physical sensors into specific platforms has been carried out, namely, function coupled textiles. Wearable platforms in the form of skin patches, wristbands, and fabrics that are conformally attached to the human body and provide long-term continuous recording of human activity and acute physical responses can be used for preventive medicine, diagnosis of disease, prognostic monitoring of patients, and elderly care.

In the past few years, a plurality of research groups had taken the first step toward merging at least two categories of sensors into a flexible device. For example, Kim and coworkers [66] demonstrated smart prosthetic skin instrumented with ultrathin, single-crystalline silicon nanoribbon strain, pressure and temperature sensor arrays, and associated humidity sensors, electroresistive heaters, and stretchable multielectrode arrays for nerve stimulation. Gao et al. [67] presented a mechanically flexible and fully integrated sensor array for multiplexed in situ perspiration analysis, which simultaneously and selectively measures sweat metabolites and electrolytes, as well as the skin temperature.

Coupled multiple wearable sensors are not only integrated circuits (ICs) fixed on wearable platforms but also, from a more modern point of view, electronic or sensing components in the form of fibers or stripes that can be directly woven into fabric or clothes. A textile integrated sensor is developed by Rooij and coworkers [68]. They presented low-power humidity and temperature sensors fabricated on Kapton polyimide sheets that are successfully woven into textile using conventional weaving machine (Figure 8.29a,b). Figure 8.29c shows the humidity sensing behavior of the woven sensors with PDMS sensing layer, measured under constant 25 °C ambient

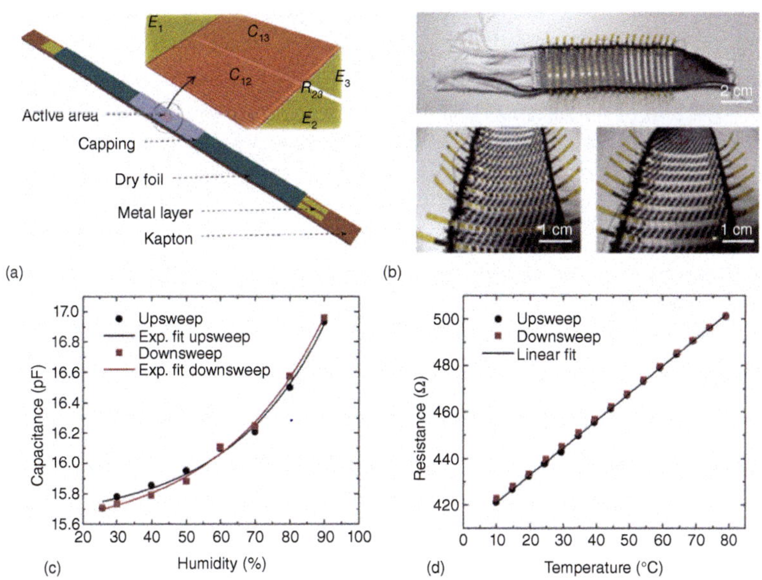

Figure 8.29 Other function coupled textiles. (a) 3D depiction of the multilayer sensor structure. (b) Images of the woven sensors. (c) Basic humidity sensing and (d) temperature sensing behavior of the woven sensors. Source: Adapted with permission from Ataman et al. [68]. Copyright (2011) by Elsevier.

temperature. The sensing capacitance has a raised exponential dependence on the ambient humidity. The integrated temperature sensors, as shown in Figure 8.29d, have a linear sensitivity of 1.175 Ω/°C, which corresponds to a temperature coefficient of resistivity (TCR) of 2701 ppm/°C.

8.5 Conclusions and Outlook

In this chapter, we have presented some kinds of emerging electronic fibers and textiles based on different sensing functionalities. Although advances are continuing and more prototypes are becoming available based on various materials and structures, there remains a large gap between the best current textile-type electronics and real applications. On the one hand, from a viewpoint of practical application, some critical challenges remain that need to be addressed. Firstly, it is difficult to fabricate highly efficient fiber- and textile-based electronic devices on a large scale, as their performances greatly decrease as their lengths are increased from centimeters to meters or even longer. Secondly, it is challenging to connect a bunch of fiber electrodes after the fiber-based electronic devices are woven into textiles. Thirdly, it is hard to decide whether or not to seal the electronic textiles. They become less stable and even soon fail without sealing, but they cannot effectively display the advantages of the textile structure after sealing. Fourthly, washing the electronic textiles is problematic as they may stop working after contact with water. On the other hand, at present, many commercial giants (such as Google) have developed very mature wearable electronic fabrics, but their high prices and the lack of corresponding industrialization standards have caused them to not be popularized in people's production and life. More efforts are needed to verify the above hypotheses. It is foreseeable that in the next 10 years, the emerging electronic fibers and textiles will usher in the peak of development.

In summary, although a lot of studies are required to optimize the structures and enhance the properties of these emerging electronic fibers and textiles, they have been demonstrated to be promising for a variety of applications. It is proposed that they may bring a technical revolution to our life in the near future.

References

1 Weng, W., Chen, P., He, S. et al. (2016). Smart electronic textiles. *Angewandte Chemie International Edition* 55: 6140–6169.

2 Lou, C., Wang, S., Liang, T. et al. (2017). A graphene-based flexible pressure sensor with applications to plantar pressure measurement and gait analysis. *Materials* 10: 1068–1079.

3 Someya, T., Sekitani, T., Iba, S. et al. (2004). A large-area, flexible pressure sensor matrix with organic field-effect transistors for artificial skin applications.

Proceedings of the National Academy of Sciences of the United States of America 101 (27): 9966–9970.

4 Fan, Z., Johnny, C.H., Jacobson, Z.A. et al. (2008). Large-scale, heterogeneous integration of nanowire arrays for image sensor circuitry. *Proceedings of the National Academy of Sciences of the United States of America* 105 (32): 11066–11070.

5 Kuniharu, T., Toshitake, T., Johnny, C.H. et al. (2010). Nanowire active-matrix circuitry for low-voltage macroscale artificial skin. *Nature Materials* 9: 821–826.

6 Yao, H., Ge, J., Wang, C. et al. (2013). A flexible and highly pressure-sensitive graphene-polyurethane sponge based on fractured microstructure design. *Advanced Materials* 25: 6692–6698.

7 Pan, L., Alex, C., Yu, G. et al. (2014). An ultra-sensitive resistive pressure sensor based on hollow-sphere microstructure induced elasticity in conducting polymer film. *Nature Communications* 5: 3002–3010.

8 Zhou, Y., He, J., Wang, H. et al. (2017). Highly sensitive, self-powered and wearable electronic skin based on pressure-sensitive nanofiber woven fabric sensor. *Scientific Reports* 7: 12949–12958.

9 Cao, M., Wang, M., Li, L. et al. (2018). Wearable rGO-Ag NW@cotton fiber piezoresistive sensor based on the fast charge transport channel provided by Ag nanowire. *Nano Energy* 50: 528–535.

10 Francisco, P., Piero, V., Daniel, Y. et al. (2018). Easy-to-build textile pressure sensor. *Sensors* 18: 1190–1203.

11 Honda, S., Zhu, Q., Satoh, S. et al. (2019). Textile-based flexible tactile force sensor sheet. *Advanced Functional Materials* 29: 1807957–1807963.

12 Chiseon, Y., Chen, K., Daisuke, K. et al. (2015). Large-area compliant tactile sensors using printed carbon nanotube active-matrix backplanes. *Advanced Materials* 27: 1561–1566.

13 Choong, C.L., Shim, M.B., Lee, B.S. et al. (2014). Highly stretchable resistive pressure sensors using a conductive elastomeric composite on a micropyramid array. *Advanced Materials* 26: 3451–3458.

14 Wang, X., Gu, Y., Xiong, Z. et al. (2014). Silk-molded flexible, ultrasensitive, and highly stable electronic skin for monitoring human physiological signals. *Advanced Materials* 26: 1336–1342.

15 Su, B., Gong, S., Ma, Z. et al. (2015). Mimosa-inspired design of a flexible pressure sensor with touch sensitivity. *Small* 11 (16): 1886–1891.

16 Gong, S., Willem, S., Wang, Y. et al. (2014). A wearable and highly sensitive pressure sensor with ultrathin gold nanowires. *Nature Communications* 5: 3132–3140.

17 Yin, F., Yang, J., Peng, H., and Yuan, W. (2018). Flexible and highly sensitive artificial electronic skin based on graphene/polyamide interlocking fabric. *Journal of Materials Chemistry C* 6: 6840–6846.

18 Chen, H., Su, Z., Song, Y. et al. (2017). Omnidirectional bending and pressure sensor based on stretchable CNT-PU sponge. *Advanced Functional Materials* 27: 1604434–1604443.

19 Lee, J., Kwon, H., Seo, J. et al. (2015). Conductive fiber-based ultrasensitive textile pressure sensor for wearable electronics. *Advanced Materials* 27: 2433–2439.

20 Benjamin, C.K., Alex, C., Roger, R.D. et al. (2014). Tunable flexible pressure sensors using microstructured elastomer geometries for intuitive electronics. *Advanced Functional Materials* 24: 5427–5434.

21 Stefan, C.B., Benjamin, C.K., Randall, M.S. et al. (2010). Highly sensitive flexible pressure sensors with microstructured rubber dielectric layers. *Nature Materials* 9: 859–864.

22 Gregor, S., Benjamin, C.K., Mei, J. et al. (2013). Flexible polymer transistors with high pressure sensitivity for application in electronic skin and health monitoring. *Nature Communications* 4: 1859–1867.

23 Lucie, V., Alessandro, L., Massimo, T. et al. (2014). Flexible three-axial force sensor for soft and highly sensitive artificial touch. *Advanced Materials* 26: 2659–2664.

24 Zang, Y., Zhang, F., Huang, D. et al. (2015). Flexible suspended gate organic thin-film transistors for ultra-sensitive pressure detection. *Nature Communications* 6: 6269–6278.

25 Minjeong, K., Hyoungjun, K., Jinwoo, P. et al. (2017). Real-time sitting posture correction system based on highly durable and washable electronic textile pressure sensors. *Sensors and Actuators* 924 (17): 31281–31285.

26 Kapoor, A., McKnight, M., Chatterjee, K. et al. (2018). Toward fully manufacturable, fiber assembly-based concurrent multimodal and multifunctional sensors for e-textiles. *Advanced Materials Technologies* 4: 1800281–1800286.

27 Steve, P., Kim, H., Michael, V. et al. (2014). Stretchable energy-harvesting tactile electronic skin capable of differentiating multiple mechanical stimuli modes. *Advanced Materials* 26: 7324–7332.

28 Chen, W., Gui, X., Liang, B. et al. (2017). Structural engineering for high sensitivity, ultrathin pressure sensors based on wrinkled graphene and anodic aluminum oxide membrane. *ACS Applied Materials & Interfaces* 9: 24111–24117.

29 Xu, W., Huang, M., Navid, A. et al. (2013). eCushion: a textile pressure sensor array design and calibration for sitting posture analysis. *IEEE Sensors Journal* 13 (10): 3926–3934.

30 Kechichea, M.B., Bauerb, F., Harzallaha, O., and Drean, J.Y. (2013). Development of piezoelectric coaxial filament sensors P(VDF-TrFE)/copper for textile structure instrumentation. *Sensors and Actuators A* 204: 122–130.

31 Luana, P., Canan, D., Su, Y. et al. (2013). High performance piezoelectric devices based on aligned arrays of nanofibers of poly(vinylidenefluoride-co-trifluoroethylene). *Nature Communications* 4: 1633–1643.

32 Cao, R., Pu, X., Du, X. et al. (2018). Screen-printed washable electronic textiles as self-powered touch/gesture tribo-sensors for intelligent human-machine interaction. *ACS Nano* 12: 5190–5196.

33 Joseph, D.S., Marie, M., Clark, J.K. et al. (2018). Stretchable optomechanical fiber sensors for pressure determination in compressive medical textiles. *Advanced Healthcare Materials* 7 (15): 800293–1800300.

34 Heikenfeld, J., Jajack, A., Rogers, J. et al. (2018). Wearable sensors: modalities, challenges, and prospects. *Lab on a Chip* 18: 217–248.

35 Lee, Y., Bae, S., Jang, H. et al. (2010). Wafer-scale synthesis and transfer of graphene films. *Nano Letter* 10: 490–493.

36 Takeo, Y., Hayamizu, Y., Yuki, Y. et al. (2011). A stretchable carbon nanotube strain sensor for human-motion detection. *Nature Nanotechnology* 6: 296–301.

37 Lu, N., Lu, C., Yang, S., and John, R. (2012). Highly sensitive skin-mountable strain gauges based entirely on elastomers. *Advanced Functional Materials* 22: 4044–4050.

38 Sun, J., Li, Y., Peng, Q. et al. (2013). Macroscopic, flexible, high-performance graphene ribbons. *ACS Nano* 7 (11): 10225–10232.

39 Shayan, S., Joselito, M.R., Peter, C.I. et al. (2015). Knitted strain sensor textiles of highly conductive all-polymeric fibers. *ACS Applied Materials & Interfaces* 7: 21150–21158.

40 Wang, C., Li, X., Gao, E. et al. (2016). Carbonized silk fabric for ultrastretchable, highly sensitive, and wearable strain sensors. *Advanced Materials* 28: 6640–6648.

41 Cai, G., Yang, M., Xu, Z. et al. (2017). Flexible and wearable strain sensing fabrics. *Chemical Engineering Journal* 325: 396–403.

42 Jaehong, L., Sera, S., Sanggeun, L. et al. (2018). Highly sensitive multifilament fiber strain sensors with ultrabroad sensing range for textile electronics. *ACS Nano* 12: 4259–4268.

43 Yang, Z., Pang, Y., Han, X. et al. (2018). Graphene textile strain sensor with negative resistance variation for human motion detection. *ACS Nano* 12: 9134–9141.

44 Montazerian, H., Dalili, A., Milani, A.S., and Hoorfar, M. (2019). Piezoresistive sensing in chopped carbon fiber embedded PDMS yarns. *Composites Part B* 164: 648–658.

45 Yan, C., Wang, J., Kang, W. et al. (2014). Highly stretchable piezoresistive graphene-nanocellulose nanopaper for strain sensors. *Advanced Materials* 26: 2022–2027.

46 Lee, S., Sera, S., Sanggeun, L. et al. (2015). Ag nanowire reinforced highly stretchable conductive fibers for wearable electronics. *Advanced Functional Materials* 25: 3114–3121.

47 Li, Y., Shang, Y., He, X. et al. (2013). Overtwisted, resolvable carbon nanotube yarn entanglement as strain sensors and rotational actuators. *ACS Nano* 7 (9): 8128–8135.

48 Wang, C., Xia, K., Jian, M. et al. (2017). Carbonized silk georgette as an ultrasensitive wearable strain sensor for full-range human activity monitoring. *Journal of Materials Chemistry C* 5: 7604–7611.

49 Mohammad, Z.S., Joselito, M.R., Peter, C.I., and Gordon, G.W. (2014). Strain-responsive polyurethane/PEDOT:PSS elastomeric composite fibers with high electrical conductivity. *Advanced Functional Materials* 24: 2957–2966.

50 Darren, J.L., Michael, V., Benjamin, C.K. et al. (2011). Skin-like pressure and strain sensors based on transparent elastic films of carbon nanotubes. *Nature Nanotechnology* 6: 788–792.

51 Daniel, J.C., Debkishore, M., Kevin, P., and Michel, M.M. (2012). A highly elastic, capacitive strain gauge based on percolating nanotube networks. *Nano Letter* 12: 1821–1825.

52 Cai, L., Song, L., Luan, P. et al. (2013). Super-stretchable, transparent carbon nanotube-based capacitive strain sensors for human motion detection. *Scientific Reports* 3: 3048–3056.

53 Andreas, F., Joseph, T.M., Daniel, M.V. et al. (2015). Capacitive soft strain sensors via multicore–shell fiber printing. *Advanced Materials* 27: 2440–2446.

54 Changsoon, C., Jae, M.L., Shi, H.K., and Seon, J.K. (2016). Twistable and stretchable sandwich structured fiber for wearable sensors and supercapacitors. *Nano Letter* 16: 7677–7684.

55 Ozgur, A. (2018). Textile-based, interdigital, capacitive, soft-strain sensor for wearable applications. *Materials* 11: 768–779.

56 Yao, S. and Zhu, Y. (2014). Wearable multifunctional sensors using printed stretchable conductors made of silver nanowires. *Nanoscale* 6: 2345–2352.

57 Hyeon, J.S., Changsoon, C., Shi, H.K. et al. (2016). Stretchable triboelectric fiber for self-powered kinematic sensing textile. *Scientific Reports* 6: 35153–35160.

58 Gong, W., Hou, C., Guo, Y. et al. (2017). A wearable, fibroid, self-powered active kinematic sensor based on stretchable sheath-core structural triboelectric fibers. *Nano Energy* 39: 673–683.

59 He, T., Shi, Q., Wang, H. et al. (2019). Beyond energy harvesting-multi-functional triboelectric nanosensors on a textile. *Nano Energy* 57: 338–352.

60 Wang, L., Wang, L., Zhang, Y. et al. (2018). Weaving sensing fibers into electrochemical fabric for real-time health monitoring. *Advanced Functional Materials* 28: 1804456–1804464.

61 He, W., Wang, C., Wang, H. et al. (2019). Integrated textile sensor patch for real-time and multiplex sweat analysis. *Science Advance* 5: 0649–0657.

62 Li, B., Xiao, G., Liu, F. et al. (2013). Flexible humidity sensor based on silk fabric for human respiration monitoring. *Journal of Materials Chemistry C* 6 (16): 1–3.

63 Yong, J.Y., Won, G.H., Choi, N.J. et al. (2015). Ultrasensitive and highly selective graphene-based single yarn for use in wearable gas sensor. *Scientific Reports* 5: 10904–10911.

64 Gao, Z., Lou, Z., Chen, S. et al. (2018). Fiber gas sensor-integrated smart face mask for roomtemperature distinguishing of target gases. *Nano Research* 11 (1): 511–519.

65 Byungjin, C., Ah, R.K., Dong, J.K. et al. (2016). Two-dimensional atomic-layered alloy junctions for high-performance wearable chemical sensor. *ACS Applied Materials & Interfaces* 8: 19635–19642.

66 Jaemin, K., Mincheol, L., Hyung, J.S. et al. (2014). Stretchable silicon nanoribbon electronics for skin prosthesis. *Nature Communications* 5: 5747–5758.

67 Gao, W., Sam, E., Hnin, Y.N. et al. (2016). Fully integrated wearable sensor arrays for multiplexed in situ perspiration analysis. *Nature* 529: 509–519.

68 Ataman, C., Kinkeldei, T., Vasque-Quintero, A. et al. (2011). Humidity and temperature sensors on plastic foil for textile integration. *Procedia Engineering* 25: 136–139.

9

Towards Self-Powered Electronic Textiles

A self-powered electronic device is a device that does not require an external power supply for operation. By integrating more than one self-powered electronic function device (i.e. biological signal conduction, power/signal conditioning, or data transmission), an integrated electronic system is formed without batteries, also called a self-powered electronic system. An ideal self-powered electronic system can complete all the electronic system functions in an unrestricted time independently and sustainably, without any additional power cables or chargers. Self-powered systems have many important applications, such as implantable body monitoring, remote environment monitoring, micro-robotic system, and wearable personal health care [1].

Self-powered electronic textiles are developed with the advancement of self-powered electronic devices. Electronic textiles can adapt to a more complex environment as they are lightweight, flexible, breathable, washable, and durable, which meets many wearable applications' needs. To fabricate a self-powered electronic textile, we should first convert traditional flat-type function devices, such as sensor, energy devices, and other electronic components, into the shape of fibers, by assembling functional layers layer-by-layer on fiber substrates. These fiber-type devices are combined through embroidery, weaving, knitting, or other methods, to form a complete system with the appearance of textile.

From traditional flat-type devices to textile-type devices, the development process of self-powered electronic devices can be divided into three categories: self-powered electronic devices, flexible self-powered electronic devices, and self-powered electronic fibers, which were introduced respectively in this chapter.

9.1 Self-Powered Electronic Devices

Traditional self-powered electronic devices usually adopt rigid structures of block, plate, strip, or others. They are fabricated by assembling different functional materials on rigid substrates. A self-powered device can realize multiple functions such as energy harvesting, energy storage, and signal conversion. The fabrication methods for self-powered electronic devices can be divided into three types: independent

Textile-Based Energy Harvesting and Storage Devices for Wearable Electronics, First Edition.
Xing Fan, Nannan Zhang, and Yi Wang.

self-powered electronic devices, integrated self-powered electronic devices, and other types of self-powered electronic devices.

Independent self-powered electronic devices are a single device that can execute electronic functions and works by harvesting environmental energy. In other words, for a self-powered sensor, environmental stimulation can be converted to both electronic signals and electric energy.

Integrated self-powered electronic devices are constructed as an integrated system of different functional device units, which can accomplish functions such as energy harvest, energy storage, and sensor.

Besides, for some self-powered electronic devices, the energy harvested from the environment can be utilized to maintain its own uninterruptedly works and served as a power source for driving external electronic devices, such as a watch, light-emitting diode, and sensors.

This section will introduce the principle, structure, and application of independent self-powered electronic devices, integrated self-powered electronic devices, and other types of self-powered electronic devices.

9.1.1 Independent Self-Powered Electronic Devices

Independent self-powered electronic devices consist of only one independent device, which can uninterruptedly execute electronic functions by automatically harvesting environmental energy. With a simple structure, the independent self-powered electronic devices can be easily fabricated. Nevertheless, the device-structure optimization for electronic function and energy harvesting are sometimes conflicted with each other. Such a simple device structure could somehow become a limitation for the compatibility of multiple functions.

The most popular independent self-powered electronic devices include two types: triboelectric nanogenerators (TENGs) and piezoelectric nanogenerators (PENGs).

The first type of independent self-powered electronic device is TENG that converts the mechanical energy of friction into electrical energy.

A typical TENG device is usually composed of two types of friction materials with different electronegativity (positive and negative). Among them, positive friction materials include metal, nylon, and wool, while negative friction materials include polyvinyl chloride (PVC), poly tetra fluoroethylene (PTFE), and polyimide (Kapton). The two types of friction materials are usually coated on the surface of the two electrodes, and electricity is generated through the frictional movement of the two electrodes. The greater the difference between the electronegativity of the two electrodes' materials, the more conducive to the generation of charges. Besides increasing friction and improving power generation efficiency, it is often necessary to add nano-structures in the material design process.

The working principle of TENG relies on the coupling of contact charging and electrostatic induction. Due to the triboelectric effect, charge transfer occurs between the two layers of friction materials on different frictional electrodes in the internal circuits. A potential difference formed in the external circuit, electrons are driven by the potential difference between two electrodes to flow towards the back surface of the triboelectric material layer to balance this potential difference. Its working modes can be divided into four types: vertical contact separation mode, lateral sliding mode, single electrode mode, and independent triboelectric mode [2].

For the self-powered electronic device based on TENG, its energy sources are extensive because friction is the most common phenomenon around the human body, accompanied by almost all the mechanic movements. However, the TENG's power output is usually low and unstable, limited by the weak mechanic energy.

Such self-powered electronic devices based on TENG can be employed in many physical, environmental, biomedical applications [3, 4]. For example, Xie et al. [3] reported a triboelectric speed sensor for integrating into the bearing in physical monitoring. The system is mainly composed of bearing housing, bearing, adjusting ring, rotor, and stator (Figure 9.1a). During the rotation of the bearing, the copper grating on the rotating rotor and Polytetrafluoroethylene (PTFE) on the surface can form frictions, which will then generate periodic electrical signals, with its amplitude is proportional to the rotation speed. The analysis and comparison

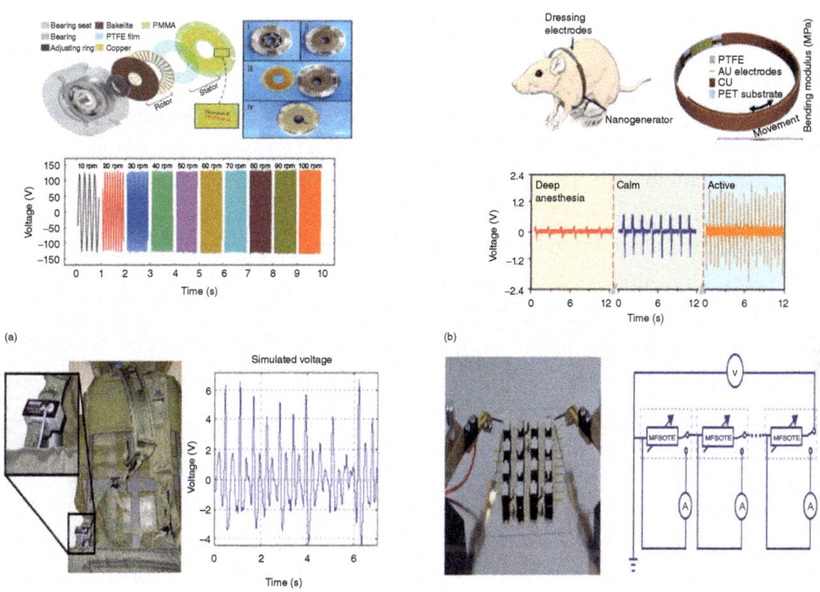

Figure 9.1 Independent self-powered electronic devices: (a) Schematic diagram of TRSS functional components, open-circuit voltage at different speeds. Source: Adapted with permission from Xie et al. [3]. Copyright (2019) by Elsevier. (b) NG-based bandages and potential wound healing applications, generating electricity from natural gas-driven by rat breathing at different frequencies. Source: Adapted with permission from Long et al. [4]. Copyright (2018) by American Chemical Society. (c) The instrument backpack is used to determine the load in the stacked amplifier experimentally. Source: Adapted with permission from Feenstra et al. [5]. Copyright (2007) by the American Chemical Society. (d) Photograph and circuit diagram of temperature/pressure sensor array with 4 × 4 pixels. Source: Adapted with permission from Zhang et al. [6]. Copyright (2015) by Springer Nature Limited.

of the output signal with commercial sensors show that the device can measure 10-1000 rpm speed. In terms of biomedical applications, Long et al. [4] reported an effective electrical bandage that can accelerate the healing of skin wounds (Figure 9.1b). A wearable nanogenerator generates alternating discrete electric fields on the bandage by converting mechanical displacement in skin motion into electricity. Studies in vitro show that skin wound healing acceleration is attributed to the migration, proliferation, and transdifferentiation of fibroblasts promoted by the electric field. This self-powered dressing method can provide a simple treatment strategy for non-healing skin wounds.

The second type of independent self-powered electronic devices is PENG. PENG is a device that uses the piezoelectric effect of special nano-materials to convert mechanical energy into electrical energy when subjected to external tension or compression.

Generally, PENG can be obtained by assembling piezoelectric materials on conductive substrates. There are three types of piezoelectric materials, including perovskite piezoelectric materials ($Pb(Zr_{1-x}Ti_x)O_3$, $KNbO_3$, $NaNbO_3$), hexagonal wurtzite piezoelectric materials (ZnO, GaN, InN), and polymer piezoelectric materials (PVDF). After assembling the piezoelectric material layer, another layer of conductive material was coated to serve as a counter electrode.

The working principle of PENG relies on the piezoelectric effect of the piezoelectric materials. Taking ZnO as an example, the most common crystallographic phase of ZnO in the ambient condition is hexagonal wurtzite structure, which causes non-centrosymmetric in ZnO crystal structure. When unstressed, the centers of mass of the positive and negative ions coincide at the tetrahedron center. Therefore, there is no net polarization in the unit cell. Upon applying pressure at an apex of the tetrahedron, the centers of mass of the cations and anions are relatively displaced, making the generation of a net polarization or charges on the crystal surface of ZnO. Under the strained condition, all the dipole moments develop an internal field within the ZnO crystal. Therefore, it exhibits a piezo-potential between two opposite electric charges of its opposite faces. Consequently, piezo-polarization charges and piezo-potential occur since the applied strain is sustained and driven the electrical charges to the external circuit [5].

The self-powered electronic device based on PENG does not need motion with a large amplitude, such as contact and separation, to generate electricity. So, the device can convert tiny pressure energy into electric energy. However, its voltage output is minimal. Many devices need to be connected in series to solve this problem. Such self-powered electronic devices based on PENG can be employed in many physical and biological applications.

For example, Feenstra et al. [5] reported a backpack that can generate electric energy from different forces between humans and the backpack. The PENG integrated at the top and bottom of the backpack belt to measure the change of tension through use. Researchers put 220 N steel plates in packages to act as loads and run on treadmills at 2–3 mph speeds. The test results show that the load impedance matches the input data very well at 2.75 Hz, representing the walking frequency. Therefore, this device can monitor the data of human walking by collecting the energy of human motion. Zhang et al. [6] reported a flexible dual-parameter temperature/pressure sensor based on organic thermoelectric materials supported by a micro-structure framework (Figure 9.1c). The temperature or pressure stimuli are effectively converted into two independent electrical signals, sensing temperature and pressure instantaneously. More importantly, these dual-parameter sensors can be self-powered and have excellent sensing performance. The excellent sensing performance of the device makes it have broad application prospects in electronic skin and health monitoring components.

9.1.2 Integrated Self-Powered Electronic Devices

Integrated self-powered electronic devices are usually constructed by integrating multiple device units such as energy harvesting, energy storage, and signal conversion. Through the cooperation of the above device units, the system can collect energy in the environment and convert it into electric energy stored in an energy storage unit for continuous electronic functions, such as sensors and signal modulation.

The advantages of integrated self-powered electronic devices include two aspects:

1) The function of integrated self-powered electronic devices can be easily expanded (such as lighting [7], alarming [8], and monitoring [9]) through the combination with different functional device units.
2) The performance of each device unit can be optimized independently, which can make function optimization very convenient. However, one integrated self-powered electronic devices need multiple device units to be integrated, so it is relatively large in volume and may interfere with each other.

The integrated self-powered electronic devices can be divided by their energy harvesting units, such as TENG, PENG, thermoelectric generator (TEG), and a photovoltaic cell.

The first type of integrated self-powered electronic devices is based on TENG. The device converts frictional energy into electrical energy and power other devices. The advantage of this type of device is that frictional energy is common in our daily life. The device can harvest energy anytime and anywhere. By combining different electronic function devices, integrated self-powered electronic devices based on TENG can implement different functions.

For example, on the aspect of integrating with a biological sensor, Zhao et al. [7] adjusted the Schottky barrier height (SBH) through the TENG and fabricated the device to realize the reversible conversion between Schottky contact and ohmic contact (Figure 9.2a). For such integrated self-powered biosensors, high-sensitivity detection of biomolecules (i.e. neurotransmitters) and neuroelectric signals can be achieved in different contact states. On integrating with a chemical sensor, Wen et al. [8] reported a blowing-driven triboelectric nanogenerator (BD-TENG). The induced voltage on the sensor is proportional to the alcohol concentration exhaled using the electrical energy generated through the mouthpiece. The device shows excellent selective anti-interference ability in alcohol detection (Figure 9.2b). On integrating

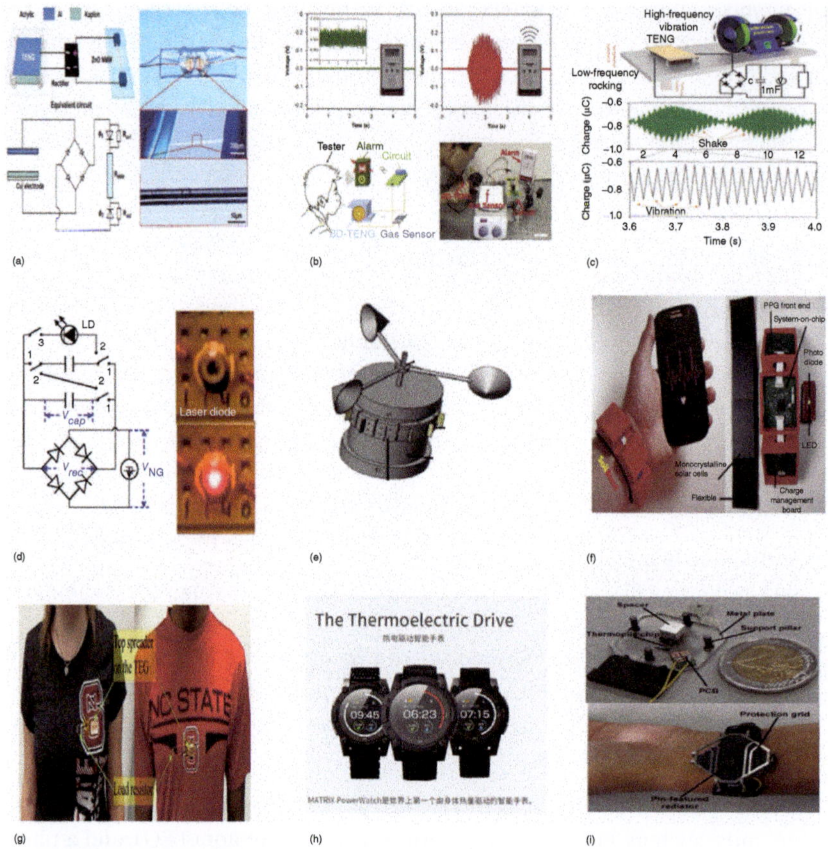

Figure 9.2 Integrated self-powered electronic devices: (a) SOR biosensor device, equivalent circuit, and basic phenomenon. Source: Zhang et al. [6]. Licenced under CC BY 4.0. (b) Demonstration of BD-TENG as an automatic breath analyzer. Source: Adapted with permission from Zhao et al. [7]. Copyright (2019) by WILEY-VCH Verlag GmbH & Co. KGaA, Weinheim. (c) System configuration of self-powered sensors that monitor the operating status of devices based on the energy collected from the grinding machine in operation. Source: Adapted with permission from Wen et al. [8]. Copyright (2015) by Elsevier. (d) Schematic design drawing of the rectifier circuit and charge storage device, using the stored charge to light up the LD snapshot before and at the moment. Source: Adapted with permission from Xu et al. [10]. Copyright (2010) by Springer Nature Limited. (e) Design of small vertical windmill, schematic diagram, and internal structure of cylindrical windmill. Source: Adapted with permission from Myers et al. [11]. Copyright (2007) by AIP Publishing. (f) Self-powered watches driven by photovoltaics. Source: Adapted with permission from Dieffenderfer et al. [9]. Copyright (2014) by IEEE. (g) T-shirt with integrated TEG devices. Source: Adapted with permission from Hyland et al. [12]. Copyright (2016) by Elsevier. (h) Thermoelectric drive smartwatch (www.cnpowerwatch.com). Source: Matrix China Guangdong Jieer Technology Co., Ltd. (i) Photograph of wearable TEG for the human body. Source: Adapted with permission from Wang et al. [13]. Copyright (2009) by Elsevier.

with a physical sensor, Wang et al. [14] introduce a self-powered electronic device based on TENG as an energy harvesting unit. The TENG can effectively convert environmental vibration energy into electric energy, with an instantaneous output power density of 102 W/m^3. It can provide continuous, direct current (DC) power with an average power of 2 mW in a regulated way. The device can continuously supply power to many electronic devices, such as temperature/humidity indicators and speedometers (Figure 9.2c).

The second type of integrated self-powered electronic devices is based on PENG. Like a TENG-based integrated device, it can also harvest various mechanical energy, such as pressure, wind, and water flow. Nevertheless, this type of device can harvest smaller mechanical energy to power the self-powered electronic system.

The self-powered electronic system can realize the different applications, such as driving the laser diode (LD) lamp [10], driving sensor [11], and others through various devices. For example, on the aspect of integrating with LD lamp, Xu et al. [10] reported a PENG. The commercial full-wave bridge rectifier composed of four diodes and eight capacitors can realize current rectification and energy storage (Figure 9.2d). By using voltage amplification technology, the stored charge is quickly released to power commercial LDs successfully. On integrating with a sensor, Myers et al. [11] reported a piezoelectric windmill uses 18 piezoelectric units. The device can solve problems related to autonomous sensor networks outdoors in inaccessible places by collecting wind energy. This research demonstrates a method of generating electricity from wind energy that is freely available on a small scale. The generated power can be stored in supercapacitors and can be used to wirelessly transmit sensor data to remote locations (Figure 9.2e).

The third type of integrated self-powered electronic device is based on photovoltaic cells, converting solar energy into electric energy.

The photovoltaic cell can be obtained by assembling photosensitive semiconductive materials on conductive substrates. There are five types of photosensitive semiconductive materials: polycrystalline film (GaAs, CdS, CdTe), organic polymer, nanocrystalline (nano-TiO_2), organic films, and photosensitive dyes. After assembling the photosensitive semiconductive material layer, another layer of conductive material was coated to serve as a counter electrode.

The photovoltaic cell is formed by joining P-type and N-type semiconductors to form a P–N junction on the interface. When irradiated by sunlight, the electrons at the interfacial P–N junction are excited to produce pairs of electrons and holes. The electric field affected these electrons and holes resulting from the polarity difference between the materials, then separate and inject into different layers of N-type or P-type semiconductors, and was finally collected by two electrodes. It will form a circuit loop for the electric current to run through by connecting the positive electrode and the negative electrode.

The energy source of sunshine is endless and will never be exhausted. Besides, the photovoltaic power generation process is simple and does not need additional mechanical movement. However, photovoltaic power generation still has some disadvantages. For example, photovoltaic generation is greatly affected by the environment, and energy can only be harvested in places with sufficient light.

After integrated with other device units, the device can implement many applications. For example, Dieffenderfer et al. [9] developed a solar-powered wireless wrist-worn platform. The device is used to monitor physiological and environmental parameters in daily activities (Figure 9.2f). It can convert light into electricity through the solar cell to provide the calculator and reach normal work.

The last type of integrated self-powered electronic device is based on TEG, converting heat into electric energy.

TEG can be obtained by coating thermoelectric materials on conductive substrates. The thermoelectric materials can be divided into three types: Bi_2Te_3 and its alloys, PbTe and its alloys, and SiGe alloy.

Because there is a constant temperature difference between the human body and the environment, TEGs are often used to harvest heat energy from the human body. The advantage of TEG is that its energy output is stable, and the human body can continuously provide an energy source for TEG. However, the power output of TEG is still small.

Small pieces of TEGs have been integrated on cloth to drive wearable electronics such as sensors and watches [12, 13].[1] For example, Hyland et al. [12] developed a kind of TEG. The TEG was made into T-shirts and recorded the power of different human activities. In experiments performed on different body parts and T-shirts, the power generated is the upper arm, wrist, chest, and T-shirt in descending order (Figure 9.2g). PowerWatch is a commercial thermoelectric watch, equipped with a thermoelectric engine powered by TEG (Figure 9.2h).[1] The patented technology is exclusive worldwide. It allows PowerWatch to be powered by its wearer's body temperature, which maximizes the flexibility of the smartwatch and is no longer subject to battery power. It is the world's first smartwatch powered in this way. Wang et al. [13] introduced a fully functional wearable micro TEG designed for human applications (Figure 9.2i). The wearable TEG worn on the human body can provide an open circuit output voltage of approximately 0.15 V, and the output power under a matched external load is approximately 0.3 nW.

9.1.3 Other Types of Self-Powered Electronic Devices

For other self-powered electronic devices, the energy harvested from the environment can be utilized to maintain its works and served as a power source for driving external electronic devices. The device consists of an energy harvesting unit, energy storage unit, and signal conversion unit. The energy harvesting unit can execute electronic functions and supply energy for other devices to achieve more functions.

This type of self-powered electronic device is usually based on TENG or PENG. By collecting the mechanical energy in the environment, it maintains its works while supplying power for the other applications. For example, Chen et al. [15] proposed a motion-triggered cantilever (MC), combined with an integrated real-time vibration monitoring sensor based on the principle of TENG (Figure 9.3a). They demonstrated a self-powered micro-scale system based on MC and a monitoring sensor. This

1 www.cnpowerwatch.com

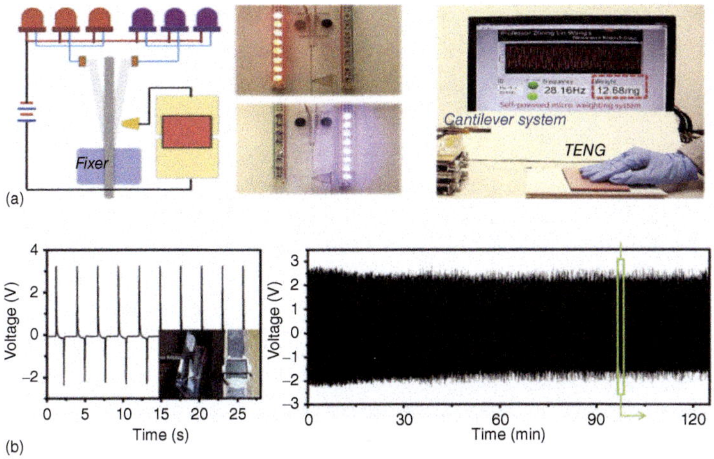

Figure 9.3 Other types of self-powered electronic devices: (a) Demonstrate the cantilever beam as a high-frequency electronic switch triggered by motion. Source: Adapted with permission from Chen et al. [15]. Copyright (2019) by Elsevier. (b) Self-powered electronic device for collecting human movement energy. Source: Adapted with permission from Fan et al. [16]. Copyright (2016) by Advanced Materials.

work demonstrates the significant advantages of using TENG technology for power and sensing in MEMS applications. Based on the single-electrode mode TENG, the integrated self-powered ROS has ultra-high sensitivity (223 mV/mm, while the laser micro-meter is 72 mV/mm). Fan et al. [16] reported a stretchable porous nanocomposite used to harvesting mechanical energy. Under the condition of stretching, an open-circuit voltage (V_{oc}) of 60 V and a short circuit current (I_{sc}) of 180 nA are generated. The deformable energy generation process makes nanocomposite suitable for various mechanical interactions, including pressing, stretching, bending, and twisting. Energy can be collected in human motion, including bending, stepping, and walking, and used to driving a watch (Figure 9.3b).

9.2 Flexible Self-Powered Electronic Devices

In general, flexible self-powered electronic devices are developed from the traditional self-powered electronic devices. Flexible self-powered electronic devices usually adopt a slice structure that can bend or fold to meet the need for human wear. There are two methods to achieve the demand for flexibility. Firstly, it can directly assemble flexible, functional materials like polymer on slice substrates to form the flexible device. Secondly, it can integrate some small rigid units through an island structure (mosaic structure) on slice substrates to form the flexible device [21].

Compared to traditional self-powered electronic devices, flexible self-powered electronic devices are more adaptable, variable, and flexible. Therefore the device can withstand many complex mechanical deformations, such as stretching, twisting, bending, and tearing. More importantly, due to the flexibility of such devices, they are more suitable for wearing on the human body.

This section will introduce the material, structure, and application of the flexible, independent self-powered electronic devices, flexible integrated self-powered electronic devices, and other types of flexible self-powered electronic devices.

9.2.1 Flexible Independent Self-Powered Electronic Devices

Like traditional independent self-powered electronic devices, flexible, independent self-powered electronic devices consist of only one independent device. The device uses flexible slices (polymer film, paper, metal film) as the substrate. The layers of functional materials were assembled on the slice by coating, deposition, chemical growth, and other methods. The flexible, independent self-powered electronic devices are softer and lighter and can adapt well to more diverse environments.

Flexible, independent self-powered electronic devices are simple, easily fabricated, and suitable for more applications. Especially in combination with smart wearable devices, it can fit well with the human body. However, the conflict between electronic function and energy harvesting remains to be a challenge.

The most commonly reported flexible, independent self-powered electronic devices also included TENG and PENG.

The first type of flexible, independent self-powered electronic devices is TENG. It converts the mechanical energy of friction into electrical energy. The two electrodes of flexible TENG are usually composed of thin, flexible slices (usually polymer or polymer-based composite material) with different electrophilic properties. Like rigid-type TENG, mechanical energy can be converted into electrical energy through the friction of the two slices. The greater the difference in electrophilicity between the two materials, the higher the efficiency of flexible TENG.

Flexible self-powered sensors based on TENG have been used to monitor human movement [19, 20]. For example, Lin et al. [19] reported smart insoles based on TENGs for real-time gait monitoring (Figure 9.4c). TENG-based sensors can be easily integrated into conventional insoles. Through the human's movement, mechanical triggers/impacts can be converted into electrical output. By analyzing these electrical signals, the smart insole can accurately monitor and distinguish various gait patterns in real time, including jumping, stepping, walking, and running. Smart insoles can also be used to monitor abnormal gait for a rehabilitation assessment. Yang et al. [20] proposed a flexible triboelectric nanogenerator (FTENG) based on a Kapton film with a wavy structure and serpentine electrodes stretchable substrate. The FTENG can collect environmental, mechanical energy through compression and tension modes (Figure 9.4d). Also, FTENG can be adaptively attached to the human skin for monitoring gentle body movements.

The second type of flexible, independent self-powered electronic devices is flexible PENG, which can convert pressure energy into electric energy. The device can be obtained by depositing or growing piezoelectric material on a flexible slice coated with conductive materials. And then, through deposited conductive material on the other side of the piezoelectric materials layer, the device can be connected with an external circuit to export either sensor signals or energy output. Compare with

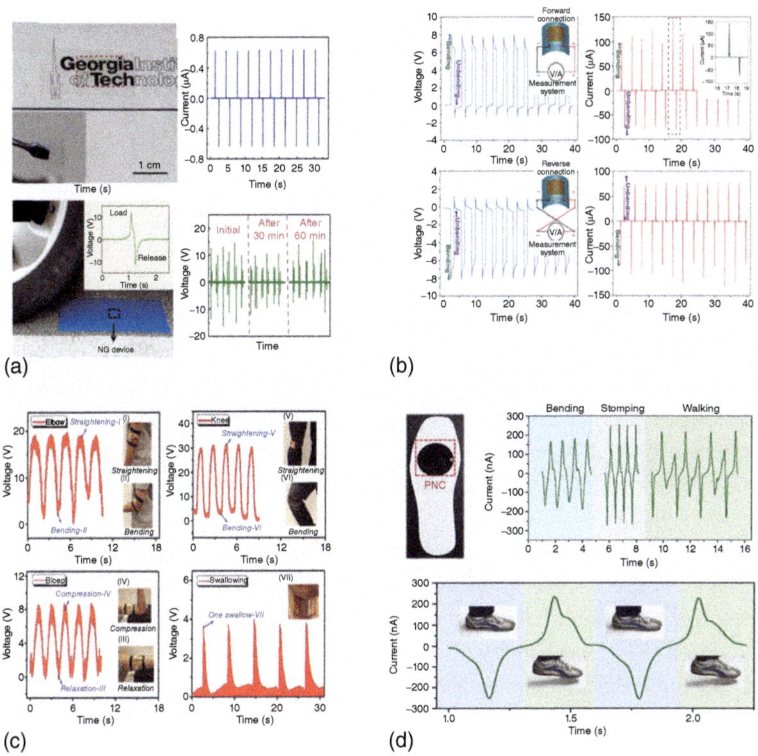

Figure 9.4 Flexible, independent self-powered electronic devices: (a) This photo shows an experimental device used to harvest energy when the tires of a vehicle are rolling. Source: Adapted with permission from Lin et al. [17]. Copyright (2016) by Elsevier. (b) The image of the flexible energy harvesting devices in the original, bent and released state, used for power generation. Source: Adapted with permission from Hwang et al. [18]. Copyright (2014) by John Wiley & Sons. (c) The application of FTENG in human motion monitoring. Monitor the relative changes of the voltage to time for the following actions. Source: Adapted with permission from Lin et al. [19]. Copyright (2018) by WILEY-VCH Verlag GmbH & Co. KGaA, Weinheim. (d) Demonstrate that the device collects energy from human movement. Source: Adapted with permission from Yang et al. [20]. Copyright (2015) by WILEY-VCH Verlag GmbH & Co. KGaA, Weinheim.

traditional PENG device, the structure of flexible PENG can adapt to more complex environments.

For example, Lin et al. [17] used flexible polydimethylsiloxane (PDMS) coated with indium tin oxide (ITO) as substrates to grow ZnO nanowires and fabricated transparent flexible PENGs. The output voltage and current are 8 V and 0.6 μA, respectively. The device can steadily remove the energy in vehicle movement. Figure 9.4a shows its application as a self-powered sensor for monitoring vehicle speed and detecting vehicle weight. Li et al. [22] reported a kind of flexible PENG. The flexible PENG allows the device to act as a microphone while still maintaining its speaker function. PENG-based microphones can effectively "hear" music or voice with high quality and high fidelity. Hwang et al. [18] demonstrated a flexible single-crystal piezoelectric energy harvester that can realize a self-powered artificial cardiac pacemaker. The short-circuit current generated by the energy harvesting device is 0.223 mA, and the open-circuit voltage is 8.2 V (Figure 9.4b). This is sufficient to meet commercial battery charging standards and stimulate the heart without an external power source.

9.2.2 Flexible Integrated Self-Powered Electronic Devices

Flexible integrated self-powered electronic devices are constructed by integrating the device units on a flexible substrate. There are two methods to achieve flexibility for the device. For one thing, by assembling flexible, functional materials (like polymer materials) layer by layer on a slice substrate, more than one flexible device with a laminated structure can be obtained in a tandem way. For another thing, by integrating small pieces of rigid-type units on a slice substrate, a flexible device with a mosaic structure can be obtained. By cooperating with each device unit, flexible integrated self-powered electronic devices can simultaneously harvest energy, storage energy, and conversion signal.

The advantage of flexible, integrated self-powered electronic devices is their wide applications and excellent wearable electronics flexibility. This type of flexible device can well meet the requirement of the human body's wear and is also convenient to be stored and carried [21]. Compared with traditional rigid-type self-powered devices, flexible self-powered devices use polymer as the substrate, which is intolerant to high temperature. Therefore, the fabrication of the device can be a challenge for some types of devices. Besides, the poor adhesion between the functional layer and polymer substrate will lead to a faster performance decaying of the flexible device.

The flexible integrated self-powered electronic devices can be divided by their energy harvesting units, such as TENG, PENG, TEG, and photovoltaic cell.

The first type of energy harvesting unit is based on TENG. This kind of device is usually combined with wearable devices to harvest friction energy generated by humans' movement.

For example, Jiang and coworkers [23] proposed a highly compact self-charging power unit by integrating TENGs into a wearable and flexible harvester-storage unit (Figure 9.5a). The device can utilize and store random energy, giving a maximum power output of 7.8 $\mu W/cm^2$. It can drive electric watches for monitoring human activities. Yang et al. [24] reported a TENG device. This device is a self-powered electronic skin composed of a light-emitting diode-triboelectric nanogenerator (LED-TENG) array, which can be used to reflect the instantaneous touch signal. When an instant touch is applied to the LED-TENG array's integrated matrix, the corresponding LED can be directly lit (Figure 9.5b). The TENG unit can provide an open circuit voltage of about −1070 V, a short circuit current density of 10 mA/m^2, and a power density of 288 mW/m^2 on a 100 MΩ external load.

The second type of energy harvesting unit is based on PENG, which converts pressure energy into electric energy. This kind of device can harvest smaller energy than TENG, and its power output is usually smaller than TENG. However, PENG can harvest mechanical energy from small movements of the human body and even organs. The device usually is used to driving low-power devices, especially for biomedical

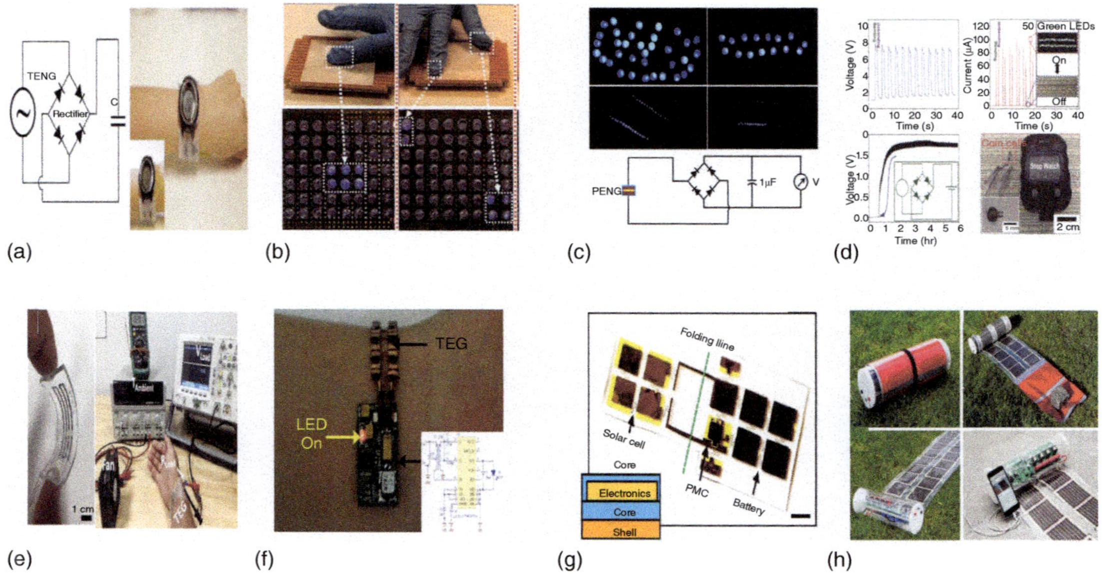

Figure 9.5 Flexible self-powered electronic devices: (a) Circuit diagram of energy supply mode, the device drives a digital watch. Source: Adapted with permission from Jiang et al. [23]. Copyright (2018) by Elsevier. (b) A map of electroluminescence signals obtained from all pixels when a hand wearing a nitrile glove touches the device from the back. Source: Adapted with permission from Yang et al. [24]. Copyright (2014) by American Chemical Society. (c) Snapshots of commercial blue LEDs driven by PENG connected in series and parallel. Source: Adapted with permission from Hoque et al. [25]. Copyright (2017) by American Chemical Society. (d) The short-circuit current measured after rectification. The illustration shows the lighting of 50 green LEDs, corresponding to the current peak value of PENG. Source: Adapted with permission from Hwang et al. [18]. Copyright (2014) by WILEY-VCH Verlag GmbH & Co. KGaA, Weinheim. (e) Devices for performance testing based on changes in air velocity. Source: Adapted with permission from Suarez et al. [26]. Copyright (2017) by Elsevier. (f) Schematic diagram of the drive circuit for wearable TEG devices and the TEG devices worn in the wrist is used to power the LED. Source: Adapted with permission from Shi et al. [27]. Copyright (2018) by IEEE. (g) Solar cell array. Source: Adapted with permission from Schubert et al. [21]. Copyright (2006) by Elsevier. (h) Photo of the portable and rollable solar cell system. Source: Adapted with permission from Garcia-Valverde et al. [28]. Copyright (2016) by Elsevier.

purposes. The device is suitable for supplying power to implantable devices, of which the flexibility brings much convenience to both installation and use.

For example, Hoque et al. [25] manufactured simple, lightweight, flexible, and cost-effective PENGs by integrating aluminum electrodes on PVDF membranes. The developed PENG has excellent energy harvesting performance, high power density, and high-speed charging capability. The PENG leads to a power density of ~160 mW/cm^3 under periodic finger imparting. The manufactured self-powered PENG can also light up 54 commercially available light-emitting diodes (Figure 9.5c). Hwang et al. [18] used the flexible PENG energy harvester as an additional energy source for the artificial pacemaker. The device harvest high electrical power from the mechanical movement of the flexible device (Figure 9.5d). The maximum rectified output current and voltage of the device were measured to be about 100 μA and 8 V, respectively. These output values are large enough to drive commercial electronic products without any auxiliary storage system directly.

The third type of energy harvesting unit is based on TEG, converting temperature energy into electric energy. We all know that the human body's temperature is higher than the ambient temperature, so this kind of device is suitable for wearable devices. More importantly, the human body temperature is constant. The output power of this kind of self-powered electronic device is more stable than the TENG and PENG.

For example, Shi et al. [27] proposed a novel wearable TEG device for powering electronic devices by collecting human body heat. The structural design of TEG devices consists of 12 thermoelectric units, which are connected by copper tapes in electrical series and thermal parallel (Figure 9.5f). After being worn on a person's wrist, the TEG device can generate an open circuit voltage of 10.5 mV, and the ability to use this TEG to power light-emitting diodes has been proven. Suarez et al. [26] proposed a method is the use of EGaIn liquid metal interconnects. The results show that this novel method can finally produce flexible TEGs, which may challenge rigid TEGs and provide a way to realize self-powered wearable electronic devices (Figure 9.5e). Under the condition of an environment temperature of 5 °C, the TEG can reach a maximum voltage and power output of 8.22 mV and 46.28 μW, respectively.

The fourth type of energy harvesting unit is based on the photovoltaic cell, which converts solar energy into electrical energy. The device can be used in many outdoor environments. However, limited by the power generation characteristics of the device types, it has higher requirements for the environment illumination.

For example, Garcia-Valverde et al. [28] proposed a flexible integrated self-powered electronic device based on a fully printed polymer solar cell with a nominal power output of at least 15 W (AM 1.5, 1000 W/m^2). The device inherently enables scaling up of power output through a single-point electrical connection (Figure 9.5g). The power supply unit is proposed for mobile applications, including mobile device charging, lighting LEDs, and pumping tiny water flow. Lee et al. [29] demonstrated concepts that allow the integration of thin millimeter-scale solid-state batteries and multi-junction solar cells with wearable wireless sensors for skin thermal imaging function. (Figure 9.5h). Thin, wearable wireless sensor for skin thermal imaging has a data recording function. The 2 × 2 array solar cells produce an open-circuit voltage, and a short-circuit current of 4.4 V and 3.4 mA, respectively, under AM 1.5 illumination. The maximum power and fill factors are 12.5 mW and 0.84, respectively.

9.2.3 Other Types of Flexible Self-Powered Electronic Devices

Other flexible self-powered electronic devices can harvest energy from the environment. The device can maintain its own uninterruptedly works and serve as a power source for driving external electronic devices. Compared with traditional self-powered electronic devices, the devices are flexible and more suitable for wearable on-body applications.

Distinguished by energy harvesting units, this type of flexible self-powered electronic devices also included TENG-based devices and PENG-based devices.

For example, Pu et al. [30] reported a soft skin-like triboelectric nanogenerator (STENG). STENG can output alternative electricity with an instantaneous peak power density of 35 mW/m^2 and energy converted by human motion to drive wearable electronic devices (such as electronic watches). The device can also be used as artificial electronic skin for touch/pressure sensing (Figure 9.6a). Lai et al. [31] reported an ultra-stretchable nanogenerator. This newly designed nanogenerator can generate electricity by touching or knocking even under various extreme mechanical deformations, even after being damaged (Figure 9.6b). The device can be used as a deformable and wearable power source and a fully autonomous and self-sufficient adaptive electronic skin system. Yang et al. [32] reported on human skin-based TENGs, which can harvest biomechanical energy and be used as a self-powered tactile sensor system for touch panel technology. The manufactured TENG can provide a short circuit current density of 8 mA/m^2 and a power density of 500 mW/m^2 under a load of 100 MΩ, which can be used to directly drive dozens of green light-emitting diodes (Figure 9.6c). Cui et al. [33] reported a new type of magnetically driven non-contact nanogenerator, which avoids direct contact between the nanogenerator and the source of mechanical motion (Figure 9.6d). Nanogenerators can collect mechanical motion energy in a non-contact manner to generate electricity. Their output voltages and currents are as high as 3.2 V and 50 nA, respectively, which are large enough to power liquid crystal displays.

9.3 Self-Powered Electronic Fibers

Flexible self-powered electronic devices are used as electronic skins, but their compatibility with the skin is not good enough to be worn continuously for 24 hours. Fiber-based textiles conform to people's long-term wearing habits. The emerging fiber/textile structure self-powered electronic devices have been regarded as an important solution to achieve 24-hour uninterrupted daily wear. To fabricate self-powered electronic fibers, researchers usually use fibrous flexible material (metal wire, cotton thread, polymer wire) as the substrate and then

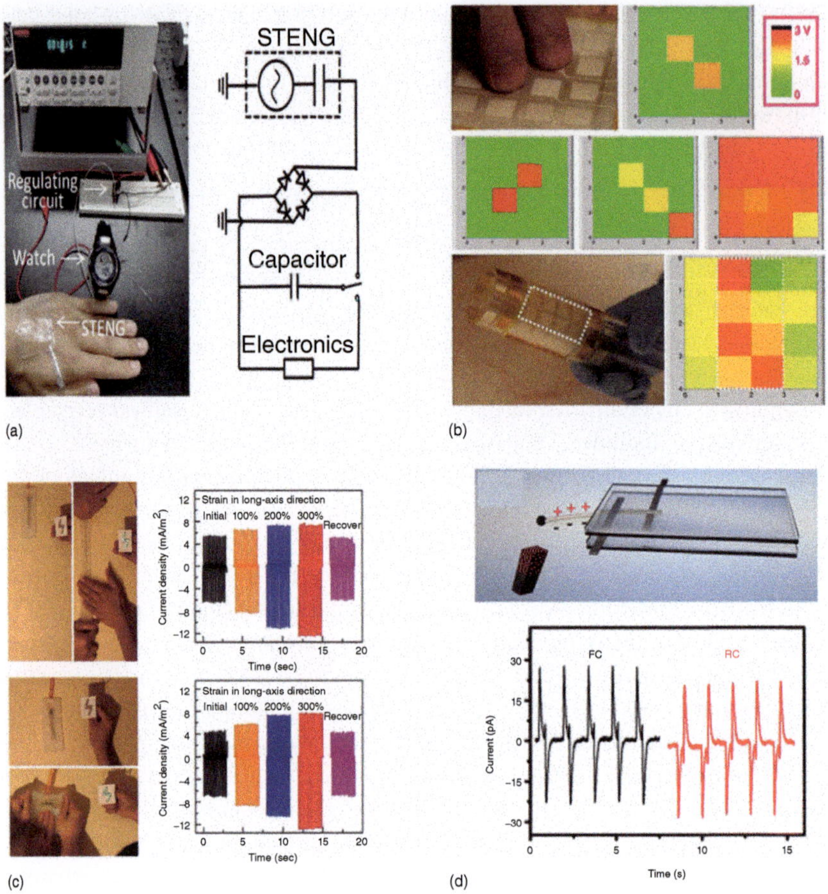

Figure 9.6 Other types of flexible self-powered electronic devices: (a) The image of 20 green LEDs lit by STENG when lightly pressed by hand. Source: Adapted with permission from Pu et al. [30]. Copyright (2017) by Science Advances. (b) The map measured when a human hand touches the device. Source: Adapted with permission from Lai et al. [31]. Copyright (2016) by WILEY-VCH Verlag GmbH & Co. KGaA. (c) Evaluation of nanogenerators under various deformations. Source: Adapted with permission from Yang et al. [32]. Copyright (2012) by the American Chemical Society. (d) The structure and performance of nanogenerators. Source: Adapted with permission from Cui et al. [33]. Copyright (2012) by American Chemical Society.

assemble functional materials layer-by-layer to make a fiber-shaped electrode. The fiber-shaped electrode can be coated with a layer of conductive materials or twisted with another conductive fiber to form a self-powered electronic fiber device unit. Finally, to make self-powered electronic textiles, fiber-type energy harvesting devices, and other fiber-type functional devices integrated into a textile through weaving, embroidery, and knit.

Self-powered electronic textiles can be adapted to the clothing system due to good air permeability [34]. The fiber device unit has higher flexibility, which means it is suitable for the wearable requirements of high curvature of the human body surface. However, in order to fabricate efficient fiber devices, we still face many challenges [35].

Self-powered electronic fibers can also be divided into independent fiber-based self-powered electronic devices and integrated fiber-based self-powered electronic devices. In this section, we respectively introduced the development of these self-powered electronic devices in recent years.

9.3.1 Fiber-Type and Textile-Type Independent Self-Powered Electronic Devices

Independent fiber-type self-powered electronic devices are usually composed of only one fiber device, which can weave into the textile to form the self-powered electronic system. The device of a single fiber can uninterruptedly execute electronic functions by automatically harvesting environmental energy.

The independent fiber-type self-powered electronic device is a miniaturized device with a simple structure, saving space for a wearable microelectronic system. By the energy harvesting method, independent fiber-type self-powered electronic devices can be divided into TENG-type, PENG-type, and TEG-type.

The first type of independent fiber-type or textile-type device is based on TENG. Triboelectric textiles can be fabricated in two methods. The first method is to weave fiber-type fibers into textile-type fibers. Different types of fiber-shaped TENG electrodes can be fabricated by coating functional materials with different electrophilic properties on the fiber-conductive substrate, respectively. Two electrodes with different electrophilic properties are then woven as warp and weft together with yarns to make a single-layered textile. The second method is to directly coat functional materials with different electrophilic properties on a different textile substrate, stacked together to make a multiple-layered textile.

When the self-powered textile was under external impact, both contact separation and relative sliding could occur between two fiber-type electrodes. The relative-movement-induced friction would contribute a large portion to the total electrical output. At larger impact momentum, the air gaps between two fiber-type electrodes decreased, contributing to a larger electrical output [36].

TENG is often combined with wearable devices to monitor human activities. For example, Fan et al. [37] reported a triboelectric all-textile sensor array for precise epidermal physiological signal monitoring. It shows high-pressure sensitivity, fast response time, stability, wide operating frequency bandwidth, and machine washability. Besides, the manufactured device is sewn to different parts of the clothes to simultaneously monitor arterial pulse waves and respiratory signals, which can be comfortably worn (Figure 9.7a).

The second type of independent fiber-based self-powered electronic devices is based on PENG. By integrating with textiles, piezoelectric fibers can monitor human behaviors (walking, finger bending, arm swinging). Therefore, the self-powered electronic fabric based on piezoelectric materials can be applied to the medical and health fields. For example, Zhang et al. [39] reported a two-dimensional fabric nanogenerator based on flexible hybrid piezoelectric fibers, which is expected to

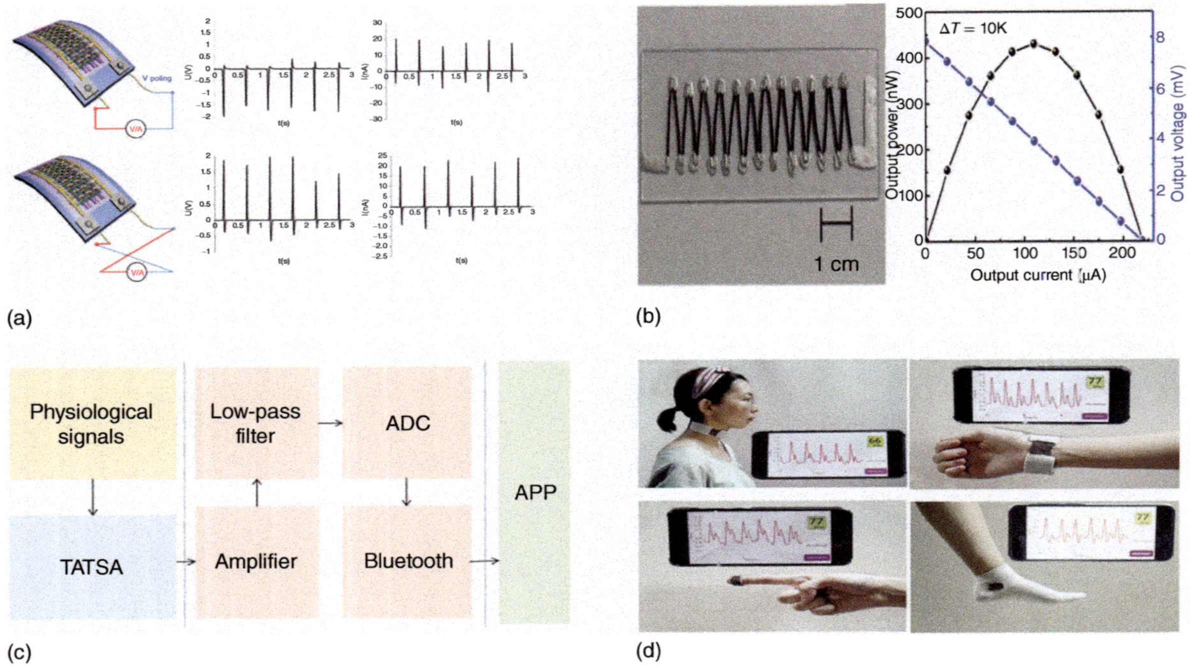

Figure 9.7 Fiber-based independent self-powered electronic devices: (a) On-off evaluation of open-circuit voltage and short circuit current measured from FNG in forward and reverse connections, showing that the output signal is based on piezoelectric response. Source: Modified from Fan et al. [37]. (b) High-performance organic fiber thermoelectric generator. Source: Adapted with permission from Kim et al. [38]. Copyright (2018) by Elsevier. (c) Schematic diagram of the structure of the device. [39] (d) Actual application scenarios of device. Source: Adapted with permission from Zhang et al. [39]. Copyright (2015) by Elsevier.

be easily integrated with clothing and convert the mechanical energy of human motion into electrical energy (Figure 9.7c,d). By installing the fabric nanogenerator on the elbow pad bent by a person, the nanogenerator can generate 1.9 V output voltage and 24 nA output current. The output is large enough to power LCD.

The third type of independent fiber-based self-powered electronic devices is based on TEG. TEG can use the temperature difference between the human body and the surrounding environment to generate electricity without any transmission parts, an ideal energy source for intelligent textiles.

Thermoelectric fibers are woven into the textile structure, which can be combined with the human body to collect or reflect the human body's heat changes. For example, Kim et al. [38] reported wet spinning and post-processing methods of single-walled carbon nanotube (CNT) composite fibers used in high-performance organic fiber TEGs. Moreover, manufactured organic fiber TEGs with P-type and N-type thermoelectric fibers confirmed their power generation capabilities (Figure 9.7b). Its voltage output can be potentially used as a sensor signal for temperature.

9.3.2 Textile-Type Integrated Self-Powered Electronic Devices

Fiber units usually combine integrated fiber-based self-powered electronic devices with different functions. These fiber units usually included fiber TENG, fiber dye-sensitized solar cells (DSSCs), fiber battery, fiber sensor, and fiber transistor. The device can realize more applications by combining the above units, such as wireless health monitoring, motion capture, and other aspects.

The power supply unit for a textile-type integrated self-powered electronic device can consist of different energy harvest units (such as TENG, TEG, and DSSCs) and energy storage units (such as supercapacitor (SC), lithium-ion battery [LIB], and $Zn\text{-}MnO_2$ battery).

For textile-type integrated self-powered electronic devices using TENG as the energy harvest unit, Park et al. [40] proposed a new structure of fiber-based single-electrode triboelectric nanogenerator (FSTENG). The electrical output of FSTENG is produced by continuous contact and separation between human skin and silicone rubber. The prototype of the proposed FSTENG shows an electrical output of 28 V and 0.56 μA when in contact with human skin and shows a high strain as high as 100% (Figure 9.8a). We combined FSTENG to produce a woven structure with 45 mm × 45 mm and confirmed its power generation capacity using LEDs and electronic watches. Yang et al. [41] proposed a liquid metal-based triboelectric nanogenerator (LM-TENG), which uses Galinstan as the electrode and silicone rubber as the triboelectric and encapsulation layer (Figure 9.8b). Also, LM-TENG maintains stable performance under various deformations (such as stretching, folding, and twisting). Various forms of LM-TENG (such as large blocks, bracelets, and fabrics) can collect mechanical energy from people's walking, arm shaking, or hand pat to sustainably drive wearable electronic devices such as electric watches and LED screen.

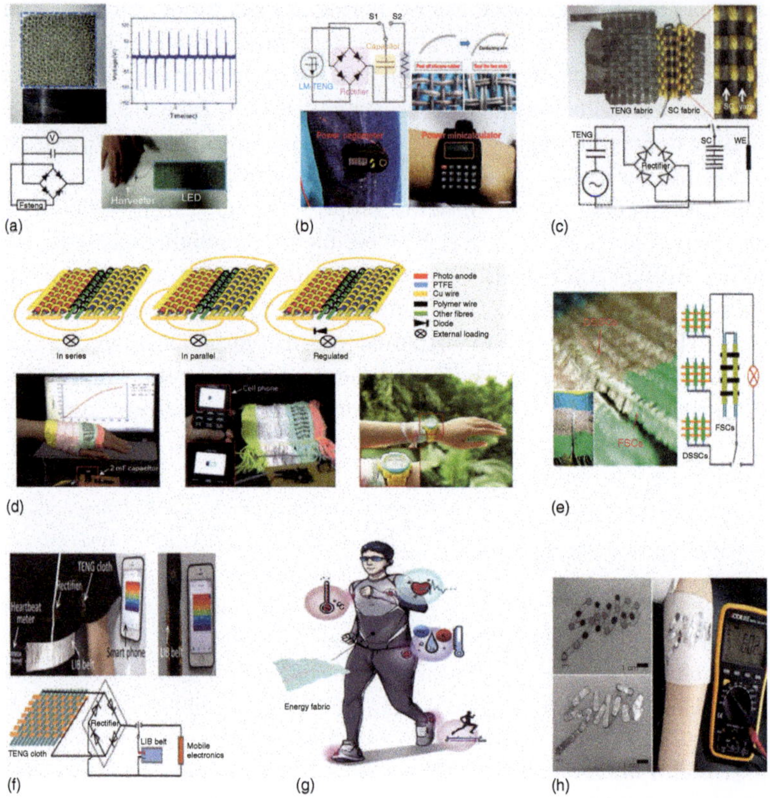

Figure 9.8 Fiber-based integrated self-powered electronic devices: (a) Photograph of woven structure based on FSTENG. The light of 82 white LEDs is visible in the dark environment. Source: Adapted with permission from Park et al. [40]. Copyright (2017) by RSC Adv. (b) The working circuit of a self-charging system based on LM-TENG, including LM-TENG, rectifiers, commercial capacitors, and wearable electronic devices. Source: Adapted with permission from Yang et al. [41]. Copyright (2018) by American Chemical Society. (c) Photograph of self-charging power textile woven with TENG fabric and yarn supercapacitor (SC) fabric. Source: Adapted with permission from Pu et al. [42]. Copyright (2015) by the American Chemical Society. (d) The power output of wearable all-solid hybrid textiles under various conditions. Source: Adapted with permission from Chen et al. [36]. Copyright (2016) by Springer Nature Limited. (e) Integrated energy textiles, showing the customizability of textiles with DSSC and FSC functions. Source: Adapted with permission from Chai et al. [43]. Copyright (2016) by American Chemical Society. (f) Use TENG-cloth to collect the mechanical energy of human movement, use the LIB belt to store energy, and then power the heartbeat strap. Source: Adapted with permission from Pu et al. [44]. Copyright (2015) by WILEY-VCH Verlag GmbH & Co. KGaA. (g) Photo-rechargeable fabric powering a body area sensor network for personalized healthcare. Source: Adapted with permission from Zhang et al. [45]. Copyright (2020) by Elsevier. (h) Power generation of "power shirts". Source: Adapted with permission from Lu et al. [47]. Copyright (2020) by Elsevier.

For textile-type integrated self-powered electronic device using TENG as the energy harvest unit and supercapacitors as the energy storage unit, Pu et al. [42] reported a self-rechargeable power fabric that integrates three functional units in a single cloth: a friction nanogenerator (TENG) fabric for energy collection, a supercapacitor fabric for energy storage, and wearable electronic devices or sensors (such as buttons) (Figure 9.8c). The intermittent energy generated by human movement is stored in the textile energy storage unit to stably supply power to smart electronic devices in the form of textiles or accessories.

Pu et al. [44] designed a wearable power supply by integrating TENG cloth full textiles and a flexible LIB belt for textile-type integrated self-powered electronic devices using TENG as the energy harvest unit and LIB as the energy storage unit. When TENG-cloth is worn in different positions on the human body, it can convert the mechanical energy of various human movements into electrical energy. The LIB belt shows good electrochemical performance even if it is violently folded 30 times at 180°. The LIB belt is charged by TENG-cloth for three cycles and powers the heartbeat strap to communicate with a smartphone remotely (Figure 9.8e).

For textile-type integrated self-powered electronic device using DSSCs as the energy harvest unit and Zn–MnO_2 battery as the energy storage unit, Zhang et al. [45] made a photo-rechargeable fabric using an industrial shuttle-flying weaving process (Figure 9.8g). The photo-rechargeable fabric consists of a photovoltaic energy-harvesting component and a rechargeable fabric battery. With decent air stability, the energy fabric can even function normally under mechanical twisting and rainfall. Given such features as lightweight, robustness, and cotton-like softness, the photo-rechargeable fabric is a critical component in realizing an autonomous textile sensor network for continuous human physiological, biomechanical, and ambient environmental monitoring, with the possibility of succeeding as a future medical application in the era of the Internet of Things.

For textile-type integrated self-powered electronic device using TENG and DSSCs as the multi-sourced energy harvest unit, Fan and coworkers [43] proposed an all-solid customizable energy textile with comprehensive functions of solar energy collection and storage (Figure 9.8d). This technology can cut multifunctional textiles into any design shape without affecting its performance, and produce fashionable smart energy clothing for wearable automatic power supply systems, thereby enhancing the user experience and leaving more space for fashionable designs. The optical fiber SC is customizable, ultra-fast charging capability [46]. The energy storage unit uses ultra-high bending resistance, and the solar energy collection unit uses all solid-state dye-sensitized solar cell fabric. Chen et al. [36] used a micro-cable powered textile to collect energy from the surrounding movement simultaneously. Solar cells made of micro-cables made of lightweight polymer fibers are then woven with fiber-based TENGs through a shuttle flight process to form a smart fabric (Figure 9.8f). This kind of textile can continuously power the electronic watch, directly charge the mobile phone, and promote the water decomposition reaction.

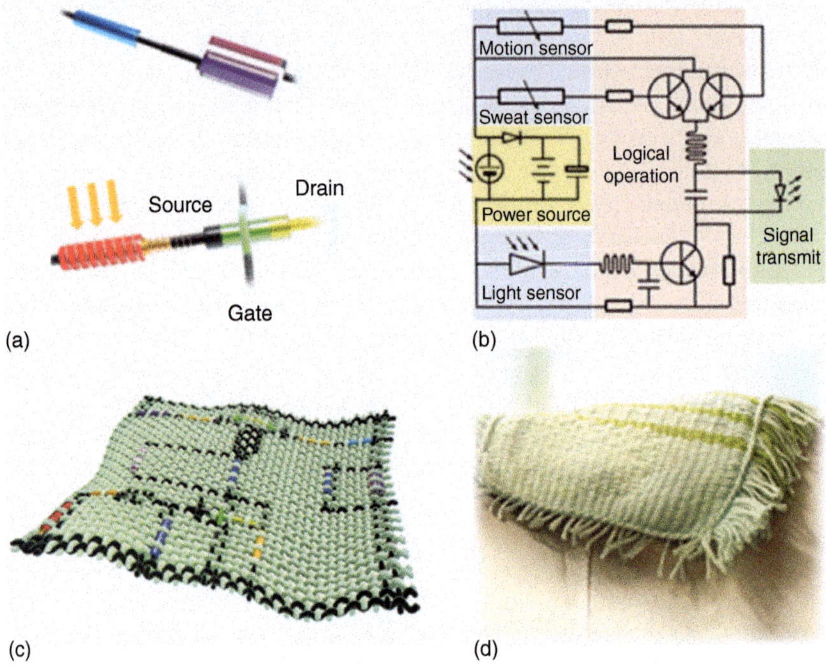

Figure 9.9 Integration of fiber-shaped electronic devices into textiles by interweaving IC: (a) Fiber-shaped field-effect transistor. (b) Design drawing of IC. (c) A designed interwoven IC textile. (d) The practical interwoven IC textile. Source: Adapted with permission from Fan et al. [37].

For textile-type integrated self-powered electronic device using TEG as the energy harvest unit, Lu et al. [47] synthesized nano-structured Bi_2Te_3 and Sb_2Te_3 and deposited them on both sides of the silk fabric to form thermoelectricity (TE) pillars. These TE pillars are connected with silver foil to make a prototype that integrates 12 thermocouple arrays. The generator can effectively convert thermal energy into electrical energy within a temperature difference of 5–35 K. In 100 bending, and twisting cycles, about 15 nW have no significant changes, respectively (Figure 9.8h). Before and after walking for 30 minutes, different voltage output curves were collected from the arm generator to highlight the great potential of the silk fabric-based TE generator. This research provides a new method for developing fabric-based TE generators for practical applications in wearable electronic devices.

Fan et al. have already presented a cloth-like 24/7 private AI nursing system based on a fully interwoven integrated circuit (IC) fabric. Both devices (Figure 9.9a) assembling and circuit wiring (Figure 9.9b) were completely accomplished along polymer wires or at their cross-nodes during weaving, rather than on a printed circuit board.

During the weaving process, there are two main assembling strategies for device units. The first strategy is to make complete fiber-shaped electronic devices, such as sensors, solar cells, lithium-ion batteries, and light-emitting devices, and then integrate them into textiles through weaving or embroidery technologies. The second strategy is to weave fiber-shaped cathode and anode with an interlaced structure to construct textile-type electronic devices directly. Compared with the first strategy, the second strategy has been verified by successfully fabricating solar cell textiles. The warp and weft yarns serve as the anode and modified cathode, respectively [48].

Figures 9.9c and d show a designed interwoven IC textile and the practical interwoven IC textile, respectively. Such a body-fitted cloth has demonstrated its capabilities of uninterrupted physiological monitoring, signal amplifying, logic computing, and wireless communication as an independent mobile electronic system for routine monitoring and emergency assistance.

9.4 Summary

This chapter provides an overview of the development of self-powered electronic textiles. Nowadays, self-powered electronic devices have evolved from the original self-powered electronic devices to flexible self-powered electronic devices and even textile-type or fiber-type self-powered electronic devices. They can adapt to more complex environments and collect even smaller amounts of energy. The self-powered electronic textiles overcome the shortcomings of traditional flexible self-powered electronic devices, such as poor air permeability [53–56]. It is more suitable for human wear and can be better used in various fields such as human health monitoring and human motion capture.

The self-powered electronic fabric can also be formed by integrating multiple fiber-shaped electronic function fibers with fiber-type or textile-type energy devices into one fabric through different weaving methods. Among them, the power output of piezoelectric devices can reach 64.8 $\mu W/cm^2$ [49]; the power output of TENGs can reach 50 $\mu W/cm^2$ [50]; the power output of photovoltaic devices can reach 1–10 mW/cm^3; the power output of thermoelectric devices can reach 15.79 $\mu W/cm^2$ [51]. Although the research on the fiberization of signal processing and conversion units is still limited, fully woven wearable self-powered integrated systems with fiberized diodes, transistors, and field-effect transistors (FETs) are a step close to real life.

The textile covers the surface of the human body and is exposed to a large number of acoustic, optical, electrical, biochemical, and biological signals from the human body. These signals can theoretically become useful data reflecting the human body. For ordinary textiles, we will clear these useful data every time we wash and then lose the data reflecting the physical condition, which is a pity. We collect, store, analyze, and feedback the above data for self-powered electronic textiles through energy collection and energy storage. We can realize the close connection between human health and wearable smart devices. Therefore, the application prospects of wearable self-powered fabrics are extensive, which will play a very important role in future human health monitoring [52].

References

1 Khoshnoud, F., Sundar, D.B., Badi, M.N.M. et al. (2013). Energy harvesting from suspension system using regenerative force actuators. *International Journal of Vehicle Noise & Vibration* 9 (3/4): 294–311.

2 Ahmed, A., Hassan, I., El-Kady, M.F. et al. (2019). Integrated triboelectric nanogenerators in the era of the internet of things. *Advanced Science* 6 (24): 1802230.

3 Xie, Z., Dong, J., Li, Y. et al. (2020). Triboelectric rotational speed sensor integrated into a bearing: a solid step to industrial application. *Extreme Mechanics Letters* 34: 100595.

4 Long, Y., Wei, H., Li, J. et al. (2018). Effective wound healing enabled by discrete alternative electric fields from wearable nanogenerators. *ACS Nano* 12 (12): 12533–12540.

5 Feenstra, J., Granstrom, J., and Sodano, H. (2008). Energy harvesting through a backpack employing a mechanically amplified piezoelectric stack. *Mechanical Systems and Signal Processing* 22 (3): 721–734.

6 Zhang, F., Zang, Y., Huang, D. et al. (2015). Flexible and self-powered temperature-pressure dual-parameter sensors using microstructure-frame-supported organic thermoelectric materials. *Nature Communications* 6 (1): 1–10.

7 Zhao, L., Li, H., Meng, J. et al. (2020). Reversible conversion between schottky and ohmic contacts for highly sensitive, multifunctional biosensors. *Advanced Functional Materials* 30 (5): 1907999.

8 Wen, Z., Chen, J., Yeh, M.H. et al. (2015). Blow-driven triboelectric nanogenerator as an integrated alcohol breath analyzer. *Nano Energy* 16: 38–46.

9 Dieffenderfer, J.P., Beppler, E., Novak, T., Whitmire, E., Jayakumar, R., Randall, C., and Bozkurt, A. (2014). Solar powered wrist worn acquisition system for continuous photoplethysmogram monitoring. *In 2014 36th annual international conference of the ieee engineering in medicine and biology society*. IEEE, 3142–3145.

10 Xu, S., Hansen, B.J., and Wang, Z.L. (2010). Piezoelectric-nanowire-enabled power source for driving wireless microelectronics. *Nature Communications* 1 (1): 1–5.

11 Myers, R., Vickers, M., Kim, H., and Priya, S. (2007). Small scale windmill. *Applied Physics Letters* 90 (5): 054106.

12 Hyland, M., Hunter, H., Liu, J. et al. (2016). Wearable thermoelectric generators for human body heat harvesting. *Applied Energy* 182: 518–524.

13 Wang, Z., Leonov, V., Fiorini, P., and Van Hoof, C. (2009). Realization of a wearable miniaturized thermoelectric generator for human body applications. *Sensors and Actuators A: Physical* 156 (1): 95–102.

14 Wang, X., Niu, S., Yi, F. et al. (2017). Harvesting ambient vibration energy over a wide frequency range for self-powered electronics. *ACS Nano* 11 (2): 1728–1735.

15 Chen, J., Guo, H., Wu, Z. et al. (2019). Actuation and sensor integrated self-powered cantilever system based on TENG technology. *Nano Energy* 64: 103920.

16 Fan, Y.J., Meng, X.S., Li, H.Y. et al. (2017). Stretchable porous carbon nanotube-elastomer hybrid nanocomposite for harvesting mechanical energy. *Advanced Materials* 29 (2): 1603115.

17 Lin, L., Hu, Y., Xu, C. et al. (2013). Transparent flexible nanogenerator as self-powered sensor for transportation monitoring. *Nano Energy* 2 (1): 75–81.

18 Hwang, G.T., Park, H., Lee, J.H. et al. (2014). Self-powered cardiac pacemaker enabled by flexible single crystalline PMN-PT piezoelectric energy harvester. *Advanced Materials* 26 (28): 4880–4887.

19 Lin, Z., Wu, Z., Zhang, B. et al. (2019). A triboelectric nanogenerator-based smart insole for multifunctional gait monitoring. *Advanced Materials Technologies* 4 (2): 1800360.

20 Yang, P.K., Lin, L., Yi, F. et al. (2015). A flexible, stretchable and shape-adaptive approach for versatile energy conversion and self-powered biomedical monitoring. *Advanced Materials* 27 (25): 3817–3824.

21 Schubert, M.B. and Werner, J.H. (2006). Flexible solar cells for clothing. *Materials Today* 9 (6): 42–50.

22 Li, W., Torres, D., Díaz, R. et al. (2017). Nanogenerator-based dual-functional and self-powered thin patch loudspeaker or microphone for flexible electronics. *Nature Communications* 8: 15310.

23 Jiang, Q., Wu, C., Wang, Z. et al. (2018). MXene electrochemical microsupercapacitor integrated with triboelectric nanogenerator as a wearable self-charging power unit. *Nano Energy* 45: 266–272.

24 Yang, Y., Zhang, H., Zhong, X. et al. (2014). Electret film-enhanced triboelectric nanogenerator matrix for self-powered instantaneous tactile imaging. *ACS Applied Materials & Interfaces* 6 (5): 3680–3688.

25 Hoque, N.A., Thakur, P., Roy, S. et al. (2017). Er^{3+}/Fe^{3+} stimulated electrointegrated, visible light emitting, and high dielectric flexible PVDF film based piezoelectric nanogenerators: a simple and superior self-powered energy harvester with remarkable power density. *ACS Applied Materials & Interfaces* 9 (27): 23048–23059.

26 Suarez, F., Parekh, D.P., Ladd, C. et al. (2017). Flexible thermoelectric generator using bulk legs and liquid metal interconnects for wearable electronics. *Applied Energy* 202: 736–745.

27 Shi, Y., Wang, Y., Mei, D. et al. (2017). Design and fabrication of wearable thermoelectric generator device for heat harvesting. *IEEE Robotics and Automation Letters* 3 (1): 373–378.

28 Garcia-Valverde, R., Villarejo, J.A., Hösel, M. et al. (2016). Scalable single point power extraction for compact mobile and stand-alone solar harvesting power sources based on fully printed organic photovoltaic units and efficient high voltage DC/DC conversion. *Solar Energy Materials and Solar Cells* 144: 48–54.

29 Lee, J.W., Xu, R., Lee, S. et al. (2016). Soft, thin skin-mounted power management systems and their use in wireless thermography. *Proceedings of* *the National Academy of Sciences of the United States of America* 113 (22): 6131–6136.

30 Pu, X., Liu, M., Chen, X. et al. (2017). Ultrastretchable, transparent triboelectric nanogenerator as electronic skin for biomechanical energy harvesting and tactile sensing. *Science Advances* 3 (5): e1700015.

31 Lai, Y., Deng, J., Niu, S. et al. (2016). Electric eel-skin-inspired mechanically durable and super-stretchable nanogenerator for deformable power source and fully autonomous conformable electronic-skin applications. *Advanced Materials* 28 (45): 10024–10032.

32 Yang, Y., Zhang, H., and Lin, Z.H. (2013). Human skin based triboelectric nanogenerators for harvesting biomechanical energy and as self-powered integrated tactile sensor system. *ACS Nano* 7 (10): 9213–9222.

33 Cui, N., Wu, W., Zhao, Y. et al. (2012). Magnetic force driven nanogenerators as a noncontact energy harvester and sensor. *Nano Letters* 12 (7): 3701–3705.

34 Castano, L.M. and Flatau, A.B. (2014). Smart fabric sensors and e-textile technologies: a review. *Smart Materials and Structures* 23 (5): 053001.

35 Stoppa, M. and Chiolerio, A. (2014). Wearable electronics and smart textiles: a critical review. *Sensors* 14 (7): 11957–11992.

36 Chen, J., Huang, Y., Zhang, N. et al. (2016). Micro-cable structured textile for simultaneously harvesting solar and mechanical energy. *Nature Energy* 1 (10): 16138.

37 Fan, W., He, Q., Meng, K. et al. (2020). Machine-knitted washable sensor array textile for precise epidermal physiological signal monitoring. *Science Advances* 6 (11): –eaay2840.

38 Kim, J., Lee, W., Kang, Y.H. et al. (2018). Wet-spinning and post-treatment of CNT/PEDOT: PSS composites for use in organic fiber-based thermoelectric generators. *Carbon* 133: 293–299.

39 Zhang, M., Gao, T., Wang, J. et al. (2015). A hybrid fibers based wearable fabric piezoelectric nanogenerator for energy harvesting application. *Nano Energy*: 298–305.

40 Park, J., Choi, A.Y., Lee, C.J. et al. (2017). Highly stretchable fiber-based single-electrode triboelectric nanogenerator for wearable devices. *RSC Advances* 7 (86): 54829–54834.

41 Yang, Y., Sun, N., Wen, Z. et al. (2018). Liquid-metal-based super-stretchable and structure-designable triboelectric nanogenerator for wearable electronics. *ACS Nano* 12 (2): 2027–2034.

42 Pu, X., Li, L., Liu, M. et al. (2016). Wearable self-charging power textile based on flexible yarn supercapacitors and fabric nanogenerators. *Advanced Materials* 28 (1): 98–105.

43 Chai, Z., Zhang, N., Sun, P. et al. (2016). Tailorable and wearable textile devices for solar energy harvesting and simultaneous storage. *ACS Nano* 10 (10): 9201–9207.

44 Pu, X., Li, L., Song, H. et al. (2015). A self-charging power unit by integration of a textile triboelectric nanogenerator and a flexible lithium-ion battery for wearable electronics. *Advanced Materials* 27 (15): 2472–2478.

45 Zhang, N., Huang, F., Zhao, S. et al. (2020). Photo-rechargeable fabrics as sustainable and robust power sources for wearable bioelectronics. *Matter* 2 (5): 1260–1269.

46 Zhong, J., Zhang, Y., Zhong, Q. et al. (2014). Fiber-based generator for wearable electronics and mobile medication. *ACS Nano* 8 (6): 6273–6280.

47 Lu, Z., Zhang, H., Mao, C., and Li, C.M. (2016). Silk fabric-based wearable thermoelectric generator for energy harvesting from the human body. *Applied Energy* 164: 57–63.

48 Wang, L., Fu, X., He, J. et al. (2020). Application challenges in fiber and textile electronics. *Advanced Materials* 32 (5): 1901971.

49 Yeo, H.G., Xue, T., Roundy, S. et al. (2018). Strongly (001) oriented bimorph PZT film on metal foils grown by rf-sputtering for wrist-worn piezoelectric energy harvesters. *Advanced Functional Materials* 28 (36): 1801327.

50 Zhu, G., Zhou, Y.S., Bai, P. et al. (2014). A shape-adaptive thin-film-based approach for 50% high-efficiency energy generation through micro-grating sliding electrification. *Advanced Materials* 26 (23): 3788–3796.

51 Potnuru, A. and Tadesse, Y. (2014). Characterization of pyroelectric materials for energy harvesting from human body. *Integrated Ferroelectrics* 150 (1): 23–50.

52 Loke, G., Alain, J., Yan, W. et al. (2020). Computing fabrics. *Matter* 2 (4): 786–788.

53 Kwak, S.S., Yoon, H.J., and Kim, S.W. (2019). Textile-based triboelectric nanogenerators for self-powered wearable electronics. *Advanced Functional Materials* 29 (2): 1804533.

54 Von Buren, T., Mitcheson, P.D., Green, T.C. et al. (2006). Optimization of inertial micropower generators for human walking motion. *IEEE Sensors Journal* 6 (1): 28–38.

55 Roundy, S. (2005). On the effectiveness of vibration-based energy harvesting. *Journal of Intelligent Material Systems and Structures* 16 (10): 809–823.

56 Wang, Z.L. and Wu, W. (2014). Piezotronics and piezo-phototronics: fundamentals and applications. *National Science Review* 1 (1): 62–90.

10

The Future of Electronic Textiles

Textiles have played an important role in the human civilization for thousands of years. With the rapid development of wearable electronics, there is a tremendous growth of research on fiber-type or textile-type energy, sensor, or other electronic devices, which has been pushing research to explore sustainable, eco-friendly, and pervasive solutions of distributed electronics for a smart way of on-body applications. In this book, we have comprehensively introduced the development of fiber-type or textile-type electronic devices not only on the function evolution from traditional fiber and textile materials with basic wearing functions to textile devices with various electronic functionalities of energy harvesting, energy storage, sensor, and their integrated system but also on the challenges for fiber electrode preparation and textile device assembling technologies, beyond classical fiber and textile materials synthesis and modification technologies.

In the evolution history of textile function, functional fiber and textile materials have been endowed with flame retardancy, static resistance, ultraviolet resistance, etc., in addition to basic needs of covering human body for beauty adorning, warm keeping, and body protection. However, people's pursuit of new textiles is not satisfied but comes with increasing demand for merging with electronic devices, which not only supply fundamental electronic functionalities of energy harvesting and storage but also aim to realize more electronic functionalities or even serve as integrated electronic system. Thus, the idea of the intelligent textile has been explored by weaving with fiber-shaped devices or electrodes based on functional fiber substrates, which can be defined as textile devices rather than textile materials. As discussed in this book, different from the classical functional textiles, intelligent textiles can generate two main types of electric responses to thermal, mechanical, or electrical stimulations: (i) electricity can be generated by energy harvesting units of photovoltaic fibers, piezoelectric fibers, triboelectric fibers, and thermoelectric fibers, etc. and then stored in supercapacitor fibers, battery fibers, or phase-transit fibers or even maintains a self-powered electronic textile system without an external power supply, and (ii) electric signals can be generated by stress sensing textile, strain sensing textile, chemical sensing textiles, and other function coupled textiles. Meanwhile, except for the electric responses, other related properties were also discussed in this book, including stretchability, hydrophobic, stable, air permeability, and color-changing properties.

Textile-Based Energy Harvesting and Storage Devices for Wearable Electronics, First Edition.
Xing Fan, Nannan Zhang, and Yi Wang.

Meanwhile, the function evolution has pushed forward the advancement of fabrication technologies for textiles. Traditionally, functions were endowed to fiber and textile materials by several methods: (i) surface coating or modifying with special agents, (ii) co-spinning of ordinary polymer with functional polymer and nanoparticles, and (iii) copolymer with functional monomer. For more functions and applications, intelligent electronic fibers and textiles have been developed by nano/micro-assembly of layers of functional materials, on the basis of functional fibers and textiles. During which, fiber-shaped electrodes were fabricated by coaxially coating conductor and semiconductor layers on fiber, and then, fiber-shaped or textile-shaped devices can be assembled from fiber-shaped electrodes in ways of winding, side-by-side placing, or interweaving. Many film deposition technologies for nano/micro-assembly have been employed, which can be divided into three categories, namely, dry-process film deposition, wet-process film deposition, and patterning technique. In response to the challenges for devices assembling on highflyer curved fiber surface, the film structure should be optimized with a higher precision or even endowed with a newly designed micro/nanostructure, while film deposition technologies with new principles and new setups should be designed to realize a well-controlled fluid field and electro field. Besides, for increasing the power output and integrating multiple electronic functionalities, intelligent fibers are weaved, knitted, or embroidered into textiles toward large-scale production, rather than setting on printing circuit boards of the current electronic industry.

In the future, integrated electronic fabric systems will have significant application on the wearable electronics market by not only promoting energy harvesting and conversion efficiency but also meeting with other commercialization requirements beyond energy supply. Thus, we have provided a prospective discussion on intelligent electronic textiles to present a critical analysis and insights into remaining challenges and future directions of electronic textiles.

10.1 Commercialization Requirements Beyond Energy Efficiency

Emerging intelligent fibers or textiles can be defined as fiber or textile devices that are able to give active response to changes in the environment, rather than a type of fiber or textile materials. As a matter of fact, these textiles are a new form of electronic equipment, which combines various electronic function devices but in similar ways to fiber drawing and textile weaving. Thus, the commercialization requirements of smart fabrics should mainly concentrate on the demand of energy supply, function expansion, durability, and wearing comfort, as that in electronics industry. For the electronic devices, energy supply is the most fundamental character, and the development of different electronic functions is also an urgent need to meet on the commercialization. In additional, the smart electronic textile still should meet with basic wearing requirements, including mechanical durability and wearing comfort.

10.1.1 Energy Supply

Energy supply is the fundamental of all the electronic devices. Energy fabrics were a hot research topic in recent years, which has succeeded in harvesting and storage multi-sourced renewable energy from environment around the human body.

As previously reported, for the energy harvesting devices [1–3], various types of environmental energy have been harvested including solar energy, mechanical energy, and thermal energy. On the solar energy harvesting, the power efficiency of bendable slice-type devices is about 18.4%, and the textile type is about 10.0%. In terms of the wearable area of the human body, if the surface area of the human body that can be directly irradiated by light is about $0.5\,m^2$, the power can reach up to 50 W under one sunlight intensity. On the mechanical energy harvesting, including human movement, walking, environmental wind energy, vibration, etc., the power performance of film-type device can reach to $25\,mW/cm^2$, and the fabric type can also reach to $3.5\,mW/cm^2$. On the thermal power harvesting, the thermoelectric textile is about $7\,\mu W/cm^2$ calculated according to temperature difference of the body temperature ambient (9 °C). Moreover, the hybrid harvesting efficiency of mechanical energy and solar energy is up to $15\,mW/cm^2$.

For the energy storage devices, it mainly includes supercapacitors and secondary batteries [4, 5]. The power density of supercapacitors is larger and suitable for starting and fuze power supply, while secondary batteries with larger energy density is suitable for long-term power supply. The efficiency of supercapacitor fabric can reach up to $6.7\,F/cm^2$, and the efficiency of the secondary battery is more than $15\,mWh/cm^2$. If the wearing area can reach to $1.3\,m^2$, the secondary energy storage battery can provide energy of 216 Wh, and the power density of supercapacitors can reach up to 301 Wh.

Furthermore, for the hybrid energy harvesting and storage, various types of textile devices with higher energy efficiency and stable energy output were proposed, such as solar cell and supercapacitor, solar cell and second battery, triboelectric nanogenerator (TENG), and supercapacitor [6–8]. Among them, the output of solar cell and supercapacitor hybrid fabric can reach up to $20\,mWh/cm^2$, and the discharge voltage of film-shaped solar cell and lithium-ion battery (LIB)-integrated devices is about 3.4 V, and the discharge capacity is close to 100 mAh/g. There is even the hybrid of three fiber devices like solar cell, TENG, and supercapacitor. The energy harvesting power of the integrated textile is close to $6\,mW/cm^2$, and the energy density can reach to $14\,mWh/cm^2$.

Considering the energy demands of common integrated electronic system, by taking an on-body integrated circuit (IC) system for personalized physiological signal monitoring and healthcare as an example, to realize efficient and in-time sensory responses on the body movements and sweating, both the motion and sweat monitoring modules require no more than $45\,\mu W$ of power consumption, for both cases of day and night. Moreover, the signal transmission power consumption of the low-power wireless chip is about 10–100 mW, the average power consumption of emerging smartwatches is about 30 mW, which can be supplied by the hybrid energy harvesting and storage textiles or the film-shaped devices. As more and more attention concentrated on the promising future of electronic textiles, more textiles with different functions have been developed, which also demand the fabrics to promote the efficiency for energy supply.

As a matter of fact, energy harvesting around the human body usually exhibits uncontrollable fluctuations in electric outputs due to the variation of environmental energy input over time. It produces a practical problem in driving on-body electronics, which normally requires a constant and stable power supply. The first viable method is to integrate textile generators with textile-based energy storage devices, like a supercapacitor or battery, to form a textile-based self-charging power unit. Textile generators will first charge the energy storage devices, which can produce a steady and continuous electric output at its discharging plateau. Another possible approach to solve this problem is utilizing power management circuits, such as a rectifier or a buffer, and then integrating them with textile generators to suppress electric output fluctuations. Developing sewable or weaveable power management circuits that can be merged with smart textiles will be a promising research direction in the future.

10.1.2 Electronic Function Expansion

Recently, flexible electronic devices have implemented functions like a self-powered integrated system for body-health monitoring, wireless communication, or even the health data analysis. However, the function of intelligent textiles still mainly focuses on the power supply and sensing, which should be expanded to satisfy the demands increasing with the prominent progress in wearable electronics.

With ideas of artificial intelligence (AI)-assisted healthcare and human-computer interface, the emergence of daily big-data monitoring can define a healthier and smarter way of lifestyle. Global clothing industry giants are eager for a better way to merge electronics into common textile in addition to the accessories and are chasing an ultimate dream of "fabric computer" or "smart personal assistant hidden in the clothes," which everybody likes to wear in every ordinary day. In order to realize the "fabric computer," on one hand, more functional fiber or fabric devices should be developed, including signal transduction, conditioning, logic judgment, and wireless transmission, which also challenge for the advancing structural design. In another word, emerging wearable electronic intelligent textiles refer to highly intelligent textiles that can also function as the integrated electronic systems for information processing, computing, and other electronic functions.

10.1.3 Mechanical Durability

Except for the electronic demand, the wearing demand of mechanical durability is also an essential requirement for commercial applications. In practical wearing scenarios, various and ceaseless mechanical stresses can routinely subject smart textiles to damage, leading to the cracking of the adhesive conductive layer or even peeling off from the textile substrates. On the one hand, due to small diameter and high curvature of the fiber substrate, it is more challenging for function layers to be coated on the electrodes firmly. On the other hand, for a decent mechanical strength and durability, a good adhesion between functional coatings and soft textile materials should be achieved.

Technically, to enhance the surface roughness of yarn/textile substrates is an effective way to strengthen the adhesion and yield large loading areas, which is especially favorable for coating functional material via solution process. Meanwhile, it is also an important strategy to prepare functional materials with stronger coating adhesion on the fiber substrate. For example, recently developed polymer-assisted metal deposition technologies could also shed some light on this factor [9], and more research on enhancing interface adhesion and stability is desired for improving the mechanical robustness of smart textiles.

10.1.4 Wearability

Wearability, a key feature for the commercialization, has not been well developed in the area of intelligent electronic textiles so far. Fabric is the most intimate medium in contact with the human body. On the one hand, most of the current fiber and fabric structure devices have only initially achieved the basic goals of being bendable and stretchable while there is still a long way to the ultimate goal of "soft and comfortable clothes." On the other hand, the wearing comfort of fabrics depends on the combination of the thermohumid comfort, tactile comfort, and pressure comfort, where the contribution to the wearing comfort is 61.5%, 11.5%, and 9%, respectively [10].

Among them, the contribution of thermohumid comfort is the highest. Traditionally, in order to improve the thermohumid comfort of clothing, the moisture permeability of the fabric is generally strengthened to maintain the human body's moisture and heat balance in a thermal environment. At present, researches on wearable functional fabrics are mainly focused on improving electrical performance, but there are relatively few studies on wear comfort such as breathability, tactile sensations, biocompatibility, and even esthetic properties in the wearing applications. Wearabilities are the important commercialization requirements of the smart electronic textile. How to effectively improve the wearability of fabric devices still remains to be a key research content for the commercialization of electronic textile.

10.2 Challenges for Smart Electronic Textiles

To meet with the commercial requirements listed above, there are many technical challenges for the development of smart electronic textiles to be overcome. Firstly, energy supply is an important part for the wearable electronics; thus the improvement of energy efficiency will be always a challenge for the smart electronic textiles. Secondly, electronic fabrics with more functions should be manufactured, in addition to stable and sustainable supply of energy, to meet the emerging demand of on-body electronics. Thirdly, wearing comfort is another essential requirement of the smart textiles contacting with our body. How to improve the wearing comfort is also considered to be the future directions of smart electronic textiles. Moreover, the fabrication of smart textiles reported was mainly completed on the laboratory, and the large-scale fabrication technology will also be a challenge for smart textiles to realize the commercial applications. The challenge of energy efficiency, diversity of functions, wearing comfort, and fabrication technology will be summarized as follows.

10.2.1 Energy Efficiency

Renewable energy fabrics have attracted intensive attention in the community as a sustainable and wearable power solution. However, the energy efficiency of the fabric-shaped devices is lower than other flexible types of devices with the same mechanism, at the current stage. Therefore, it is still a challenge to increase the energy output of the smart energy fabrics, from aspects of materials innovation, textile structure, and harvesting more environment energies.

First of all, fundamental materials innovation will be the direct method to push forward the boundaries of energy conversion efficiency. For example, graphene is a good choice of lightweight capacitive materials for supercapacitor applications, which has favorable properties such as good electrical conductivity even at room temperature, good mechanical flexibility, high surface area (2675 m^2/g), and low preparation cost. Furthermore, many attempts have been made to get higher-performance fibrous LIBs; an aligned carbon nanotube (CNT)/MoS_2 nanosheet hybrid fiber was synthesized to display combined remarkable mechanical, electronic, and electrochemical properties. It was used to fabricate LIBs with high specific capacitance and specific capacity.

Secondly, it is also an efficiency way to improve the output performance within a limited wearing area by designing different textile structure according to the different energy harvesting or storage mechanism. For example, integrated energy textile with regular-shaped device patterns cannot maximize the space utilization on the complex-shaped human body, which limited the overall energy outputs. Couching embroidery has been proposed to assemble various types of cable electrodes and device units with a large size and shape difference directly and reliably on a light tulle, which can be further integrated following any irregular pattern and any electrical connection design.

Besides, manufacturing hybrid energy harvesting devices to capture more energy from the environment would also be an effective way for the increasing of energy efficiency. In terms of environmental multi-energy generation, wearable energy harvesting technologies have been reported for biochemical energy, mechanical energy generation, mechanical energy–thermal energy generation, light energy–mechanical energy–thermal energy generation technology, etc. Taking the mechanical and biochemical energy harvestings as an example, the output performance of the integrated fiber devices can reach about 6 $\mu W/cm^2$, which can be an important complement for solar energy harvesting.

In fact, the average power consumption of our cell phones is about 300 mW; if the energy density of a hybrid energy textile can reach up to 72 mWh/cm^2, an energy fabric with 10 cm × 10 cm size can supply a phone for 24 h use, which only takes a small area of the cloth on the chest or back of a human body.

10.2.2 Diversity of Functions

With the prominent progress in wearable electronics, the functions of the smart electronic textile are no longer limited to the energy supply and sensor signals. With the ideas of "fabric computer" or "smart personal assistant hidden in the clothes" proposed, the integrated smart fabric system with diversity electronic functions of signal transduction, conditioning, logic judgment, and wireless transmission is in urgent need.

In order to realize the "fabric computer," on the one hand, more functional fiber or fabric devices should be developed, including signal transduction, conditioning, logic judgment, and wireless transmission, which also challenge for the advancement of structural design. For example, different form the electronic devices with two electrodes, transistor, the most common electronic components, has three electrodes, which will increase the possibility of fabrication and connection process. Furthermore, fabric-type voltage-stabilizing and charging–discharging management circuits will be designed for the process of energy collection and storage, based on newly invented wire-shaped switching and Zener diodes. On the other hand, the weaving process for the integrated electronic fabric is also a significant portion. Different devices have different electrical properties. With more and more devices integrated on the same fabric, how to connect the diverse function electronic components and how to manage the IC will also be a big challenge for the future work on the smart textile.

Ideally, an all-woven self-powered integrated wireless electronic system in the form of common-looking fabric for intelligent big-data analysis will be needed, which can accomplish a complete set of tasks, including signal transduction, conditioning, logic judgment, and wireless transmission as an integrated smart system. Such a system should be based on but far beyond single-fiber-type or textile-type electronic devices like sensors, diodes, and power sources, etc., which have emerged in recent years and gained wide attentions. Covering the surface of our bodies, fabrics are exposed to a vast trove of optical, mechanical, chemical, and biological information from daily environment, which could form a massive dataset and make sense for the human health.

10.2.3 Wearing Comfort

Recently, flexible electronic devices with a flat structure of the mosaic, tattoo, and bandages have developed rapidly. Their electronic efficiency has been evidently improved, though they still follow the adorn-and-patch paradigm. Therefore, they still have been regarded as medical therapeutic and monitoring measures in the case of specific environmental and physical conditions, rather than lifestyles or habits, because they simply appear different from everyday wears. Thus, many people would psychologically consider the big-data acquiring as additional and burdensome for daily use and dislike to be adorned or patched when they subjectively think they are in a state of health, which usually and objectively increases the chance for missing timely early diagnosis of recessive diseases in many clinic cases. However, to promote the electricity generation of the smart textile, most textile devices are woven with metal or metal-coated wires, which to a certain extent has a negative impact on wearing comfort of smart textiles. Therefore, except for the electronic performance, the wearing properties of the smart textile directly contacted on our body will bring along many challenges, including lightweight, tactilc comfort, breathability, biocompatibility, and even the esthetic characteristics.

The lightweight of smart textile is important for human wear. Metal-based electrodes with good electrical conductivity in smart textiles may be too heavy for human daily-wear applications. Replacing metal coatings or metal wires with conductive carbon-based materials (such as CNTs and graphene) is a viable way to obtain lightweight smart textiles. For example, the density of CNT-coated polyethylene terephthalate (PET) conductive substrates used in Zn/MnO_2 battery is about 1.402 g/cm^3. Furthermore, if the PET fiber was replaced by hollow tube, the weight of the CNT-coated fiber electrode can be further reduced. Compared with common carbon fiber electrode, the weight of CNT-coated fiber electrode is much lower, which is good for fabricating lighter energy textile. It is highly desirable to optimize the conductivity of the carbon-based electrode without sacrificing the lightweight of the carbon-based electrode. In addition, ultralight materials (such as aerogels, copper–iron ore, metal foam, polymer foam, and metal microlattices) can also be developed to reduce the weight of smart textiles.

Tactile comfort is the direct sensation of the skin when it comes into contact with smart textiles. Someone tested the stretchability, bending, and even twisting properties to support the tactile comfort of electronic textiles, but comfortability of smart textile has almost never been discussed in the literature. Now Zhi and coworkers [11] proposed a softness parameter to evaluate the softness of flexible electronic device, referring to the comfortability evaluation of textile. With his fabric softness tester, the softness of commercially available leather and Nylon textile was about 6–7 mm, while softness of the battery with the most flexible CNT paper substrate and less mass loading of coated active materials was only 3–4 mm. With the increased mass of loaded active materials, the softness decreases remarkably while the areal electronic performance increases, signifying that the compromise between softness and areal electronic properties should be considered when designing smart fabrics. Therefore, improving the softness of textiles may be a solution that can be achieved through material innovation and structural design. On the one hand, we can introduce more soft products (such as silk, wool, and cotton) into smart textiles. On the other hand, a woven/knitted smart textile with a porous structure is more desirable to increase the soft touch. In addition, replacing metal conductors with carbon-based materials or conductive polymers can improve the touch of smart textiles.

Breathability refers to the ability to allow moisture and air to pass through. It is another key parameter to measure the abrasion resistance of textiles because it is essential for regulating human body temperature and humidity. However, embedding functional materials (such as the thermoelectric material legs printed by the dispenser) on the textile substrate will affect the transmission path of air and moisture, thereby giving the wearer a feeling of moisture and humidity. For instance, the air permeability of hemp textile with good breathability can reach up to 535 mm/s, while the air permeability of an electronic textile fabricated with conductive CNTs and screen-printing technology is about 100 mm/s. Solving the volume between the interwoven yarns to obtain a proper pore structure and controlling the size of the embedded functional material may be a feasible solution. In addition, natural products that are hygroscopic, such as wool and cotton, can absorb sweat and release heat at the same time. Thus, combining these hygroscopic materials with smart textiles may enhance the ventilation of the skin and the surrounding environment.

The biocompatibility of smart textiles is another serious issue that needs to be considered for long-term skin applications. The leakage of potential toxic chemicals is one risk for widespread applications. For current device manufacturers, toxic chemicals include the harmful metal elements in lead zirconate titanate (PZT), the most commonly used N719 dye in dye-sensitized solar cells (DSSCs), which were traditionally not involved in textile industry [12]. Therefore, finding nontoxic substitutes suitable for textile device is an urgent problem for human application. For example, natural products have aroused great research interest in the field of energy harvesting, and blackberries can be used as natural dyes for DSSCs without toxic effects. In addition, the packaging polymer materials graft biologically side chains or modified with hydrophilic group could also make a significant contribution to enhance the biocompatibility of smart textiles.

Esthetic characteristics of the smart textile, as color or clothing style, would also be key factors that can determine the success or failure of the wearable electronics market. However, esthetic requirements of smart textile have almost rarely been considered in the current research. With the development of tailorable electronic fabrics, more and more fabrics have the possibility to be restored by cutting out the damaged part and reconnecting the intact part, like a delicate sewing process. Furthermore, as a reliable and industrially scalable electrode assembling strategy, the couching embroidery may be another solution for the diversity fashion design of electronic fabrics.

Moreover, wearability is a key feature in the design of electronic textiles, and there is currently no evaluation standard. The textile industry has developed comprehensive specifications and standards to test different textile material products; in order to standardize textile device manufacturers, it may be necessary to establish similar testing procedures for electronic textile devices.

10.2.4 Fabrication Technology

Developing the large-scale fabrication technology is a key challenge for the commercialization of smart textiles. Different from the planar devices and traditional textile, the smart fabrics were woven by the fiber assembled functional layers. On the one hand, compared with the typical film-forming technology, the diameter of the fiber substrate is much smaller than the typical film substrate while the curvature of the fiber substrate is also much higher, which leads to the difficulty of continuously producing fiber substrates. On the other hand, compared with the traditional textile, smart fabric with conductive substrates, functional layers, and electrical connection has much more demand for the woven machine. As is known, there is still no large-scale machine for the fabrication of the smart electronic textile.

Smart textile devices must be massively produced. The challenges mainly include the following three aspects: (i) how to produce functional electrodes suitable for mass production, (ii) how to develop large-scale woven technology, and (iii) how to ensure large-scale good encapsulation of wearable electronic textiles.

Firstly, due to the bending structure of fiber, uneven surface morphology, and complex material properties, it is difficult to achieve a good compatibility between function fiber materials and textile systems. The challenge in the large-scale preparation of function fiber materials is to provide a convenient and economical way to achieve good compatibility with existing textile technologies. Coating materials onto miles of fibers/yarns or several square meters of fabrics by chemical vapor deposition (CVD) is very costly while many researchers can only demonstrate an electronic textile device of a small size. Film deposition technologies with new principles and new setups should be designed to realize a well-controlled fluid field and electro field during film deposition process. Many promising laboratory coating techniques such as dip coating, screen printing, and spinning should be combined with machines used in industry that have large-scale manufacture ability. Besides, to help with the scalability of their designs, researchers will need to understand both how their proposed techniques can be scaled up and how the increases in size affect the performances of their devices, which could be highly challenging.

Secondly, to improve the efficiency of large-scale woven technology, functional electrode yarns as the building blocks of smart textiles should be more endurable for tensile or friction, even with increasing length. Many functional inorganic coating materials for mechanic enhancement would usually enlarge the diameter of the yarns, which may exceed the yarn count range (0.5–50 denier) of commercial looms. In addition, many of these electronic yarns with brittle and stiff coatings are not stretchable enough to withstand the average tensile strength of 50 MPa during the shedding and insertion motions of machine weaving procedures [13], which might easily cause crack propagation or delamination to these coatings and lead to degradation. Moreover, most fabrication of currently reported smart fabrics is in the laboratory-scale manual production stage; the traditional weaving machine may not be suitable for the woven of smart fabrics directly. Specific machine for the integration of wearable electronic textiles into full textiles should also be developed to be compatible with conventional textile fabrication techniques. Hence, large-scale weaving technology including weaving method and weaving machine is highly desired to advance in the smart fabric field.

Thirdly, to ensure large-scale good encapsulation of wearable electronic textiles, more attention need to be paid to the safety and esthetical issue of device encapsulation. On the one hand, for many materials in wearable electronic textiles proposed in the literature, their potentials dangerous for human health are unclear. Electrode materials and electrolytes should be properly encapsulated to prevent leakage or exposure to the external environment. It is worth noting that in most literature regarding wearable electronic textiles, polymeric materials including silicone rubber, PET thin films, etc. have been widely employed as flexible encapsulation layers. However, their poor barrier capability to water vapor and oxygen could result in high risks of active material corrosion, performance loss, and solution leakage. Aluminum/plastic composite films, which are widely used as the encapsulation material in LIB industry, are also far from ideal for the encapsulation of wearable electronic textiles due to the bulky and opaque nature. More efforts toward advanced encapsulation materials with flexible, lightweight, and transparent features and the corresponding packaging technologies are still needed in the development of future wearable electronic textiles. On the other hand, from the aspect of fashion and daily-life use, clothing and electronic devices show different life cycles, where many electronic components may have life span of more than two years but a typical item of clothing may be disposed after several seasons. The encapsulation is also an important solution for ensuring the esthetical appearance from fiber to wearable electronic textiles and toward the final garments.

In summary, the scale-up fabrication for weaving fiber-based devices into textile, especially with reasonable compatibility to the current textile technology, is very important while challenging for the state of the art of this field.

10.3 A Prospective Discussion on Smart Electronic Textiles

As is discussed in this book, with the development of planar devices and the birth of the fiber-shaped function devices, the smart electronic textiles were proposed. In fact, there are many similarities between making the fabric patterned and the electronic circuits.

From view of electronics, traditional electronics were fabricated on a flat plastic board, just like the printed circuit board (PCB) process. In such process, electric device components are assembled by stacking conductive and semiconductive materials layer by layer, forming a flat or cylindrical structure. They were then electrically connected by a single-layered or laminated conductive tracks, which are printed or etched in advance on a plastic substrate like a pattern of artwork. In this way, different function modules can also be assembled and arranged together to form a complete integrated device (Figure 10.1) [14].

From the view of textile, the device assembling and integrating have many similarities with the pattern making process in textile industry. For a typical textile device fabricating process, fiber-shaped electrodes were fabricated by coaxially coating conductor and semiconductor layers on fiber. Then, device components can be assembled from fiber-shaped electrodes in ways of twisting or interweaving. During which, both the devices assembling from fiber-shaped electrodes and the device connecting for circuit wiring are almost the same as the pattern making in textile industry.

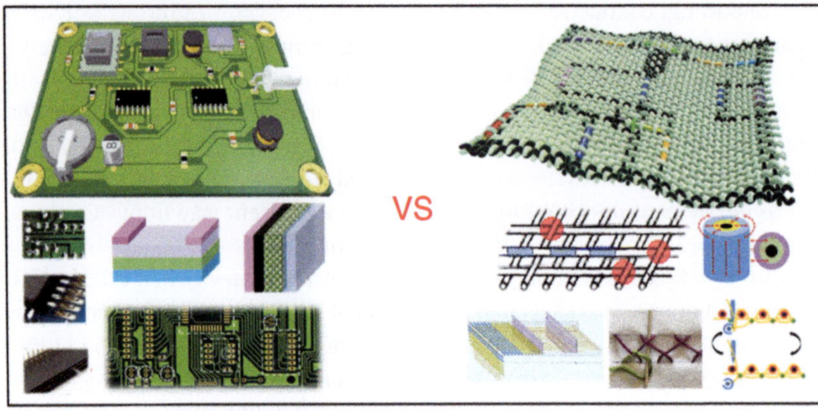

Figure 10.1 A comparison between PCB and smart electronic textiles. Source: Adapted with permission from Huang et al. [14]. Copyright (2020) by American Chemical Society.

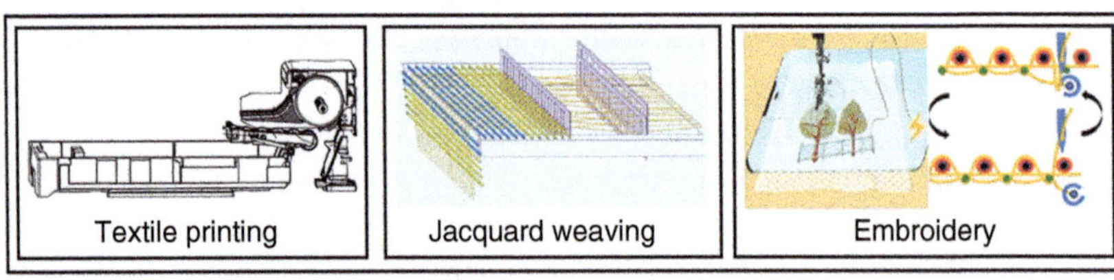

Figure 10.2 Schematic illustration of pattern making process in textile industry. Source: (center image) Zhang et al. [15]. Copyright (2020) by Elsevier.

Current pattern making process in textile industry includes textile printing, jacquard weaving, embroidery (Figure 10.2) (http://www.rayizhan.com/service info.asp?id=37, http://www.360doc.com/content/18/0527/15/33604672_757429801 .shtml, https://www.sohu.com/a/254614996_465156), etc. Textile printing is the process of applying color to fabric in definite patterns or designs, which is similar to printing on a paper or conventional dyeing process for textile, while dyeing would usually form a uniformly colored surface. A jacquard weave is created through a loom process, which is programmed to raise each warp thread independently of the other threads. The design of the textile is incorporated into the weave, instead of being printed or dyed onto the fabric. Embroidery is the craft of decorating fabric or other materials using a needle to apply thread or yarn. Either colored fibers or other materials, such as pearls, beads, quills, and sequins, can be incorporated via embroidery for caps, hats, coats, blankets, dress shirts, denim, dresses, stockings, and golf shirts.

For one thing, conductive pattern can be construct on a textile with a similar appearance to the PCB circuit by textile printing, jacquard weaving, embroidery, or even simply sewing conductive wires to connect different components into a fabric. For another thing, fiber-shaped anodes can be embroidered on textile cathodes or directly interwoven with cathode wrap as wefts in jacquard weaving, which actually formed textile-like devices as previously reported [14]. In a word, patterns but functions as device modules could be constructed in a different way from PCB process but similar way to pattern making process in textile industry.

With the increasing attention on the textile electronic devices and their related interwoven systems, the traditional impression of integrated electronic circuits system for ordinary people may be changed [16]. For one thing, the function of smart textiles can be expanded to power supply and management, transistor, logic operation, wireless communication, etc. and act as corresponding substitutes for current wearable electronic devices or even as the fabric computer. For another thing, owing to their flexible shape, prominent wearable comfort, and beautiful appearance, wearable hardware can be presented as comfortable cloth that can smartly change color, heat/cool, emit light or sound, etc., not only for more possibilities of fashion or function design but also for impressive innovations by acting as a cloth-like 24/7 private AI nurse, or curtain-like restless housekeeper, or even a softer but powerful Tony Stark's smart Ironman suit to become a superhero.

As a promising AI technology, the smart electronic textiles with ubiquitous and endless applications will play an important role in our future life, especially in two representative aspects. One is body healthcare, and the other is human–machine interface.

For the body healthcare, the smart electronic fabric can be constantly worn by a person as common clothing to provide intelligent assistance for human body care such as monitoring, sensing, communicating, and interacting. For example, an outdoor walker comfortably puts on a leisurewear, for which the IC fabric can be merged and distributed arbitrarily on the shoulder, joint, or any part of the body. Day or night can be determined by detecting the light illumination. His sweating and motion profile can be sent out wirelessly to mobile terminal or cloud servers. Consequently, doctors or caregiver can make suitable medical diagnosis and customized healthy plans for different population (infants, elders, rehabilitation people) based on the intelligent big-data analysis. Meanwhile, analysis results can be fed back to their friends or family members. In the case of safety and health emergency, especially for some common but potentially fatal emergencies for elders, such as dehydration and slipping down, the extent of casualty can be effectively reduced by giving timely assistance. As an integrated smart system, such IC fabric can make intelligent judgments by itself, after accomplishing a set of tasks including signal transduction, logic operation, and wireless transmission.

For human–machine interface, intelligent textile can be worn on human body for the recognition of gesture and even unique human physiological characteristics, such as face, voice, odor, fingerprint, and heartbeat but also function as computers in an intelligent manner to assist in planning, learning, reasoning, motioning, manipulating, problem solving, knowledge representation, language programming, and so on. For example, it can be worn on the skin as a platform for programmable control of large arrays of miniaturized haptic actuators in wireless modes of operation and with lightweight, battery-free designs. It can serve as integrated electronically programmable interfaces, not only for a broad spectrum of potential applications, including social media interactions, prosthetic feedback, and video gaming, but also for hardware systems to assist personalized rehabilitation, surgical training, educational feedback, and multimedia entertainment experiences. The development and application of smart electronic textiles will certainly contribute to the advancement of human–machine interface. Many opportunities exist to improve the performance of these systems, which can one day revolutionize and reshape our physical and spiritual world.

Moreover, the smart electronic textile can be fabricated for the wearing applications of the spacemen, firemen, and policemen to protect them and enhance their work efficiency. It also provides a solution for the smart home system, for which curtain can serve as energy supply, carpet can serve as safety monitoring, and beddings can serve as body-health sensing (Figure 10.3).

In summary, with a very bright future, the emerging smart textiles still need the continuous effort to achieve commercialization requirements, which is still challenging in many aspects, including energy efficiency, diversity of functions, wearing

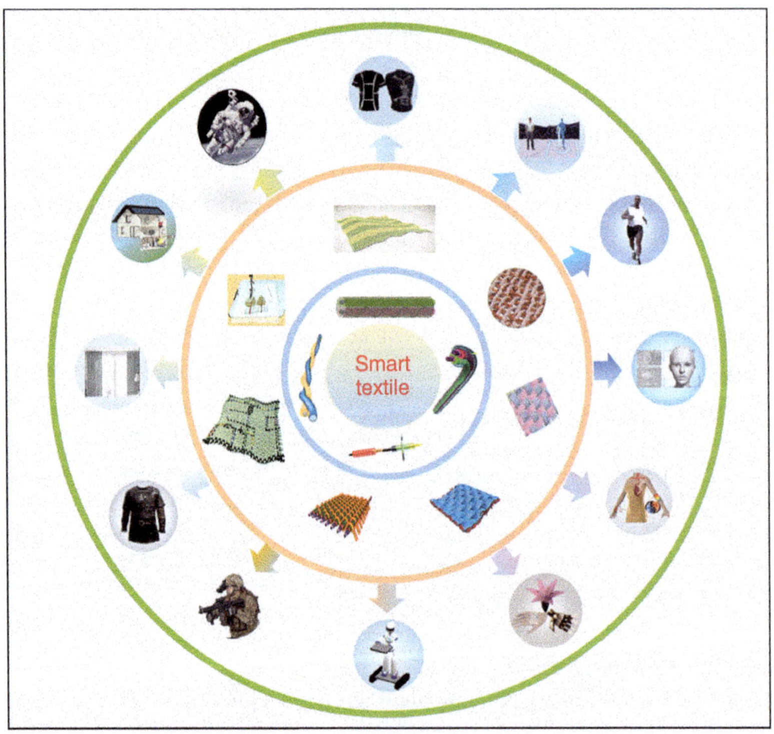

Figure 10.3 The potential application of the smart electronic textiles.

comfort, and fabrication technology. The completion of these challenges requires multidisciplinary efforts from materials science, chemistry, machine engineering, etc. across the textile and electronics industries. We believe that the success of the smart electronic fabrics is not merely a dream, but a solid step forward to realize the new lifestyle of our future or even our childhood dream of becoming a superhero like Tony Stark or Spiderman.

References

1 Zhang, N., Chen, J., Huang, Y. et al. (2016). A wearable all-solid photovoltaic textile. *Advanced Materials* 28 (2): 263–269.

2 Zhu, G., Zhou, Y., Bai, P. et al. (2014). A shape-adaptive thin-film-based approach for 50% high-efficiency energy generation through micro-grating sliding electrification. *Advanced Materials* 26: 3788–3796.

3 Lee, J.A., Bykova, J.S., Kim, D. et al. (2016). Woven-yarn thermoelectric textiles. *Advanced Materials* 28 (25): 5038–5044.

4 Yang, Y., Zhang, N., Zhang, B. et al. (2017). Highly-efficient dendritic cable electrodes for flexible supercapacitive fabric. *ACS Applied Materials & Interfaces* 9: 40207–40214.

5 Wang, Y., Chen, C., Xie, H. et al. (2017). 3D-printed all-fiber Li-ion battery toward wearable energy storage. *Advanced Functional Materials* 27 (43): 1703140.

6 Chen, J., Huang, Y., Zhang, N. et al. (2016). Micro-cable structured textile for simultaneously harvesting solar and mechanical energy. *Nature Energy* 1: 16138.

7 Lee, S., Bae, S.H., Lin, L. et al. (2013). Flexible hybrid cell for simultaneously harvesting thermal and mechanical energies. *Nano Energy* 2: 817–825.

8 Chai, Z., Zhang, N., Sun, P. et al. (2016). Tailorable and wearable textile devices for solar energy harvesting and simultaneous storage. *ACS Nano* 10 (10): 9201–9207.

9 Li, P., Zhang, Y., and Zheng, Z. (2019). Polymer-assisted metal deposition (PAMD) for flexible and wearable electronics: principle, materials, printing, and devices. *Advanced Materials* 31 (37): 1902987.

10 Yang, X., Che, Y., and Zhao, X. (2014). Comparison of test methods for textile moisture permeability. *Dyeing and Finishing Technology* 10: 47.

11 Li, H., Tang, Z., Liu, Z., and Zhi, C. (2019). Evaluating flexibility and wearability of flexible energy storage devices. *Joule* 3: 613–619.

12 Chen, G., Li, Y., Bick, M., and Chen, J. (2020). Smart textiles for electricity generation. *Chemical Reviews* 120 (8): 3668–3720.

13 Khan, G., Terano, M., Gafur, M., and Alam, M. (2016). Studies on the mechanical properties of woven jute fabric reinforced poly (L-lactic acid) composites. *Journal of King Saud University-Engineering Sciences* 28: 69–74.

14 Huang, F., Yu, H., Xiang, S. et al. (2020). Embroidering a filmsy photo-rechargeable energy fabric with wide weather adaptability. *ACS Applied Materials & Interfaces* 12 (3): 3654–3660.

15 Zhang, N., Huang, F., Zhao, S., et al. (2020). Photo-rechargeable fabrics as sustainable and robust power sources for wearable bioelectronics. *Matter* 2: 1260–1269.

16 Han, M., Kwak, J., and Rogers, J.A. (2020). Soft sign language interpreter on your skin. *Matter* 3: 337–338.

Index

Textile-Based Energy Harvesting and Storage Devices for Wearable Electronics, First Edition.
Xing Fan, Nannan Zhang, and Yi Wang.

d

e

t

u

v